21st International Scientific and Research Conference on Metallurgy: Technologies, Innovation, Quality (Metallurgy-2019)

AA007006

IOP Conference Series: Materials Science and Engineering
Volume 866

Novokuznetsk, Russia
23 - 24 October 2019

ISBN: 978-1-7138-3618-6
ISSN: 1757-8981

Printed from e-media with permission by:

Curran Associates, Inc.
57 Morehouse Lane
Red Hook, NY 12571

Some format issues inherent in the e-media version may also appear in this print version.

Additional copies of this publication are available from:

Curran Associates, Inc.
57 Morehouse Lane
Red Hook, NY 12571 USA
Phone: 845-758-0400
Fax: 845-758-2633
Email: curran@proceedings.com
Web: www.proceedings.com

TABLE OF CONTENTS

Preface

From 23 October to 24 October 2019 the XIX International Scientific Conference "Metallurgy: technology, innovation, quality – Metallurgy 2019" was held in Novokuznetsk.

The conference has been held since 1997 at Siberian State Industrial University. It has become a successful forum of the international level, in which leading scientists in the field of metallurgy and materials science, experts in the steel industry and metalworking from Russia, China, Japan, Great Britain, Germany, Brazil, Austria, Israel, Poland, Uzbekistan, Kyrgyzstan, Kazakhstan, Tajikistan, Ukraine, Belarus, Latvia take part. The conference is a platform for professional discussions and implementation of the latest research results and technological solutions in the field of production and processing of materials, energy and resources, the environment and industrial waste processing by the scientific and industrial communities. This allows the dominant current trends and forecasts for the future to be determined.

This issue contains 58 specially selected research papers divided into four topics:
- Basic research, theory, modeling and technology of metallurgical processes;
- Theory and technology of processing metal materials: foundry, metal forming, heat treatment;
- Theory and technology of welding processes, powder metallurgy and the production of composite materials and coatings;
- Heat and mass transfer in metallurgical processes and units; resource and energy conservation, ecology and waste management.

The Organizing Committee would like to thank all participants for their high activity and appreciate the contribution of all authors and research groups to the development of technologies. We hope that the exchange of experience and thoughtful discussion of relevant scientific issues will be a powerful incentive for their effective solution, and the published conference materials will help to advance the achievements of the world industry.

Professor Dr. Evgeny Protopopov
Co-Chair of the Conference Metallurgy 2019
Novokuznetsk, Russia

Peer review statement

All papers published in this volume of *IOP Conference Series: Materials Science and Engineering* have been peer reviewed through processes administered by the proceedings Editors. Reviews were conducted by expert referees to the professional and scientific standards expected of a proceedings journal published by IOP Publishing.

IOP Conf. Series: Materials Science and Engineering **866** (2020) 012001 doi:10.1088/1757-899X/866/1/012001

Analysis of the status and development prospects of Siberian metallurgical cluster in Russia

E V Protopopov, M V Temlyantsev, G V Galevsky, N A Kozyrev, S G Korotkov and A R Fastykovskii

Siberian State Industrial University, 42 Kirova Street, Novokuznetsk, 654007, Russia

E-mail: kafcmet@sibsiu.ru

Abstract. The characteristics of metallurgical production and metal products of Kuzbass enterprises, their heat and energy supply, environmental safety, training of specialist for their needs, directions of their innovative development are considered. It was confirmed that the metallurgy of Kuzbass forms the basis of Siberian metallurgical cluster in Russia and includes 7 enterprises, of which ZSMK is the undisputed leader. Metallurgy takes the 2nd place with the volume of 26.6% in the structure of the industrial sector of Kuzbass.

1. Introduction

Historical traditions, the ore base, the safeguarding of national interests, the length of the territory predetermined the formation and development of three metallurgical clusters in Russia: central, Ural and Siberian. The basis of the Siberian cluster is the metallurgy of Kuzbass [1, 2].

In the structure of the industrial regional sector, metallurgy takes the 2nd place after coal mining and processing with a volume of 26.6% of PS ratio. These two sectors ensure all-Russian priorities of Kemerovo Region in the production of such important industrial products as tram rails – 100%, main rails – 65%, coal and coal concentrate – 58.5%, ferrosilicon – 52.6%, rolled metal – 10.7%, steel – 9.9%. The metallurgical complex includes seven enterprises and is geographically located in three cities: Novokuznetsk, Guryevsk, Yurga. The metallurgical capital of Kuzbass is Novokuznetsk. There are 4 metallurgical enterprises in the city, among which ZSMK is the undisputed leader.

2. Characteristics of metallurgical industries and metal products

JSC "EVRAZ West-Siberian Metallurgical Plant" is a full-cycle enterprise that includes coke-agglomeration, steelmaking, rail and rolling production [3].

The coke-chemical production of the plant, one of the largest coke-chemical enterprises in Russia, was launched in 1963 and annually produces more than 3.5 million tonnes of coke, the main consumer of which (over 80%) is blast-furnace production. The composition of the by-products includes: ammonium sulfate, naphthalene, coal tar pitch, coal oil, benzene, toluene, solvent, anthracene fraction and others (more than 20 items).

The **blast furnace production** includes 3 blast furnaces with a total working volume of 8000 m^3.

In 2019, pig iron production is planned at the level of 6 – 6.1 million tonnes. Thus, 9.5 – 10 million tonnes of iron ore materials are needed for that – 75% of it is the sinter of own production, and 25% – is pellets, mainly purchased on the Russian market of iron ore raw materials. The demand for coke is 2.6 – 3 million tonnes. Coke is produced from coal mines and open pits of the "Coal" division. To

IOP Conf. Series: Materials Science and Engineering **866** (2020) 012001 doi:10.1088/1757-899X/866/1/012001

increase the share of own iron ore raw materials, the Tashtagol mine is being reconstructed with an investment of 6.8 billion rubles.

Steelmaking includes 2 oxygen-converter shops.

In the oxygen-converter shop No. 1, steel is smelted in three converters with a capacity of 160 tonnes. The production capacity of the workshop in 2018 amounted to 2 057 000 tonnes of steel, steel is cast into molds. In the oxygen-converter shop No. 2, steel is smelted in two 350 t converters. The workshop capacity in 2018 amounted to 3 919 000 tonnes of steel. Steel is casted in molds and on two continuous casting machines to produce billets with a cross-section of 150-150 mm, 150-200 mm, 200 (250) 1050-1750 mm.

Rail production includes electric steelmaking and rail and beam shops.

The electric steel-smelting shop provides balk and ball-rolling mills with billets. Two modern arc electric arc furnaces with a capacity of 95 MVA are equipped with gas-oxygen burner systems and foaming slag in the furnace, operate according to the technology with cutting off the furnace slag and leaving some of the metal and furnace slag in the furnace, provide low energy consumption, electrodes and metal charge. The metal discharged from the furnace is processed on two aggregates of complex processing of steel "ladle furnace" and a high-performance ladle vacuum generator. The prepared liquid steel is poured into two modernized bloom casters equipped with a metal protection system against secondary oxidation and an electromagnetic stirring system. The workshop ensures the production of 1.4 million tonnes of continuously cast billets of high-quality and high-quality steels per year.

Rail and beam shop produces products of rail and construction assortment. The reconstruction of the rail and beam workshop completed in 2013 made it possible to produce all types of rails in one workshop: trunk, tram, crane, wit and rails for the subway. The rail and beam workshop was the first in Russia to master the technology of differentiated hardening and received a certificate for differentiated heat-strengthened rails up to 100 meters long. These rails have increased wear resistance and improved performance. Also, the capabilities of the new rolling mill make it possible to produce products of the construction range: beams, channels, piling bar, square and round billets. The volume of production is about 1 million tonnes of rolled metal per year.

Rolling production includes blooming, a continuous billet mill, two continuous small-section mills, a continuous wire mill, a ball rolling mill and is focused on the production of construction rolled products (reinforcement, beams, channels, angels, etc.) in volumes up to 4.2 million tonnes per year.

JSC "Guryevsk Metallurgical Plant" the oldest plant in Siberia and Kuzbass has been operating since 1816, first as a producer of copper and silver, and since 1840 – iron and its alloys [4]. The steelmaking capacities of the plant allow producing up to 210 thousand tonnes of steel per year in two open-hearth furnaces. Section rolling production differs with its rolling mill from other similar ones, and the ball rolling line is equipped with complexes with a capacity of more than 200 thousand tonnes per year. Today from 16 to 23% of mineral raw materials in Russia and the CIS are crushed by balls made in Guryevsk.

JSC "Kuznetskie Ferrosplavy" – the enterprise started its operation in 1942 and is one of the largest producers of ferrosilicon in Russia [5]. Currently, the enterprise accounts for 52% of the total Russian production of ferrosilicon and 100% of the production of high-quality compacted silica fume. In 2011, the "Yurginsky Ferrosplavny Zavod", a branch of JSC "Kuznetskie Ferrosplavy", was put into operation with four furnace units, full cycle of production and waste processing.

LLC "West Siberian Electrometallurgical Plant" was commissioned in 2014. The company operates four ore-thermal furnaces of RKO-9MVA and RKO-8.5MVA makes. The plant's capacity for smelting high-quality ferrosilicon manganese with a low phosphorus content is 53 thousand tonnes per year. The main consumer of products (75% of the total) is JSC "EVRAZ ZSMK".

JSC "RUSAL Novokuznetsk Aluminum Smelter", the first part of which was commissioned in 1943, the second in 1961 [6]. Currently, the plant operates as part of the electrolysis and foundries. The production of raw aluminum is carried out on top-worked cells, as well as on the cells with

IOP Conf. Series: Materials Science and Engineering **866** (2020) 012001 doi:10.1088/1757-899X/866/1/012001

prebaked anodes. A total of 531 cells were installed in 3 series of electrolysis. The annual production of raw aluminum is 210 thousand tonnes.

The foundry equipment includes: 4 IAT 6M2 induction furnaces, 8 electric reflective mixers, 2 semi-continuous casting machines, 6 casting conveyors, "Brochot" automatic line, "Wagstaff" cylindrical casting machine, continuous homogenization furnace and cylindrical bars cutting "Hertwich". The main products of foundry are cylindrical ingots and small ingots from 6 to 22 kg. About 70% of the products are alloys, including multicomponent ones. About 52.7% of the plant's production is supplied to the domestic market and to the CIS countries.

3. Heat and power supply for metallurgy of Kuzbass

The leading group includes products of electrothermal production – aluminum and ferroalloys. Electricity consumption by metallurgical enterprises accounts for 28.5% of total electricity consumption in the Kuzbass and is distributed among enterprises as follows – from 10% for the Novokuznetsk aluminum plant, to 0.15% for Guryevsk metallurgical plant.

The analysis of the structure and generating capacities of Kuzbass fuel and energy complex shows that 8 thermal power plants and 2 block stations with a total installed capacity of 4979 MW operate on its territory, which confirms the guaranteed provision of metallurgical enterprises with heat and energy resources at the required level.

4. Improvement of the the environmental safety of metallurgy in Kuzbass

In 2017, emissions of the main polluting substances by enterprises of the Kemerovo region amounted to 1 million 487 thousand tonnes, including solids 146.5 thousand tonnes [1]. The total contribution of metallurgical enterprises amounted to 19.6%. In 2017, the volume of fresh water withdrawn was reduced to 142.27 million m^3, which is 7.2% of the volume of recycled re-sequential water supply (1965.15 million m^3).

5. Training of specialists for the needs of Kuzbass metallurgy

The share of the labor force in Kuzbass is 55.4%, and the share of those receiving professional education is 21.7% of the population younger than the working age. At present, almost 29 thousand people are employed at metallurgical enterprises, which is ~ 2% of the able-bodied population in Kuzbass [1]. A multilevel system for training metallurgical personnel is currently being implemented.

6. Directions of innovative development of metallurgical production in Kuzbass

The immediate plans of the Kuzbass metallurgical enterprises include reconstruction and modernization of production in the following directions [2]: in the conditions of JSC "EVRAZ ZSMK", improvement of rail steel production in order to improve the quality of metal products; mastering the production of sheet metal products at the casting and rolling complex; long drawing profiles for use as reinforcement; technologies for flash butt welding of differentially heat-strengthened railway rails into long-length (up to 800 m) strings; in terms of JSC "RUSAL Novokuznetsk Aluminum Plant", a conversion of technology implemented by JSC "RUSAL Novokuznetsk" with the installation of cells RA-167 and S-8BME EcoSoderberg.

7. Conclusion

In 2018, the following industry structure was implemented in the industrial sector of Kuzbass (% of PS ration): coal industry – 51.5, ferrous metallurgy – 24.3, energy – 8.5, ferrous metallurgy (scrap and metal processing) – 4.1 chemical and petrochemical industry – 3.1, food industry – 2.8, non-ferrous metallurgy – 2.3, construction materials and metal structures – 2.0, engineering – 1.4. The metallurgical complex includes seven enterprises and is geographically located in the cities of Novokuznetsk, Guryevsk, Yurga. The range of metal products includes cast iron, steel, all types of rails, rolled products (construction, balls, metalware), silicon, manganese-containing ferroalloys, aluminum and aluminum alloys. The total number of employees at metallurgical enterprises is 29

IOP Conf. Series: Materials Science and Engineering **866** (2020) 012001 doi:10.1088/1757-899X/866/1/012001

thousand, which is 2% of the able-bodied population of Kuzbass, amounting to 1481 thousand (55.4% of the total population of Kuzbass). The total electricity consumption by metallurgical enterprises is 10 299 million kWh or 28.51% of the total electricity consumption in Kuzbass. At the same time, electricity consumption by metallurgical enterprises is, %: JSC "RUSAL Novokuznetsk Aluminum Plant" – 9.97, JSC "EVRAZ ZSMK" – 9.4, JSC "Kuznetskie Ferrosplavy" – 9.0, JSC "Guryevsk Metallurgical Plant" – 0.14. Over the past 15 years, diversified environmental programs have been implemented at metallurgical enterprises, which led to a 70% reduction in emissions into the air basin and fresh water intake to 7.2% of the volume of recycled re-sequential water supply, amounting to 1.97 billion m^3.

Priority areas of modernization and reconstruction of metallurgical enterprises include improving the production and welding of main rails, mastering the production of sheet metal products at the casting and rolling complex, the production of long profiles by drawing, and the transition to modern technology for the electrolysis of aluminum.

References

[1] Ryabov V A and Stoblova O B 2017 *Bulletin of KemSU. Series: Biological, Technical, and Earth Sciences* **3** 41–45

[2] *Program of Kemerovo Region – Kuzbass "Development of the industry of Kuzbass" for 2019–2024* http://docs.cntd.ru/document/550201933

[3] *Encyclopedia of Zapsib* http://www.rusmet.ru/promnews/show/12223/Enciklopediya_Zapsiba

[4] *Guryevsky Metallurgical Plant* http://gmzkem.net/

[5] Korennayat K A 2017 *Metallurgy: Technology, Innovation, Quality: Proc. of the XX Int. Sci. and Practical Conference* (Novokuznetsk: SibSIU) part 1 pp 18–20

[6] Zhirnakov V S, Bolshakov D G et al *2017 Proc. of the XX Int. Sci. and Practical Conference* (Novokuznetsk: SibSIU) part 1 pp 11–17

IOP Conf. Series: Materials Science and Engineering **866** (2020) 012002 doi:10.1088/1757-899X/866/1/012002

Steelmaking: current state and development directions

E V Protopopov, A N Kalinogorskiy and L A Ganzer

Siberian State Industrial University, 42 Kirova Street, Novokuznetsk, 654007, Russia

E-mail: protopopov@sibsiu.ru

Abstract. The analysis of the current state and development directions of steelmaking was performed. The main trends of changes in production, technical and economic indicators of steelmaking in Russian metallurgy are considered. The main directions for the improvement of technological processes are determined.

1. Introduction

Metallurgical engineering in Russia is at the forefront of global steel production, which in 2018 according to the World Steel Association (WSA), increased by 4.6% in comparison to 2017 [1]. At the same time, China, with the production volume of 928.3 million tonnes, remains the leading country in steel production. India occupies the second position and has increased its production by 4.9% to 106.5 million tonnes, overtaking Japan where the production has dropped by 0.3%. The USA holds the fourth position – 86.7 million tonnes, the increase by 6.2%. Russia moved to the sixth position, after South Korea, with an output of 71.8 million tonnes which is 4% of global production.

2. Current state of steelmaking

The converter process is the main method of steel production (table 1) in the world. Converters produce more than 60% of the world steel production, which is due to their high productivity, sufficiently high lining resistance, relative simplicity of the design and equipment operation, and technological flexibility in the composition of cast iron. At the same time, the change in the scrap metal market has led to a decrease in the cost of converter steel production as compared to electric steel.

Currently, in the Russian metallurgy 8 oxygen-converter shops with 23 converters are operating, 10 of them are with volume of 300-350 tonnes, 13 – 160 tonnes. The steel production by the oxygen-converter method slightly decreased in 2018 and amounted to 64.1%, and the production of electric steel increased to 33.7%. With the withdrawal of capacities at the Vyksa Steel Works, smelting of open-hearth steel decreased by the end of the year to 2.2%. The volume of continuous casting has increased. Extra-furnace steel processing has been actively introduced – the volume of steel processed in ladle-furnace assemblies has increased, as well as the volume of evacuated steel [3].

Table 1. Steel production in the regions of the world, mln tonnes [2].

Region, country	2017	2018
World production (according to the data of 66 countries)	1729.8	1808.6
Including Europe (with the CIS)	311.7	311.8
EU (28 countries)	168.5	168.1

CIS:	100.9	101.3
Russia	71.3	71.8
Ukraine	21.3	21.1
North America	115.8	120.5
Latin America, Africa, Middle East	93.3	98.9
Asia:	1203.2	1271.1
China	870.9	928.3
Oceania	6.0	6.3

Change in the structure of metal processing in the industry allowed the cost of metal charge to be generally reduced by 0.8 kg/t for steel production. In the composition of the metal charge, the cast iron consumption was reduced by 10.8 kg/t, the consumption of steel scrap was slightly increased. The remaining components of the metal charge did not changed significantly. The converter shops maintain a steady tendency for reduction of cast iron consumption. The largest reduction was achieved in the converter shops of Magnitogorsk Iron and Steel Works (-10.3 kg/t), Cherepovets Steel Mill of PAO Severstal (-20.1 kg/t) and Chelyabinsk Metallurgical Plant (-21.5 kg/t). In the whole industry the reduction in cast iron consumption for converter steel production amounted to 7.8 kg/t (table 2).

As a result, the specific energy intensity of converter steel has been reduced compared to 2017 in all converter shops of the industry. EVRAZ ZSMK and Cherepovets Steel Mill of PAO Severstal still have the minimum specific energy intensity of converter steel. The high energy intensity of converter steel at EVRAZ NTMK is associated with the technology features of steel smelting by the duplex process in order to obtain vanadium slag. According to the results of 2018, the energy intensity of converter steel in the industry is 614.6 kg cf/t, which is 9.4 kg cf/t lower than in 2017.

Table 2. Change in specific consumption of cast iron in the converter steel production, kg/t [2].

Company	2017	2018	Change
In total in the industry, kg/t:	924.1	916.3	-7.8
EVRAZ ZSMK	836.8	835.8	-1.2
Magnitogorsk Iron and Steel Works	891.8	881.5	-10.3
Cherepovets Steel Mill of PAO Severstal	893.8	873.2	-20.6
NLMK	933.1	935.1	+2.0
Chelyabinsk Metallurgical Plant	1030.4	1008.9	-21.5
EVRAZ NTMK	1079.5	1078.7	-0.8

The peculiarity of the converters operation with top oxygen blowing is a relatively small fraction (20 – 45%) of smelting high-quality and alloyed steels, which is due to the low efficiency of converter bath agitation; uneven chemical composition and metal temperature; excessive oxidation of metal and slag during the smelting of low-carbon steels; some problems in controlling the bath behavior, especially during foaming and emissions generation; limited options in terms of improving the heat balance of smelting.

In the 80's of the last century the introduction of a process with top oxygen blowing and inert gas bottom blowing ensured intensification of bath agitation, control of slag formation and dephosphorization process, increased the yield of iron and manganese. However, the application of a slag skull to increase the durability of the converters lining led the manufacturers to abandon the bottom blowing in favor of the process of slagging of lining.

The integral components of modern technology of converter steel production are: the use of special designs of blowing devices; cut-off of slag to reduce its ingress from the converter into the ladle during production; process automation to ensure sustainable results; emission control system to increase yield, improve process control and ensure staff safety; the use of special models to predict various scenarios of the converter shop from the position of minimizing the cost of resources and time for the production and execution of technological operations [3, 4].

In electric furnace steelmaking at enterprises with a full metallurgical cycle, the share of cast iron in the charge remained almost unchanged – from 15 to 41%. In mini-mills the share of cast iron in the metal charge remains within the range from 0.07 to 1.4%, with the exception of Izhstal PAO and Vyksa Steel Works, where the share of cast iron was 12.3 and 11.1%, respectively. At individual enterprises of this group, there is a tendency to abandon completely the use of cast iron in the charge (table 3).

Table 3. Specific consumption of cast iron in the production of steel in electric arc furnaces, kg/t [2].

Group	Company	Cast iron consumption, kg/t		The share of cast iron in the charge, %	
		2017	2018	2017	2018
1	Magnitogorsk Iron and Steel Works	188.0	172.6	16.5	15.1
	EVRAZ ZSMK	279.8	279.8	25.1	25.1
	Ural Steel	471.3	488.2	39.1	41.1
	Cherepovets Steel Mill of PAO Severstal	408.0	376.8	35.8	33.2
	Chelyabinsk Metallurgical Plant	250.3	262.6	22.3	22.5
	Nadezhdinski Metallurgical Plant	397.1	393.0	34.4	34.0
2	Volgograd Metallurgical Plant Krasny Oktyabr	27.4	3.7	2.4	0.3
	Metallurgical Plant "Electrostal"	19.2	15.9	1.7	1.4
	Izhstal PAO	201.7	139.8	17.7	12.3
	Zlatoust Steel Mill	0.9	0.8	0.08	0.07
	Seversky Pipe Plant	1.5	0	0.13	0
	Taganrog Metallurgical Plant	0.3	0	0.02	0
	Volzhsky Pipe Plant	0	0	0	0
	Vyksa Steel Works	124.8	122.8	11.2	11.1
	NLMK-Kaluga	6.0	5.3	0.55	0.48
	Pervouralsk New Pipe Plant	0.9	5.1	0.08	0.46
	Abinsk Electrometallurgical Plant	4.3	2.9	0.38	0.26
	Ashinsky Metallurgical Plant	0	0	0	0

In the whole industry the share of cast iron in the charge is 10.7%, which corresponds to the level of 2017. In specific terms the consumption of cast iron for electric smelting decreased by 0.6 kg/t and amounted to 122.4 kg/t. The specific energy consumption of electric steel, in addition to the consumption of cast iron and metallized pellets, is greatly influenced by the consumption of electricity for smelting. At Chelyabinsk Metallurgical Plant an increase in electricity consumption by 239 kW·h/t with an unchanged proportion of cast iron in the composition of the metal charge increased the energy intensity of electric steel by 91 kg cf/t. At OEMK, due to the increase in the share of metallized pellets in the charge by 3% and energy consumption by 12.6 kW·h/t, the energy intensity of steel increased by 18.2 kg cf/t. At Volzhsky TK, a decrease in energy intensity of steel by 41 kg cf/t is associated with a decrease in consumption of energy by 105.7 kW·h/t and metallized pellets by 11.4 kg/t. Data on the specific energy consumption in the production of electric steel is given in table 4.

Table 4. Specific energy consumption in the production of electric steel, kW·h/t [2].

Company	Consumption, kW·h/t		Change
	2017	2018	
Russia	453.5	448.1	-5.4
Magnitogorsk Iron and Steel Works	345.3	413.3	+68.0
Cherepovets Steel Mill of PAO Severstal	283.6	285.1	+1.5
Ural Steel	292.2	266.2	-26.0
EVRAZ ZSMK	430.8	418.2	-12.6
Chelyabinsk Metallurgical Plant	348.4	587.4	+239.0
Nadezhdinski Metallurgical Plant	314.2	327.3	+13.1
Volgograd Metallurgical Plant Krasny Oktyabr	494.5	629.0	+134.5

Izhstal PAO	485.7	512.1	+26.4
Zlatoust Steel Mill	849.7	826.6	-23.1
OEMK	624.0	636.6	+12.6
Seversky Pipe Plant	455.1	459.8	+4.7
Taganrog Metallurgical Plant	626.3	527.6	-98.7
Volzhsky Pipe Plant	565.7	460.2	-105.7
Vyksa Steel Works	426.5	426.5	–
NLMK-Kaluga	356.5	359.4	+2.9
Pervouralsk New Pipe Plant	470.4	460.8	-9.6
Ashinsky Metallurgical Plant	581.0	592.3	+11.3
Abinsk Electrometallurgical Plant	394.2	375.8	-18.4
Rostov Electrometallurgical Plant	438.3	439.3	+1.0

Electric steel production is rapidly developing due to the fact that the electric furnace is a universal unit in terms of mobility, metal oxidation, furnace atmosphere and the proportion of the liquid charge, which makes it possible to melt any product in it.

The electric furnace, with the development of technologies and equipment, has gradually turned into a universal melting unit equipped with fuel burners and tuyeres for blowing gas and pulverized coal materials, oxygen blasting, bottom agitation devices, devices for preheating the charge, eccentric bottom tapping. The productivity of electric furnaces has more than doubled over the past 20 to 30 years. The power of electric furnace transformers reaches 1000 MVA, the melting time decreased from 120 to 40 minutes, as well as the specific consumption.

3. Results and discussion

The development of extra-furnace steel processing methods has significantly changed the technological process of smelting and casting. The electric furnace has evolved from a steelmaking unit to a charge melting unit. All further operations: refining, alloying and adjustment in composition and temperature before the continuous casting machine are carried out in extra-furnace steel processing units.

Thus, further unification of modern methods of steel production is observed. The converter and the electric furnace are turning into units for the smelting the intermediate product. In the production chain they are followed by the units of extra-furnace steel processing [3, 4]. The steel is treated after being tapped from the unit in various types of vacuum vessels to obtain low concentrations of dissolved gases and non-metallic inclusions in it. Units 'ladle-furnace' with vacuum processing are the most versatile; practically all steel processing can be carried out in them, especially if it is possible to flush slag from the ladle. The improvement of the plants is mainly aimed at intensifying the metal agitation in the ladle. It is being carried out by improving the equipment for inert gas injection, the location of the blowing devices, the period and intensity of gas injection; for units of batch and vacuum processing – by improving the design of the chamber and nozzles.

To improve the technological effectiveness of steel casting at the continuous casting machine, it is recommended to reduce the cross section of the workpieces and increase the casting speed, as well as the use of curved machines; the use of intermediate buckets of increased capacity and height; crystallizers with a taper of variable height, with a more perfect swing mode; electromagnetic braking of steel flows in the mold and soft compression of the unhardened ingot in the secondary cooling zone [3].

The recent increase in the requirements for product quality in the metal market determines the corresponding requirements for the quality of steel scrap in terms of the content of non-ferrous metal impurities, as well as physical properties. Moreover, steel scrap preparation technologies used in domestic metallurgy do not always solve these problems. It is necessary to improve the traditional composition and technological regime of the formation of solid metal charge using, in particular, iron-containing products of steelmaking slag processing [5].

The most important component of the metal charge in steelmaking is cast iron. At the same time, external desulphurization of cast iron has become widespread, as well as its desiliconization and dephosphorization. A decrease in the silicon content in cast iron to 0.30 – 0.35% provides not only the possibility for subsequent external dephosphorization and efficient low-slag technology in the unit, but also allows significant saving in the smelting of cast iron to be achieved [3].

In steelmaking, much attention is paid to reduction of costs for refractory materials. Therefore, in the smelting of steel in converters and arc steel furnaces, high-magnesia fluxes are widely used, which makes it possible to increase MgO content in slag, reduce its aggressive effect on the lining and obtain a refractory protective coating on the lining [4].

4. Conclusion

In accordance with the developed provisions, the scientists of the university scientific school together with the specialists from EVRAZ ZSMK have developed the main directions for improving steelmaking processes in modern conditions:

- improvement of steelmaking technologies to reduce the specific consumption of raw materials and materials, as well as improve product quality;
- reduction of the environmental impact from steelmaking and increase in the degree of industrial waste recycling, including slag, gases and associated energy flows;
- maintenance of the steelmaking competitiveness throughout the entire life cycle of the enterprise.

References

[1] *Website of World Steel Association http://www.worldsteel.org*

[2] Katunin V V, Zinovieva N G, Petrakova T N and Ivanova I M 2019 *Ferrous Metallurgy: Bulletin of Scientific, Technical and Economic Information* **3** 293–314

[3] Grigorovich K V 2018 *Proceedings of the XV International Congress of Steelmakers and Metal Producers* pp 42–59

[4] Semin A E and Shchukina L E 2019 *Ferrous Metallurgy: Bulletin of Scientific, Technical and Economic Information* **1** 10–13

[5] Amelin A V, Protopopov E V et al 2019 *Metallurg* **7** 18–24

IOP Conf. Series: Materials Science and Engineering **866** (2020) 012003 doi:10.1088/1757-899X/866/1/012003

Current trends in the development of metal forming

A R Fastykovskii, A N Prudnikov, A G Nikitin and T N Oskolkova

Siberian State Industrial University, 42 Kirova Street, Novokuznetsk, 654007, Russia

E-mail: omd@sibsiu.ru

Abstract. Current trends in the development of metal forming methods are considered. Along with the increase in production volumes, the tasks of quality improvement, expansion of the product mix and energy costs reduction are becoming more and more important. To solve the set tasks, the most promising methods are combined processing, such as rolling-pressing, casting-rolling-pressing. The combination of different processing methods in one process allows conditions for the production of new profiles to be created, energy consumption to be reduced. However, the lack of sophistication of the issue in theoretical terms, the few experimental data make it difficult to find optimal modes of shaping and layout of equipment when combining processing methods. To determine the energy-efficient conditions of the rolling-pressing process, the rational layout of the equipment, new dependencies are proposed. Conducted analytical studies have proved that when implementing the rolling-pressing process for each deformation condition in the roll-matrix system, there is an optimal matrix installation location that provides the maximum drawing coefficient with minimal energy consumption. The feasibility range of the combined rolling-pressing process was established, energy-efficient deformation modes were determined, taking into account the design features of the deforming unit and the shape parameters. Recommendations for the rational layout of the deforming unit equipment, carrying out the rolling-pressing process were developed.

1. Introduction

The products obtained by metal forming methods (MF) are used in almost all industries. Along with the increase in production volumes, the tasks of quality improvement, expansion of the product mix and energy costs reduction are becoming more and more important. An effective solution to the problems posed is possible when searching for new non-traditional approaches that significantly expand the capabilities of MF methods. Modern trends in the development of MF methods are combined processing methods: such as rolling – separation [1, 2], rolling and pressing [3, 4], casting and pressing [5, 6]. The combination of casting and metal forming can be considered as the most promising direction.

2. Stages of cast and roll units development

The idea of simultaneous casting of hot metal and forming a section in rotating rolls belongs to Sir Henry Bessimer, who patented the idea in 1865. The new casting–rolling process proposed by Henry Bessimer is shown schematically in figure 1.

For a long time, it was not possible to realize the processes of casting and rolling due to the problem of coordination of the continuous casting rates and metal entry into the first mill stand, which differ several times. Currently, most difficulties have been resolved, which allowed several types of cast and roll units (CRU) to be created. The structure of the unit (figure 2) includes: a continuous casting machine, a heating through furnace or an induction installation, a rolling mill with all technological equipment.

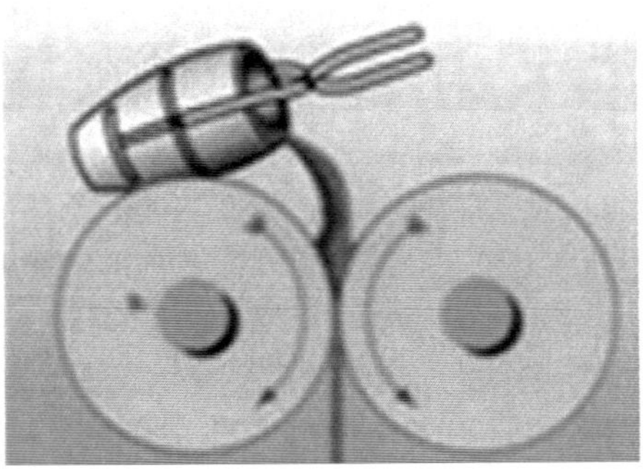

Figure 1. Design of a cast and roll unit proposed by Henry Bessimer.

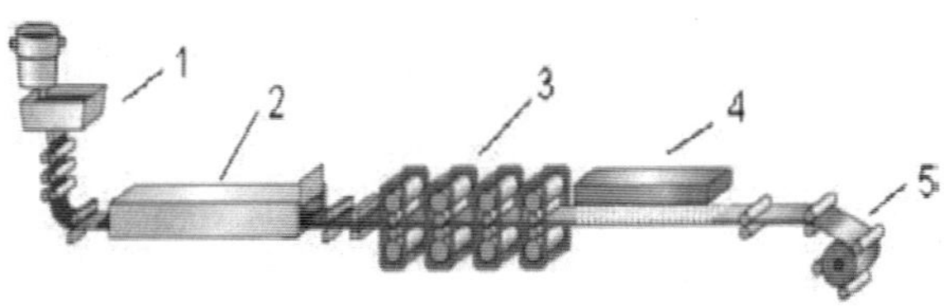

Figure 2. The CRU scheme for obtaining sheets: 1 – CC machine; 2 – through furnace; 3 – rolling mill; 4 – cooling zone; 5 – winder.

Successful experience in the use of CRU was developed in the creation of casting-pressing, casting-rolling-pressing units for manufacturing products of high complexity and precision from non-ferrous metals and alloys. In addition to the advantages listed above, this direction can significantly reduce the production cost due to conversion of the pressing process from cyclic to continuous using less metal- and energy-intensive plants operating on the principle of active friction forces.

3. Conditions for the implementation of the rolling-pressing process

Today, the process of casting, rolling, pressing remains insufficiently studied, which complicates its practical use. The combination of casting – rolling – pressing is carried out in several stages using all kinds of tools. After casting, the workpiece is deformed in the rolls of the rolling mill by the use of a reserve of friction forces and changes the shape in the matrix installed at the deformation zone exit. The options for implementing the casting – rolling – pressing process is shown in figure 3.

The process is possible in the case when the longitudinal force created by the rollers (Q) is greater than or equal to the force required for pressing through the die (Q1). Depending on the drawing coefficient in the matrix, the force Q1 can vary over a wide range. When placing the matrix on the line connecting the centers of the rolls, Q can be found by a formula that takes into account the design features of the groove and the support from the side of the matrix.

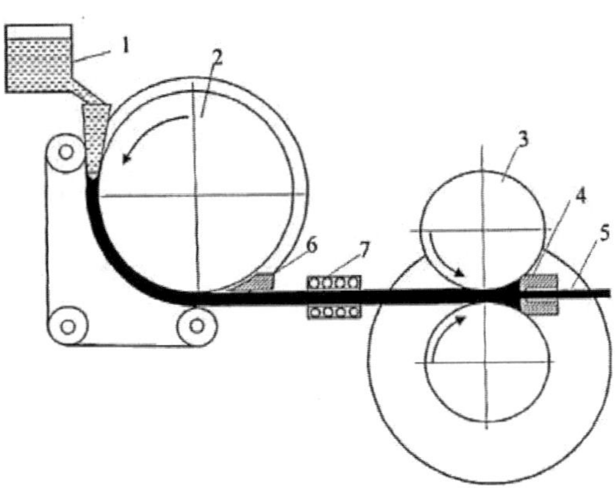

Figure 3. The scheme of the casting-rolling-pressing unit: 1 – furnace mixer; 2 – mold; 3 – rolls; 4 – matrix; 5 – a press product; 6 – a flexor; 7 – through furnace for additional heating.

$$Q = p_{ay} R_g \alpha b_g \left[2\mu_s - \alpha + \mu' \left(h_0 + h_g \right) \right], \quad (1)$$

where p_{av} – the average normal stress, MPa; b_g – groove width, mm; R_r – rolling radius, mm; μ_s – the coefficient of friction at the steady state stage of the process h_g – groove height, mm; h_0 – initial strip thickness, mm; α is the angle of capture, rad; μ´ – coefficient taking into account friction on the side walls of the groove.

Using formula (1), we can determine the pressure on the surface of the matrix (σ1):

$$\sigma_1 = \frac{p_{ay} R_g \alpha}{h_g} \left[2\mu_s - \alpha + \mu' \left(h_0 + h_g \right) \right]. \quad (2)$$

When implementing the rolling-pressing process, grooves of various designs are used [7]. If the side surfaces of the groove are part of the roll, then the friction forces on them with respect to the workpiece will be active. When using fixed side groove elements, reactive friction forces will act on the workpiece from their side. In the first case, $\mu' = \mu_s$, in the second – when using one fixed plate $\mu' = \mu_s - \mu 2$ (μ2 is the friction coefficient on the surface of the fixed plate); with two fixed plates $\mu' = -\mu 2$.

With the increase in the distance from the line connecting the centers of the rolls to the matrix (x), the position of the resultant normal and tangential friction forces changes, the angle of their application, the active area increases. Changes in the force parameters of the deformation zone with increasing distance x leads to an increase in the longitudinal force. On the other hand, increasing the distance x increases the area covered by the matrix, thereby reducing the total drawing coefficient.

4. Analytical studies of the deformation conditions, the equipment layout for energy-power parameters of the rolling-pressing process and the scope of feasibility

Let us consider the influence of the position of the matrix relative to the line connecting the centers of the rolls, for example, deformation of a 13x15 mm sample in a closed groove without release with dimensions h_r = 9 mm, b_r = 15 mm, rolling radius 150 mm, friction coefficient at the steady-state stage of the process 0.5, deformation resistance 100 MPa, speed 0.5 m/s. To establish the influence of the design features of the **grooves** on the pressing force, we consider a **groove**, the side walls of which are part of the roll, a **groove** where one side wall is formed by a fixed plate, and a caliber with two side fixed plates. The coefficient of friction on the surface of the fixed plates is 0.15. Using the results of [8], we determine the increment of the force (ΔQ) and pressure ($\Delta\sigma_1$) when the matrix is removed from the line connecting the centers of the rolls (x) for the known groove designs using the formulas

$$\Delta Q = p_{ay} x \left[2\mu_s b_g + \frac{b_g x}{R_g} + \mu' \left(2h_g + \frac{x^2}{R_g} \right) \right], \quad (3)$$

$$\Delta\sigma_1 = \frac{p_{ay} R_g x}{h_g R_g + x^2} \left[2\mu_s + \frac{x}{R_g} + \frac{\mu'\left(2h_g R_g + x^2\right)}{b_g R_g} \right]. \quad (4)$$

Figure 4 shows the calculated values of the force and pressure on the matrix obtained using formulas (3), (4) for the conditions of forming in caliber without fixed plates. According to Figure 4, an increase in the distance (x) leads to an increase in the force created by the reserve of friction forces, while the pressure on the surface of the matrix (σ1 + Δσ1) first increases and then decreases, which is associated with a nonlinear change in the area of the overlapped matrix.

In the groove, the side walls of which are fixed plates, with the increase in distance (x) the same changes in the studied parameters are observed, only the values are smaller due to the reactive action of friction forces on the side walls of the caliber. Using the formulas given in [8, 9], it is possible to determine the change in the total drawing coefficient in the roll-matrix system for different values of distance (x). The calculation results are shown in figure 5, from which it can be seen that for various conditions of the process there is a certain place for the matrix to be installed relative to the rolls, which guarantees the maximum drawing coefficient. The use of fixed elements in the caliber design reduces the efficiency of the process.

Let us consider how the distance (x) affects the power parameters of the casting-rolling-pressing process. According to figure 5, the total drawing coefficient 17 in the roll-matrix system can be obtained by setting the matrix in any place at a distance (x) from 7 to 27 mm relative to the line connecting the centers of the rolls. Let us find how in this case the energy-power parameters of the process change.

The calculation results are shown in figure 6, from which it follows that for the case under consideration, the distance x = 7 mm is optimal from the point of view of energy consumption; with an increase (x) from 7 to 27 mm, the process will be feasible, but at the same time, energy consumption will increase up to 1.7 times. Despite the fact that the distance x = 7 mm is optimal from the point of view of energy consumption, it is risky to carry out the process under these conditions, since the slightest change in the conditions can transfer the process to the area of impracticability. Therefore, the distance (x) from 8 to 10 mm can be considered optimal. In this case, energy consumption will increase by no more than 5%, but process stability will be ensured.

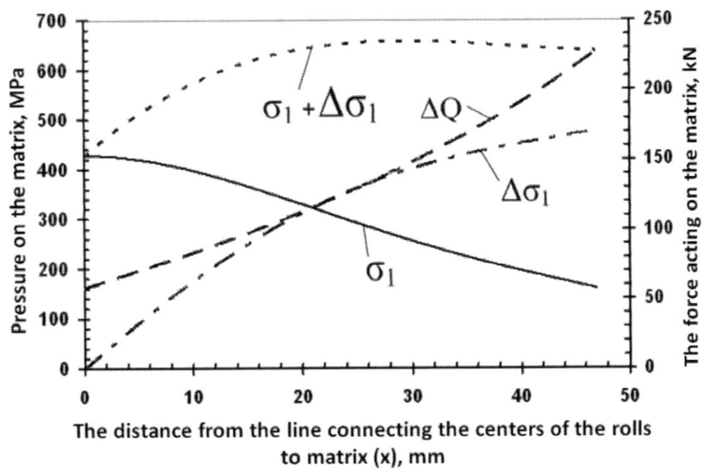

Figure 4. The effect of the matrix positioning on the magnitude of the force and pressure on the matrix.

Let us summarize the results obtained in the form of a graph of the feasibility of the rolling-pressing process shown in figure 7. Using the graph obtained, we can evaluate rationality in terms of the efficiency of the process and the location of the matrix.

The graph in figure 7 has five areas. In the upper part there is a "region of the impracticability of the process" – these are the values of the drawing coefficients unattainable under given conditions. A narrow region of "optimal conditions" of deformation allows the best combination of the total drawing coefficient – energy consumption to be obtained.

At the same time the process is carried out with maximum efficiency. "An area with unjustified high energy costs" allows the rolling-pressing process with a given drawing ratio to be implemented, but the energy costs will be significantly overestimated (up to 1.5 ... 2 times).

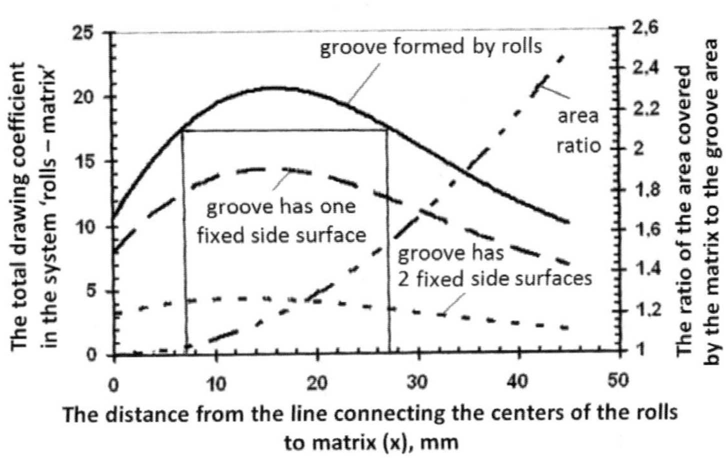

Figure 5. The effect of the installation location of the matrix (x) and the design features of the grooves on the total drawing ratio and the area ratio.

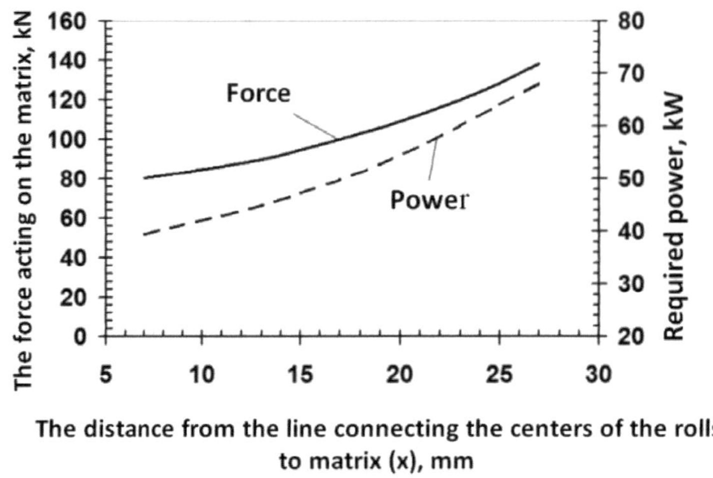

Figure 6. Change in the power parameters of the process with the increase in the distance from the matrix to the line connecting the centers of the rolls.

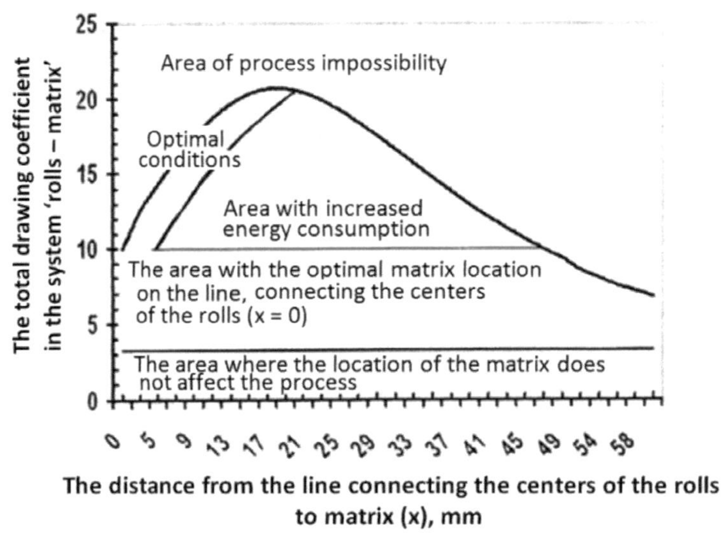

Figure 7. Area of feasibility of the rolling-pressing process.

In practice, there are cases when, having determined the feasibility domain for a particular case of shaping, it is technically impossible to fall into the zone of "optimal conditions", for example, due to the limited range of movement of the matrix relative to the rolls. In this case, according to figure 7, the process falls into the region of increased energy consumption, which is not profitable, both from an economic and technological point of view. It is possible to get out of this situation by changing the friction conditions on the contact surface of the rolls. The choice of the coefficient of friction must be commensurate with the area of feasibility of the process and equipment capabilities.

5. Conclusions

1. The prospects of using combined processes of metal forming are shown.

2. It was established that the position of the matrix relative to the rolls has a great influence on the casting-rolling-pressing process, changing such important parameters as force, pressure on the surface of the matrix, total drawing coefficient, and energy consumption for conducting the process.

3. The area of feasibility of rolling-pressing was determined and recommendations were developed for conducting the process with minimal energy consumption.

4. It is established that the use of fixed elements in the groove design reduces the efficiency of the rolling – pressing process.

References

[1] Fastikovskii A R and Fedorov A A 2016 *Production of Rolled Products* **12** 3–8
[2] Efimov N A and Fastikovskii A R 2015 *Production of Rolled Products* **4** 44–47
[3] Fastikovskii A R 2010 *Forging–stamping Production. Metal Forming* **11** 11–14
[4] Fastikovskii A R 2004 *Izv. Vuzov. Non-ferrous Metallurgy* **2** 67–70
[5] Sidelnikov S B, Dovzhenko N N et al 2015 *Production of Rolled Products* **9** 40–46
[6] Sidelnikov S B, Samchuk A P et al 2015 *Vestnik SUSU. Metallurgy* **2** 89–95
[7] Kornilov V N 1992 *Non-ferrous Metals* **5** 64–65
[8] Fastikovskii A R 2004 *Forging–stamping Production. Metal Forming* **2** 3–6
[9] Fastikovskii A R 2003 *Izv. Vuzov. Non-ferrous Metallurgy* **5** 47–51

Ferroalloy production: state and development trends in the world and Russia

I D Rozhikhina[1], O I Nokhrina[1], K S Yolkin[2] and M A Golodova[1]

[1]Siberian State Industrial University, 42 Kirova Street, Novokuznetsk, 654007, Russia
[2]JSC "UC RUSAL Engineering and Technology Center", 37 Pogranichnikov Street, Krasnoyarsk, 660111, Russia

E-mail: golodova_ma@mail.ru

Abstract. The indicators of ferroalloys production and consumption in the Russian Federation, as well as the main producers of ferroalloys are presented. The changes in the structure of ferroalloy production in the world and Russia from 2009 to 2018 are shown. The main priority tasks of the domestic ferroalloy industry are listed, the solution of which will help to increase its competitiveness.

1. Introduction

Prospects for the development of the global ferroalloy industry are determined primarily by the corresponding trends in the "large" metallurgy, especially in the production of high-quality (special) steels, which consumes the bulk of the ferroalloys produced. By 2020, global steel production will increase to 1.814 billion tonnes in the Russian Federation, steel production will be about 78 million tonnes [1-3]. The growth in steel production, including steel of special high-alloy grades, entails the need to increase the production of ferroalloys – chromium, manganese, silicon, nickel.

A growing trend in recent years in the development of steel production such as complex alloying and microalloying of not only special, but also low-alloy steels of HSLA group – high-strength low-alloy steels should be noted. New steel grades are becoming more widespread, where up to 0.5% Ni, 0.5% Mo, 0.3% Nb and 0.3% V are also included along with traditional elements (Mn, Si and Cr). The expansion of the assortment of ferroalloy production with regard to meeting the needs of the steel industry with such steel grades is primarily associated with the smelting of ferroalloys of the small tonnage group – alloys of nickel, titanium, niobium, molybdenum, vanadium and ferrovanadium, ferro-tungsten, etc.

2. Mineral resources base

By 2020, the world production of manganese ore is forecast at 58.8 million tonnes, which is caused by an increase in the consumption of manganese ferroalloys in connection with the projected increase in steel production. In South Africa, by 2020, it is planned to increase the production of manganese ore to 16 million tonnes, which corresponds to an increase of 27% compared with the existing level. In Australia, manganese production will exceed 20% of the world level. In addition to these countries, manganese ore is mined in China, Brazil, the CIS countries, Gabon, India, Ghana, Malaysia.

To ensure the resource independence of Russian metallurgy in providing manganese ferroalloys, it is necessary to carry out work to create a domestic manganese ore base.

Russia has significant balance reserves of manganese ores (more than 290 million tonnes), forecast resources are more than 1 billion tonnes [4], but manganese ores of most domestic deposits are of low quality: with a low manganese content (18 – 33%) and high specific phosphorus content (P/Mn ratio > 0.006) they have a high content of iron and silica and are difficult to reach. At the same time, about 90% of the balance reserves fall on carbonate raw materials. In Russia, manganese ores are mainly represented by three types: carbonate, oxide and oxidized. The structure of confirmed reserves is dominated (90.2%) by poor carbonate ores (19.8% Mn) with a high phosphorus content (0.2 – 0.3% or more). The reserves of oxide ores containing 23 – 26% Mn are 6% [5]. The deposits are located in Kemerovo Region (Usinskoe), in Krasnoyarsk Territory (Porozhinskoe), in the Urals (Severouralskoye), in the Komi Republic (Parnokskoye), in Irkutsk Region (Novo-Nikolaevskoye), etc.

The largest manganese ore deposit in the country, Usinskoye, accounts for 55% of the country's balance reserves. The manganese content in the ores of the deposit is 18 – 22%, the predominant type of ores is carbonate. The Usinsky ore cluster and the promising Kaygadat ore occurrence (Kemerovo region) together make up one third of the Russian forecast resources of manganese ores.

The ores of these deposits are characterized by low manganese and high phosphorus. Currently, scientific and practical research is being carried out with the aim of involving domestic manganese ores in the production and smelting of ferroalloys from them. At the same time, mining of manganese ores in Russia on a scale industrially significant for the ferroalloy industry is not currently conducted [5].

Chromium ore reserves are identified in 47 countries of the world, the largest deposits are concentrated in South Africa – 76% (1st place in the world). In the Republic of Kazakhstan, reserves are over 350 million tonnes – 9% (2nd place in the world). Next in the world in terms of reserves is Zimbabwe (with 6%) and Russia, deposits of which are concentrated in the Urals: Saranovskoye, Rai-Iz, Verblyuzhegorskoye (according to various estimates, from 0.5 to 2%). Significant reserves are also held by the USA, Greenland, Finland, India, the Philippines and Turkey – 6% in total [6].

Mining of chrome ore in Russia is at the level of 530 – 550 thousand tonnes per year with a Cr_2O_3 content of 37–39% in concentrate. New deposits are not commissioned, and the balance reserves amount to 51.2 million tonnes, forecasted – more than 540 million tonnes.

The world tungsten production volume is 85-87 thousand tonnes. The distribution of tungsten reserves in the world is shown in figure 1 [7]. About 2.6 thousand tonnes of tungsten are mined in Russia per year. In terms of tungsten production, Russia ranks third in the world.

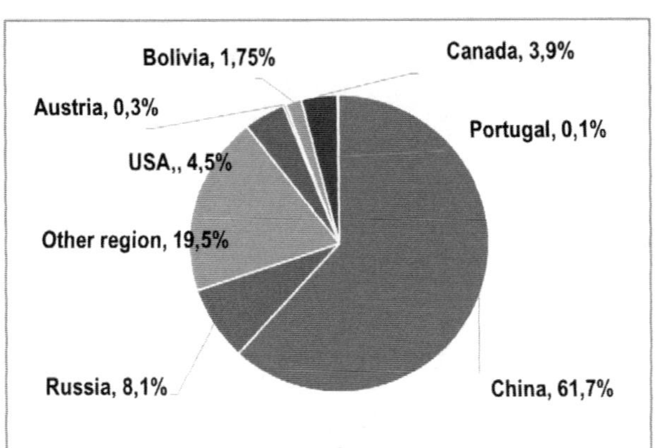

Figure 1. Worldwide distribution of tungsten reserves.

3. Dynamics of ferroalloy production

Dynamics of world production of ferroalloys for the period 2008 – 2018 is shown in figure 2 [6, 8-10]. The world structure of production of ferroalloys has not undergone significant changes in recent years and corresponds to their world consumption. In the world, manganese alloys are in first place in terms of production volumes (figure 2). In Russia (figure 3), manganese alloys also occupy the first place in production growth: in 2015, 10.8%, and in 2018, 25.8%, with more than 2 times growth (figure 3). Among the ferroalloys on the Russian market in 2018, one can distinguish ferrosilicon with a share in the production volume of 49.39% and ferrochrome – 13.92%.

IOP Conf. Series: Materials Science and Engineering **866** (2020) 012004 doi:10.1088/1757-899X/866/1/012004

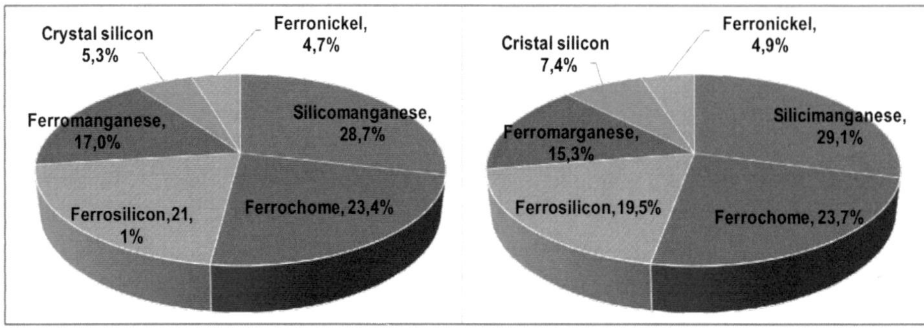

Figure 2. Change in the structure of ferroalloy production in the world.

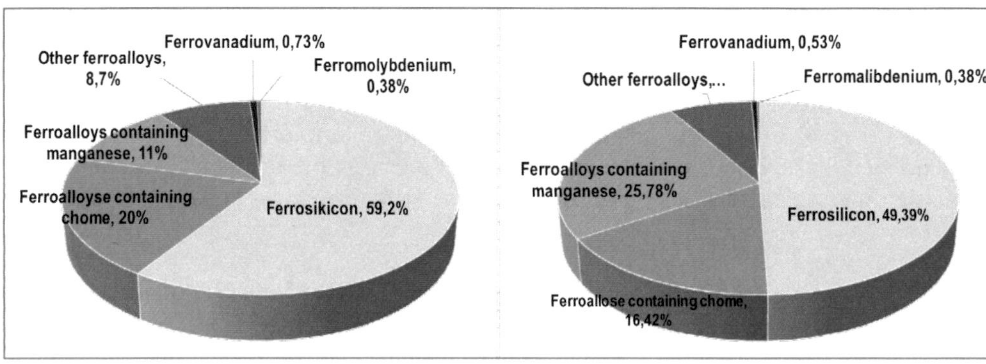

(a) 2015 (b) 2018

Figure 3. Change in the structure of ferroalloy production in the Russian Federation.

The distribution of ferroalloy production by federal districts is shown in figure 4.

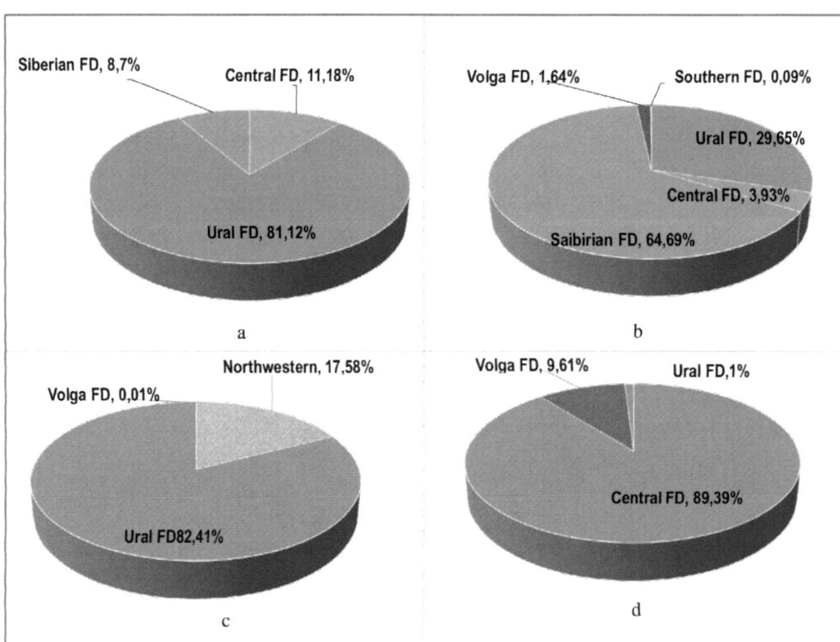

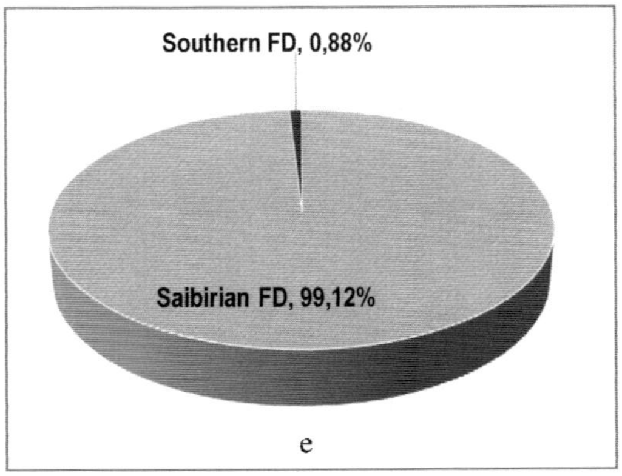

Figure 4. Distribution of ferroalloy production by federal districts: a – ferrosilicon manganese, b – ferrosilicon, c – ferrochrome, d – ferrovanadium, e – ferromolybdenum.

4. Production of certain types of ferroalloys

4.1. Production of manganese alloys

Currently, China accounts for 67% of the global smelting of silicomanganese, 53% – refined manganese alloys, and 48% – high-carbon ferromanganese. China is a leader in the production of electrolytic metal manganese. Its production is 97% of the world level.

Russia's need for manganese ferroalloys is about 600 – 650 thousand tonnes per year. Of the imported manganese ores, mainly ferromanganese is smelted carbonaceous (JSC "Satka Cast Iron Plant" and PJSC "Kosogorsk Metallurgical Plant"), silicomanganese (JSC "ChEMK" and LLC "ZSEMZ"). The dynamics of ferromanganese production in Russia since 2015 has a positive trend, in 2018 there was an increase in indicators by about 14% [4-6, 8-10].

4.2. Production of ferrosilicon

The increase in steel production determines the increase in demand for ferrosilicon. The growth of ferrosilicon consumption in the world will be determined by demand in the Asian region.

Asia continues to play a leading role in the production of ferrosilicon. Along with this, an increase in alloy production is also expected in the CIS countries, the European region and South America. China and Russia are world leaders in the production of ferrosilicon; in China, 66% of the global volume is produced.

The main producers of ferrosilicon in Russia are JSC "ChEMK"; JSC "SZF"; JSC "Kuznetsk Ferroalloys", JSC "Bratsk Ferroalloys Plant". In 2018, the production of silicon alloys in Russia reached more than 890 thousand tonnes, with the export volume of about 50%.

4.3. Production of chromium alloys

A significant amount of chromium alloys produced is primarily high-carbon ferrochrome. The world production capacities for its smelting are distributed as follows: China accounts for 39.4% of world production, South Africa – 30%, Kazakhstan and Russia – 11%, India – 9%.

Russia is one of the leading exporters of ferrochrome and one of the five world leaders. Ferrochrome production capacities account for 6% of the world, and export more than three quarters of the production volume (figure 5).

The largest producers of chromium in the Russian Federation are JSC "ChEMK", JSC "SZF", JSC "KZF" and JSC "Tikhvin Ferroalloy Plant". The dynamics of the production of alloys of ferrochrome and silicochrome is shown in figure 6.

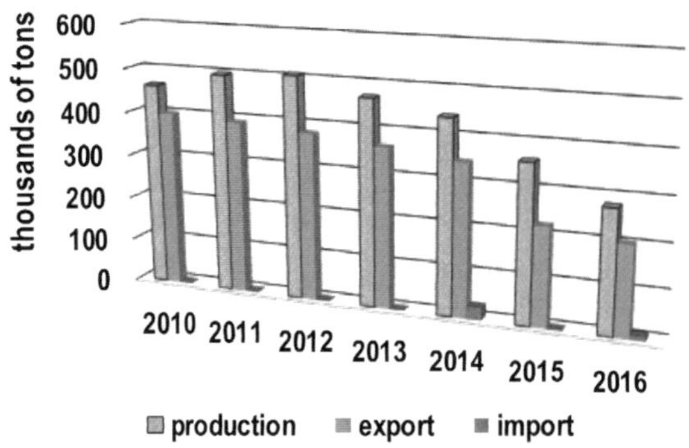

Figure 5. Dynamics of production, export and import of ferrochrome in the Russian Federation.

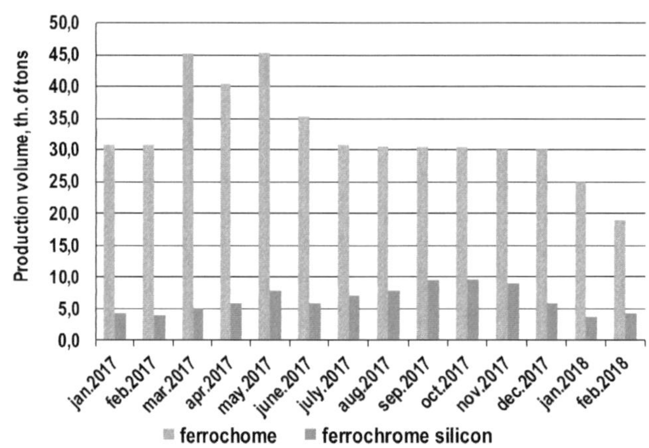

Figure 6. Dynamics of production, export and import of ferrochrome in the Russian Federation.

4.4. Production of small tonnage ferroalloys

The main producers of niobium alloys are Brazil (about 90% of world production) and Canada. In 2014, the production and consumption of ferroniobium was estimated at 83 thousand tonnes; in the past 5 years, the growth has been approximately 4.9% annually.

About 80 thousand tonnes of tungsten-containing products are sold on the world market (data from 2018), of which about 20% of the total ferro-tungsten is accounted for, i.e., about 16 thousand tonnes. The largest producers and exporters of ferro-tungsten are countries such as China (2.8 thousand tonnes), Vietnam (2.2 thousand tonnes), Sweden (0.5 thousand tonnes). The tungsten content is usually from 75 to 85%. According to the results of 2018, the total volume of imports of ferrotungsten in the Russian Federation amounted to 13.3 thousand tonnes, exports – about 880 tonnes [11].

The main producers of ferromolybdenum and ferro-tungsten in the Russian Federation are OJSC "KZF" and LLC "Ryazan Ferroalloys and Ligatures Plant". According to the results of 2018, ferromolybdenum production in Russia amounted to about 6.4 thousand tonnes.

The need for nickel is satisfied mainly due to electrolytic nickel produced at non-ferrous metallurgy enterprises. One of the largest producers of ferronickel is Indonesia, where the total ferronickel smelting capacity in Indonesia exceeded 140 thousand tonnes per year.

Ferronickel production in Russia is carried out at PJSC "Yuzhuralnickel" and LLC "Buruktalsky Nickel Plant".

The global production and consumption of vanadium in the form of ferrovanadium will increase to 140 thousand tonnes by 2020. The main producers of ferrovanadium: China – 49.7% of world production, South Africa – 13.9%, Russia – 7.8%, Europe – 6, 7%, North America – 4.5%. In the Russian Federation, the production of ferrovanadium was established at JSC "EVRAZ Vanadium Tula" and JSC "Chusovsky Metallurgical Plant" [11].

The dynamics of ferrovanadium production in Russia is shown in figure 7.

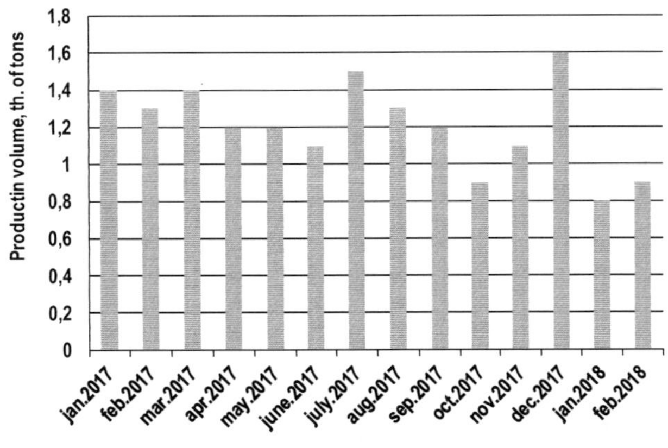

Figure 7. Dynamics of production in Russia.

5. Conclusion

In the context of the aggravation of the struggle for the domestic and world markets of ferroalloys and the strengthening of import substitution of products, the urgent task remains to increase the competitiveness of ferroalloy enterprises in Russia by reducing production costs and improving product quality. The projected increase in steel production by 2020, primarily high-alloyed and low-alloyed steel of the HSLA group, will require a corresponding increase in the output of ferroalloys and ferroalloy raw materials. To achieve these goals, it is necessary, first of all, to solve the issues of energy supply of ferroalloy enterprises and increase the energy efficiency of ferroalloy processing, as well as provision with qualified personnel, the requirements for which are constantly growing.

Acknowledgments

The study was carried out with the financial support of the Russian Federal Property Fund and the Kemerovo Region – Kuzbass in the framework of scientific project No. 20-48-420001/20.

References

[1] Budanov I A 2018 *Steel* **8** 64–66
[2] Order of the Ministry of Industry and Trade of Russia dated March 18, 2009 No. 150 on The development strategy of the metallurgical industry of Russia for the period until 2020
[3] Yuzov O V, Sedykh A M and Petrakova T M 2016 *Proceedings of the XIV International Steelmakers Congress* (Moscow: Elektrostal) pp 21–31
[4] Dashevsky V Ya 2017 *Steel* **8** 20–23
[5] Dashevsky V Ya 2019 *Steel* **1** 16–23
[6] *METALRESEARCH LLC Analytical Group "Metallurgical Research"* http://metalresearch.ru.
[7] *Metals of Russia and the CIS* http://www.metaltorg.ru/
[8] Leont'ev L I 2015 *Steel* **10** 21–25
[9] *Ferroalloy Market. Marketing research of Indexbox Marketing LLC* www.indexbox.ru
[10] *Energy efficiency and environmental friendliness - the future of the global ferroalloy industry (based on the materials of the international ferroalloy congress INFACON-14)* 2015 Steel **9** 18–22
[11] *Iron and steel industry. The current state, problems and prospects for the development of metallurgy* http://emchezgia.ru/ferrosplavy/33_proizvodstvo_Cr.php

IOP Conf. Series: Materials Science and Engineering **866** (2020) 012005 doi:10.1088/1757-899X/866/1/012005

New technology of rails welding

N A Kozyrev[1], R A Shevchenko[1], E V Protopopov[1], S N Krat'ko[2] and V E Khomicheva[1]

[1]Siberian State Industrial University, 42 Kirova Street, Novokuznetsk, 654007, Russia
[2]LLC "RSP-M" Structural unit of Rail Welding Company No. 29 (RSP-29)

E-mail: kozyrev_na@mtsp.sibsiu.ru

Abstract. Today, in the manufacture of long-length rail strings, the obligatory heat treatment is provided after the flash butt welding operation. The basis for this operation is the unsatisfactory mechanical properties of the rail joints immediately after welding. Welding heat cycles at pulsed reflow do not always allow a welded joint with a sufficient level of ductility to be obtained. In this case, the induction heat treatment used at rail welding enterprises in Russia increases the length of the heat-affected zone, which adversely affects the wear resistance of the rails during operation. This paper presents studies in industrial conditions that were previously conducted in laboratory conditions. Studies show the possibility of obtaining the necessary mechanical properties by controlling welding heat cycles without the use of induction heat treatment.

1. Introduction

Welded joints of rail strings after electric contact welding are necessarily subjected to heat treatment at induction plants. At present, the installations UIN-001-100/RT-S and UIN-001-100/RT-P (hereinafter UIN) manufactured by "Magnit M" LLC are used in the Russian Federation. Previously, heat treatment was performed by installation ITT3-250/2.4 with a power source of 250 kW, the power source of the UIN units is 100 kW at the same heating rate of the welded rail joint. The most significant change is the replacement of the air-water mixture used as a quenching medium in the ITT3-250/2.4 units with compressed air in the new generation units. The use of compressed air provided better heat treatment of welded joints, which is especially relevant in connection with the expansion of the use of alloyed steels on domestic railways, including alloyed with chromium. For these steels, quenching cooling with an air-water mixture contributes to the formation of structures with the presence of martensite, which sharply reduces the resistance of rails to fatigue and brittle fracture of rail steel [1].

With all the positive effects of heat treatment [1] using a quenching medium in the form of compressed air, modern induction plants have a number of disadvantages that are associated with:

- the emergence of new heat-affected zones during local heating of welded joints in the conditions of heat treatment,
- unilateral cooling only from the head side, which leads to the deterioration in the straightness of the welded joints after cooling,
- insufficient heating depth of welded joints during heat treatment.

Local heat treatment of the welded joint leads to an increase and the appearance of new heat-affected zones compared to zones in contact welding of rails without heat treatment. An increase in the linear magnitude of the zones after heat treatment in the rail head at a distance of about 40-47 mm from the welded joint and in its base at a distance of 70-75 mm leads to a decrease in the mechanical properties of the welded joint.

Shlatter company [2] proposed a solution to this problem using the butt welding process by flashing the rails on the equipment: the GAA 100 stationary rail welding machine. The process is divided into three stages. The first stage is the heating of the welded surfaces, the second stage is the actual welding, the third stage is the cooling of the welded joint. The first stage of heating, in turn, is divided into three stages: 1- “deliberate flashing”, 2- “pre-heating” and 3- “flashing”. The second stage of the actual welding is upsetting. During the last third stage after welding, heat treatment is applied similar to preheating to reduce the cooling rate after welding – an electric current is transmitted through the welded joint in several pulses. These current pulses are intended only to reduce the cooling rate and to eliminate the formation of quenching structures in the weld metal.

An alternative way to solve the problem, developed at SibSIU and protected by a patent of the Russian Federation, which after upsetting and cooling the welded joint at the moment the required temperature is supposed to maintain the set temperature by passing alternating electric current pulses through the welded joint. The holding temperature is selected based on obtaining a finer-grained structure of the weld metal. The exposure time is determined by the incubation period of the required structure formation and is regulated by the number of current pulses [3].

2. Materials and methods of research

The study of the influence of welding conditions with subsequent isothermal exposure of rail steel samples, produced by passing alternating electric current pulses after welding, on the weld joint quality parameters was carried out on a butt welding machine MSP-6301 in the conditions of the rail welding company LLC “RSP-M” (RSP-29). For the study the samples of full-profile rails of R65 type of DT350 category with 600 mm long were cut out. The rails were welded according to a given mode (table 1).

Table 1. P65 rail welding mode.

Sec. No.	1	2	3	4	5	6	7	8	9	10
S, mm	2	4	3	2.5	2.5	2	1.5	1.5	1.5	1.5
U, %	75	70	55	60	60	65	70	88	88	98
Vu, mm/s	0.6	1.3	0.8	1	1.1	1.2	0.7	0.9	1.1	1.2
Vd, mm/s	0.5	0.6	0.5	0.4	0.4	0.5	0.4	0.3	0.2	
Icor, A	200	300	200	200	250	250	250	300	400	500
Ist, A	300	450	350	350	400	400	460	500	600	700
Irev, A	400	500	400	400	450	450	550	600	700	800
Isc, A	1200	1200	1200	1200	1200	1200	1200	1200	1200	1200

Sups = 14 mm – upsetting path;
Vups = 100 mm/s – upsetting rate;
Si = 6 mm – upsetting under current.

The welding mode mainly consists in controlling the flashing during butt welding, which is carried out by adjusting the set flashing speed depending on the current value of the current. This is done by changing the settings of the beginning of the speed corrections (correction current Icor), the current Ist, which stops the supply, and the current Irev, which gives the command for reverse.

For this, the programmable values of the speed and reflow current, which implement feedback Icor, Ist, Irev, are selected so that the actual average value of the current on the waveform is 2 times less than the short circuit current. This corresponds to the maximum electric power developed in contact between the melted ends of the workpieces. Inadmissible current deviations in the form of spasmodic changes in its magnitude are eliminated due to high-speed feedback. The same feedback stabilizes the current value of the welding current.

To obtain information about the heat effect on the metal structure, the temperature measurements in the heat-affected zone during welding were performed using the HotFind-D thermal imager. The

HotFind-D thermal imager allows you to measure temperatures up to 1500°C. The thermal imager is equipped with an uncooled microbolometric matrix in the focal plane of the lens with a resolution of 160×120 pixels. The video image of thermograms is transmitted to a PC using a video capture board of an analog signal in NTSC format with a frequency of 60 Hz. The thermal imager was mounted on a tripod at a distance of 1 meter from the welded joint of the rails.

To search for optimal isothermal exposure conditions in the first series of experiments, a full factorial experiment $N = 2^k$ was carried out (table 2). The studied parameters were: X_1 – cooling time after upsetting (characterized by the cooling rate (degree of austenite overcooling) and the temperature T_1 to which cooling occurs); X_2 – heating time (characterized by temperature T_2, to which heating occurs); X_3 – cooling time after heating (characterized by the temperature T_1 to which cooling occurs); X_4 is the number of heating pulses (characterized by the incubation period of austenite into perlite transformation).

The exposure time (X_1) must be selected so that the weld joint cools down to the temperature at which the required weld metal structure is formed. The current transmission pulses were set at a certain interval. The pulse duration (X_2) is determined by the temperature of the welded joint, which should not rise above the temperatures required for the formation of the necessary structure. The duration of the interval (X_3) is selected so that the temperature of the weld joint does not fall below the temperature values at which the necessary structure of the weld metal is formed. The number of pulses (X_4) sets the time during which the average temperature of the weld joint is maintained, which is necessary for the formation of the required structure during welding.

Table 2. Matrix of experiment planning $N = 2 \cdot 3^k$

Sample No.	Factor values		
	X_1, s	X_2, s	X_3, s
1	60	2	30
2	160	2	30
3	60	6	30
4	160	6	30
5	60	2	15
6	160	2	15
7	60	6	15
8	160	6	15

For comparison, the sample No. 0 was also welded without isothermal exposure.
The number of pulses (X_4), to reduce the number of experiments, was set equal to 4 in all modes of isothermal exposure.

The isothermal exposure control after welding is performed using a personal computer, changing the program of the industrial controller SIMATIC S7-300. Using the Simatic Step 7 software, a program was written to control isothermal time modes. Timers were used to set the necessary time parameters (figure 1a). To activate the welding transformer, a trigger was used, to the inputs S and R of which the contacts, activated by timers, are connected in series (figure 1b). Subsequently, the heat-treated joints were tested for a three-point static bend according to STO RZD 1.08.002-2009 "Railway rails, welded by electric contact method". Static bending tests were carried out on the press of PMS-320 type. The control samples were tested after welding and removal of burr without additional processing joints. In tests for static transverse bending, the control sample had a length of at least 1200 mm with a welded joint in the middle.

The load was applied in the middle of the span of the control sample in the place of the welded joint with a distance between the supports of 1 m. The test of the control sample was performed with the application of a load on the head (figure 2). The test results are the values of the force arising during bending P_{bend}, kN and the values of the deflection f_{def}, mm at which the control sample is destroyed, or the maximum values of these indicators if the sample did not break during the test.

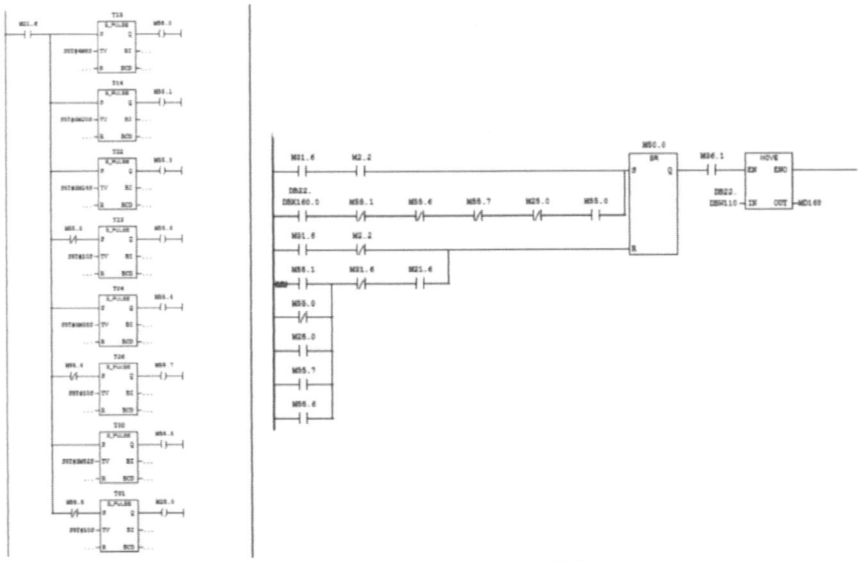

(a) (b)

Figure 1. The program responsible for the isothermal exposure.

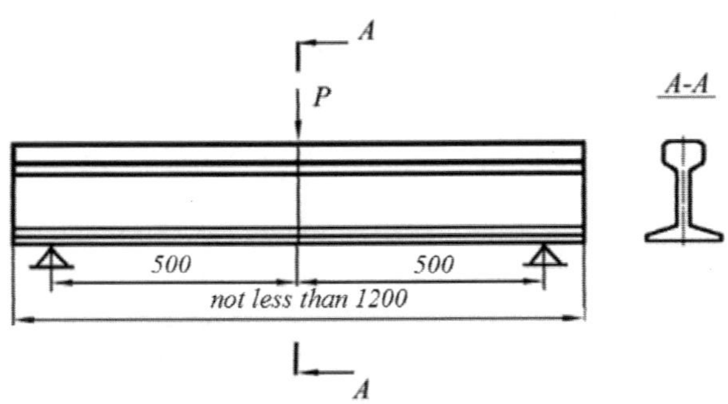

Figure 2. Test pattern for static lateral bending.

3. Results and discussion

As a result of the experiment, samples 3 (P_{bend} = 2364 kN, f_{def} = 35.4 mm) and 8 (P_{bend} = 2244 kN, f_{def} = 31.2 mm) have higher test values (P_{bend} = 2000 kN, f_{def} = 27 mm) than the requirements of STO RZD 1.08.002-2009. According to the data, the regression models were obtained for the output force parameters of bending and deflection: $P_{bend} = 1926.5 - 2.05\ X_1 + 64.38\ X_2 + 3.2\ X_3$; $R_2 = 0.61$, $f_{def} = 15.72 - 0.04\ X_1 + 2.37\ X_2 + 0.07\ X_3$; $R_2 = 0.51$. Table 3 shows the test results.

Table 3. Test results on static three-point bending.

Sample No.	Factor values	
	P_{bend}, kN	f_{def}, mm
1	2002	17.4
2	1882	17.9
3	2364	35.4
4	1970	20.1
5	2107	23.4
6	1616	11
7	2059	20.9
8	2244	31.2
0	2179	24.4

At the same time, having examined the cooling process (figures 3 and 4) of the best results, it can be concluded that the formation of favorable mechanical properties of the welded joint occurred to some extent due to a slowdown in the cooling rate, since the modes of isothermal exposure were not fully implemented.

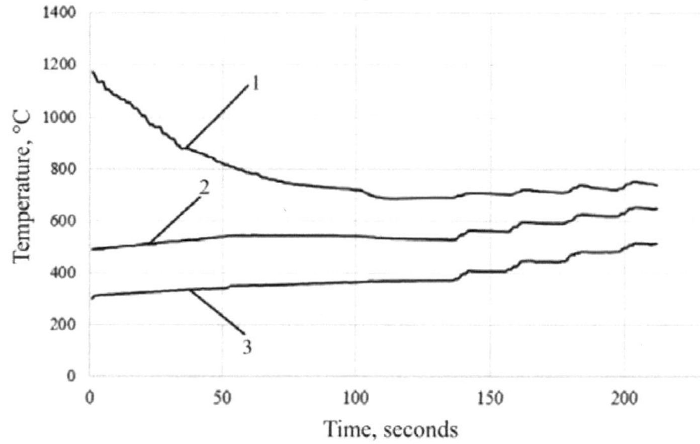

Figure 3. Pattern of cooling after welding of sample 9: 1 – the seam center, 2 – 20 mm from the seam center, 3 – 28 mm from the seam center.

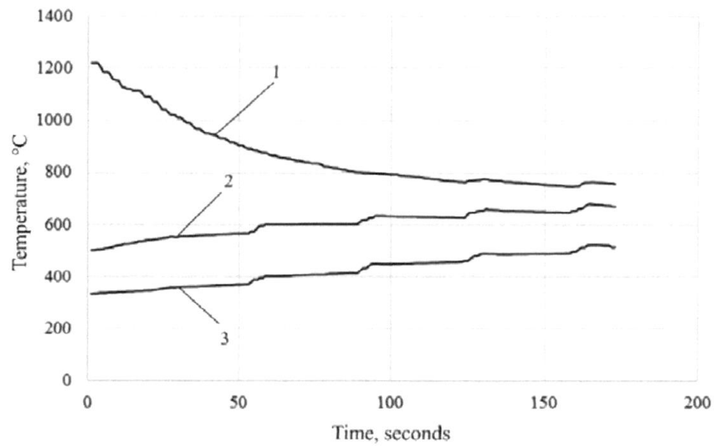

Figure 4. Pattern of cooling after welding of sample 10: 1 – the seam center, 2 – 20 mm from the seam center, 3 – 28 mm from the seam center.

To confirm the results, the second series of experiments was carried out. The welding was repeated in the modes that showed the best test results, and the isothermal holding mode was selected that fully satisfies the temperature conditions. The modes and results of the repeated series of experiments are given in table 4.

The results of the second series of experiments confirmed the reproducibility of parallel experiments. In the case of sample 9 (8), the instability of the rail welding machine played a role, since this sample was welded first after a long downtime of the equipment. One of the factors influencing instability is insufficient heating of the hydraulic fluid, which directly affects the parameters of the welding process (the movement speed of the movable bed during flashing and upsetting).

IOP Conf. Series: Materials Science and Engineering **866** (2020) 012005 doi:10.1088/1757-899X/866/1/012005

Table 4. The results of the second series of experiments.

Sample No.	Factor values			Test results for static bend	
	X_1, s	X_2, s	X_3, s	P_{bend}, kN	f_{def}, mm
9 (8*)	160	6	15	1962	21.5
10 (3*)	60	6	30	2276	34.2
11	200	4	10	2493	40.2

*Indicates the number of the repeated sample.

The number of pulses (X_4), to reduce the number of experiments, was set equal to 4 in all modes of isothermal exposure.

4. Conclusion

Based on the conducted experiments a method of contact butt welding was developed and protected by the patent of the Russian Federation [4]. It makes possible to obtain a welded joint of products from rail steel with the necessary technical requirements of STO RZD 1.08.002-2009 surpassing the mechanical properties of welded joints of rails of P65 type of DT350 category.

Acknowledgments

The study was carried out with the financial support of the Russian Foundation for Basic Research and Kemerovo Region in the framework of scientific project No. 20-48-420003 r_a "Development of physico-chemical and technological fundamentals of creating fundamentally a new method for welding differentially heat-strengthened railway rails".

References

[1] Rezanov V A, Fedin V M et al 2013 *Vestnik VNIIZhT* **2** 28–34

[2] Weingrill L, Krutzler J and Enzinger N 2016 *Materials Science Forum* vol 879 pp 2088–93

[3] Kozyrev N A, Shevchenko R A et al 2018 *Ferrous Metallurgy. Bulletin of Scientific, Technical and Economic Information* 8(1424) 50–57

[4] Protopopov E V, Kozyrev N A et al Pat. 2641586 RF, IPC B 23 K 11/04

IOP Conf. Series: Materials Science and Engineering **866** (2020) 012006 doi:10.1088/1757-899X/866/1/012006

Formation of the iron ore pellets structure with differentiated properties

V M Pavlovets

Siberian State Industrial University, 42 Kirova Street, Novokuznetsk, 654007, Russia

E-mail: pawlowets.victor@yandex.ru

Abstract. The probable mechanism of the porosity appearance in pellets formed by the technology of forced nucleation is analyzed. A review on the problem of increasing the reactivity of a sintered metallurgical product is carried out. The conditions for optimizing the pore structure of the agglomerated dispersed iron-containing materials are formulated.

1. Introduction

Agglomeration of ore materials is necessary for obtaining a solid lump product, which is a raw material for the extraction of primary metal (cast iron) in blast furnaces or for metallization of raw materials in mine units [1, 2]. To form the porosity character of the agglomerated products in various branches of technology (metallurgy, construction and refractory production), various technical means and methods are used [3]. In metallurgy, pore-forming additives must have high surface-active properties and provide the necessary cold and hot strength at a minimum consumption of binder material. One of the structural problems of agglomerated metallurgical raw materials is the diffusion difficulties in the central part of the piece, due to kinetic constraints arising during transfer of gases from the surface, for example, pellets to their center. With such structure, at the stage of pellets drying, the central part of the pellet contains a certain amount of moisture; upon firing, a zonal structure is formed with underfiring and insufficient amount of hardening melt; during recovery, a long metallization is required [1, 2].

One of the ways to overcome these drawbacks is to obtain pellets using forced nucleation (nucleation by spraying and after-pelletizing NSA) technology, in which the germinal part and its pore structure are formed by heat-spraying a wet mixture onto the bottom crust of a pelletizer [1, 2, 4]. The scheme for the production of pellets according to the NSA technology is shown in figure 1.

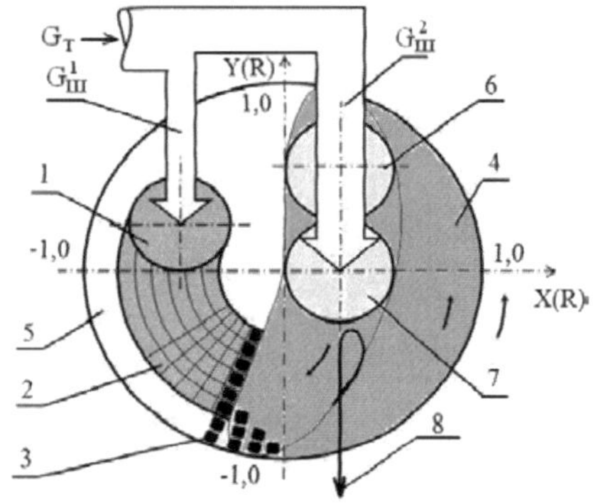

Figure 1. The scheme for producing pellets by forced nucleation: 1 – spraying area (sprayed layer); 2 – zone of nucleation; 3 – embryos; 4, 5 – working and idle zones of a pelletizer; 6, 7 – areas of moistening and after-pelletizing; 8 – pellets.

IOP Conf. Series: Materials Science and Engineering **866** (2020) 012006 doi:10.1088/1757-899X/866/1/012006

Formation of pellets according to the simplified NSA scheme begins with forced nucleation by spraying a moist mixture by compressed air in the blank area of the plate [2]. The structural diagram of such pellets is shown in figure 2.

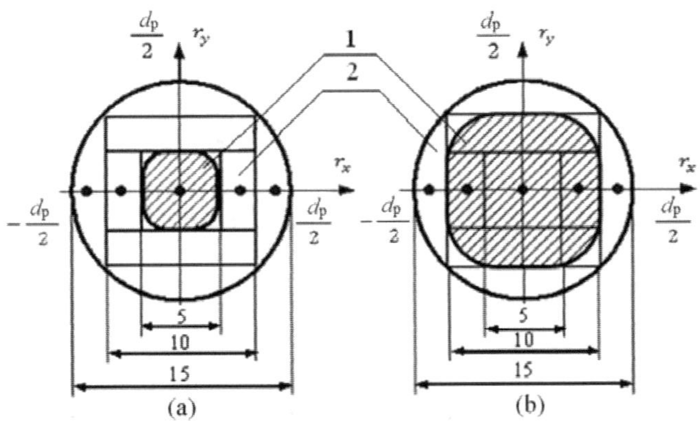

Figure 2. The structure of the pellets formed by the NSA technology: a – scheme of the pellet with the embryo size a_{ES} = 5 mm; b – also with an embryo size a_{ES} = 10 mm. The shaded area 1 belongs to the embryos; the unshaded area 2 belongs to the shell of the pellets.

There is a scheme for the pellets production in which the wet mixture is sprayed onto the seeds obtained in the rerolling mode (nucleation, pelletizing, spraying NPS) [2]. The interaction scheme of the reducing gas with the ore skeleton of iron ore pellets obtained by the NSA and NPS technologies is shown in figure 3.

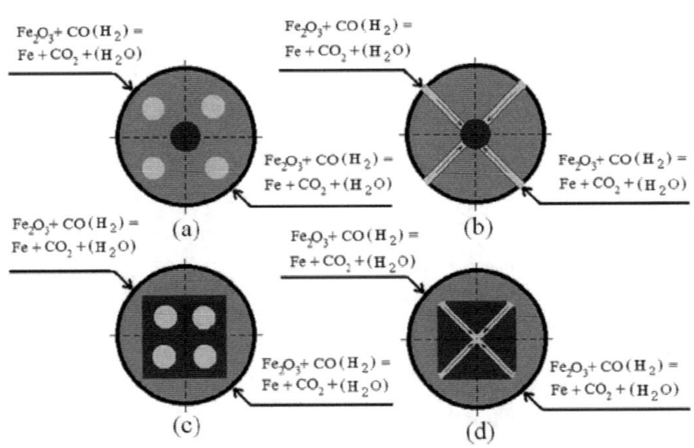

Figure 3. Scheme of the interaction of the reducing gas with the ore skeleton of iron ore pellets obtained by NSA and NPS technologies – a, b and NSA – c, d: in position: a, c – scheme of pellets with closed pores; b, d – scheme of pellets with open pores.

These technologies have been successfully tested in laboratory conditions and have shown high practical efficiency. However, the laws governing the formation of porosity of iron ore pellets and many other processes associated with the new technology remain poorly understood.

The aim of this work was to study the mechanism of nuclei porosity formation in the technology for the production of pellets based on the technique of forced nucleation.

2. Description of the experimental setup and methods of research

The experimental setup for video shooting the process of wet charge spraying onto a contrasting surface is shown in figure 4.

The sprayed charge with a moisture content of 8.4% contained iron ore concentrate of the Teysk deposit (d_{ch} = 0.068 mm) and 1% of bentonite. A wet mixture with a moisture content of 8.4% was sprayed by the compressed air at a pressure of 0.2 MPa and a flow rate of 0.6 m^3/min. In each zone of the SL limited by a relative diameter δ equal to 0 ± 0.2, 10–15 samples were taken with samplers (cutting ring) with a diameter of 10 mm. The compressive strength of wet – P_w and dry – P_d, kPa, samples was determined according to GOST 17245-79 and 26447-85. Some samples were used to determine the density ρ_w, kg/m^3, and the moisture content of the samples W, %. The relative distance L/d $_{sa}$ = 1.5 was used in the work; 2.5; 5.0; 10.0; 15.0; 20.0 (where L is the distance between the nozzle SA and ShG, m; d $_{sa}$ is the diameter of the nozzle SA, m). The angle of attack of the SA to the skull was 90 degrees. Using the nomogram developed in [2] and the experimental conditions, we determined the pressure of the internal combustion engine depending on the parameters of compressed air and the characteristics of the jet apparatus.

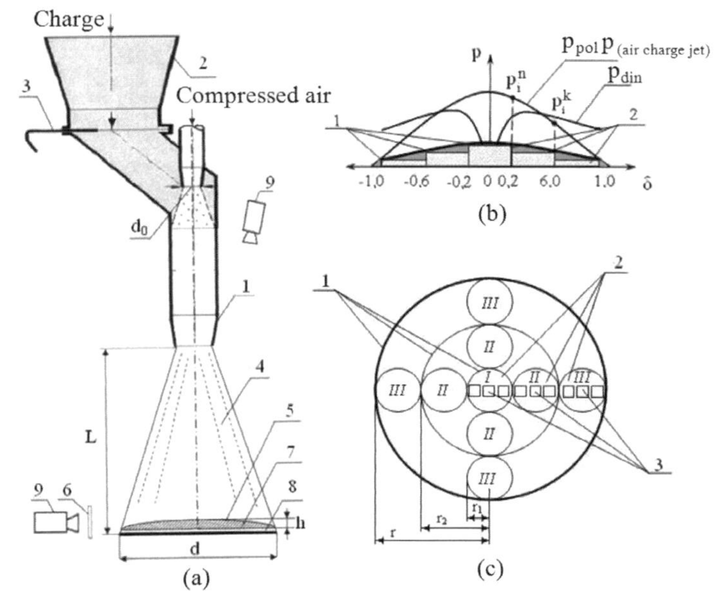

Figure 4. Scheme of the experimental setup for video recording of the spraying process of wet charge onto a contrasting surface – a, the field of jet pressures on the surface of the sprayed layer – b and the sampling scheme in zones I - III – c. In position a: 1 – inkjet apparatus; 2 – charge hopper; 3 – gate; 4 – air charge jet; 5 – sprayed charge layer; 6 – protective glass screen; 7 – contrast basis; 8 – support; 9 – video cameras. In position b and c: 1 – the boundaries of the zones; 2 – location of samples; 3 – places of photography.

3. Research results and discussion of the porogenesis mechanism

The analysis of the samples macrostructure (figure 5) showed that in each SL zone there are markedly pronounced structural features.

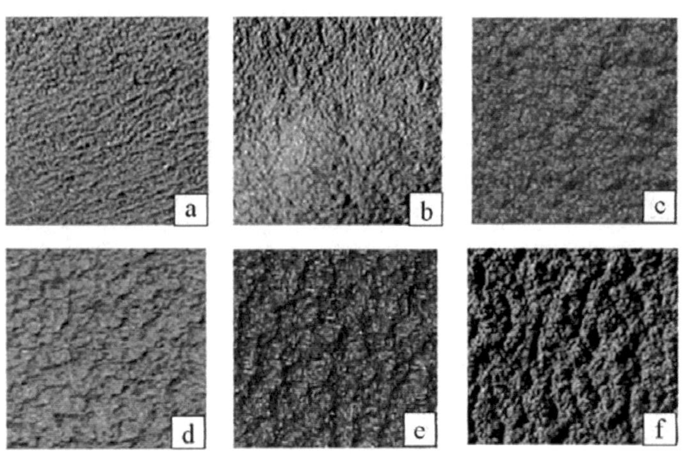

Figure 5. The macrostructure of the charge sprayed layer: a – charge skull; b – central zone NS δ=0; β=90°; c – δ=0,2; β=90°; d – δ=0,5; β=90°; e – δ=0,5; β=45°; f – δ=0.6; β=45°.

A feature of the macrostructure is the structural cavitations of NS, which are concentric, slightly sinuous channels formed by charge bursts and located along a circular path around the axis of the circular NS. This feature of sprayed coatings is noted by the authors in [4, 5]. As the parameters of the SL macrostructure, we used the relative value of the structural cavitations of the SL θ_{ho} and the relative number of structural cavities θ_N, amount/^{m2} (1/m^2) on its surface.

The relative value of the structural recesses of the SL θ_{ho} on its surface was determined by the expression: $\theta_{ho} = h_o/h$, where h_o is the average value of the structural cavitations, mm; h is the average height of the sprayed layer on its axis, mm. The relative number of structural grooves θ_N was calculated by the expression: $\theta_N = N/f_{sl}$, where N is the numerical value of the structural cavitations N, determined by the number of concentric shadow channels in each zone of SL; f_{sl} – the area of the sprayed layer with a diameter of d, m^2.

The parameters θ_{ho} and θ_N were estimated depending on the pressure of the NSA, the relative distance L/d $_{sa}$, and the humidity of the sprayed charge (figure 6).

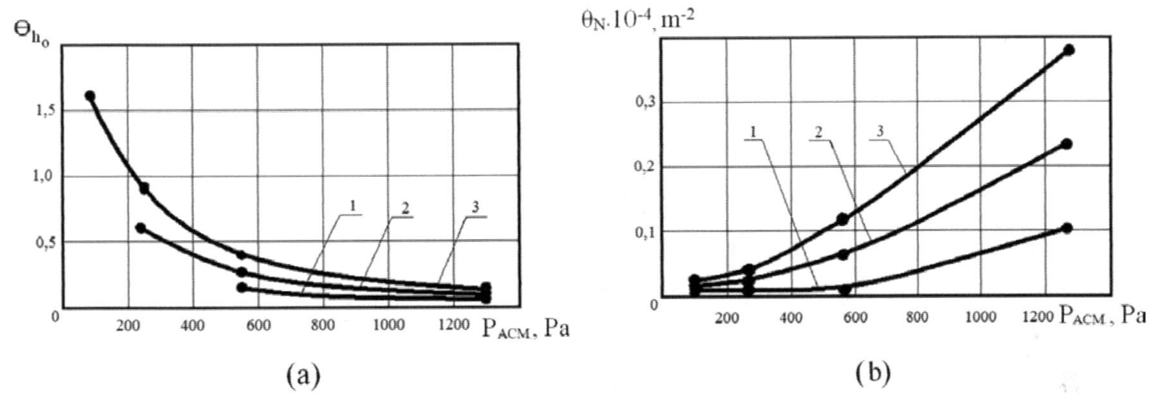

Figure 6. Dependences of the relative magnitude of structural recesses – a and the relative number of structural recesses of the SL mixture – b on the pressure of ACM: mixture the moisture content: 1 – 5.5 %; 2 – 7.5%; 3 – 9.5%.

It was found that in the general case, the parameter θ_{ho} sharply decreases with an increase in the pressure of the ACM to 800 Pa, after which there is a slow decrease in its value (figure 6). A large role on the parameter θ_{ho} is exerted by the moisture content of the sprayed charge. When W_c = 5.5% and the pressure of the ACM less than 800-1000 Pa, low flows and small structural cavitations (h_o <0.1-0.2 mm), inaccessible for instrumental measurement, are formed on the surface. At W_c = 7.5 and 9.5%, large charge flows of sufficient height and structural cavitations are formed, which are accessible for visual observation and instrumental measurement. The parameter θ_N characterizes the number of concentric structural recesses on the surface of the sprayed charge layer, depending on the P_{acm} Up to a pressure of 500-600 Pa, the parameter θ_N grows at a slower rate than at P_{acm} > 600 Pa (at W_c = 7.5 and 9.5%). This is explained by the fact that, with an increase in P_{acm}, the growth rate of the number of structural depressions significantly exceeds the increase in the area of the deposited layer f_{sl} and its diameter d. Moreover, with an increase in the charge moisture from 5.5 to 9.5%, the θ_N parameter increases almost 4 times at P_{ACM} = 1280 Pa. During the formation of SL obtained at a moisture content of the mixture W_c = 5.5% and a pressure of ACM less than 800-1000 Pa, the height of the structural recesses is practically impossible to measure and it was arbitrarily assumed to be 0.1 mm.

In the central part of the SL, the ACM pressure is maximal; in zone III, the ACM pressure is minimal. The dynamic pressure on the axis of the SL at an angle of incidence of the ACM of 90 degrees is zero, after which it increases to a maximum value at about half the radius of the SL (zone II), and then gradually decreases to a minimum in zone III [6] (figure 4).

A moving flow of gases and particles experiences resistance from the side of the surface of the SL due to friction when moving. For this reason, the surface zone of the NS experiences shear loads [2]. In the center of zone I of SL, with an ACM average pressure (in the range of 1000-2000 Pa), the maximum pressure is exerted on the charge, which squeezes moisture from the sprayed charge onto the surface of the SL, especially at W_c> 7.5%.

As a result, a mobile charge pulp is formed with a moisture content W = (1.25-1.50) W_c, which impregnates the charge and actively participates in structure formation. In zone II of the SL, the

IOP Conf. Series: Materials Science and Engineering **866** (2020) 012006 doi:10.1088/1757-899X/866/1/012006

dynamic pressure of the carrier gas is maximal, and the static component of the total pressure is minimal (figure 4). In zone III of the SL, all components of the total pressure of the ACM have a minimum value. The sprayed layer of charge in this zone consists mainly of their loosely coupled charge conglomerates of low humidity W = (0.85-0.95) W_c, which are formed from partially destroyed crest ridges. The cross-sectional microstructures in different zones of the SL, annealed at 800 °C, are presented in figure 7.

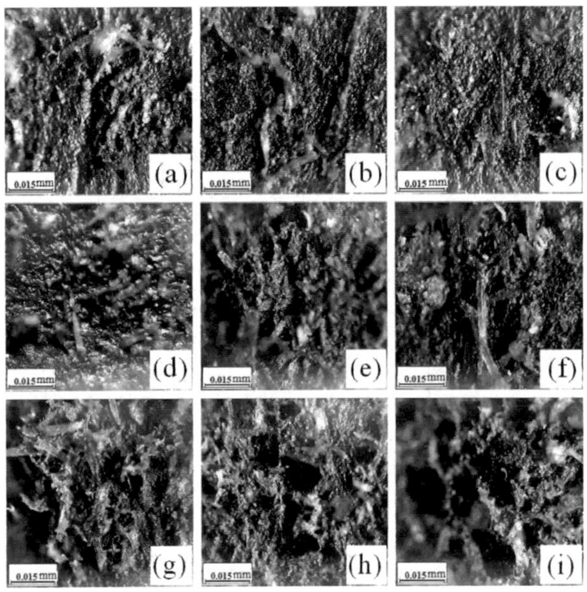

Figure 7. The microstructure of fired SL (T=800 °C, τ= 20 min); W= 10 %; Magnification 120^x: Positions: a, b, c – zone 1; d, e, f – zone 2; g, h, i – zone 3; Coordinates : h = 0.5, δ = 0.15 (a); h = 0.7, δ = 0.1 (b); h = 0.3, δ = 0,15 (c) h = 0.3, δ= 0.5 (d); h = 0.2: δ = 0.4 (e); h = 0.2, δ = 0.5 (f) h = 0.2, δ = 0.65 (g); h = 0.1, δ = 0.7 (h); h = 0.2, δ = 0.9 (i).

Analysis of the SL microstructure in different zones of the thin section showed the presence of extended, small, sinuous pore channels of various depths and densities, both in the SL height and in its diameter. Moreover, the channels are located mainly vertically or with a slight inclination towards the attack of the jet. Since the entire SL array experiences power loads from the dynamic pressure of the air flow, a characteristic inclination of the pores towards the jet attack appears.

In zone 2 (d = 0.2-0.6), the width and depth of the pore channels increase, the length decreases significantly, the tortuosity and density of the channels increase. In this zone, a small number of closed-type pores of irregular shape appear. In zone 3 (d = 0.6–1.0), the porosity of SL is noticeably higher, the number of channel-type pores sharply decreases, and they are observed only at the beginning of the zone.

The mechanism of porosity formation is based on the overlapping of charge bursts at each other under the dynamic pressure of the ACM. The air cavity (pore) is most likely to form at the base of the structural cavity, where the strongest adhesion of the charge flow to the base is. The forces that prevent separation of the influx from the base are capillary and interparticle interaction forces, viscous friction forces [1, 2]. The influx crest, which is under the ACM dynamic pressure, on the contrary, has higher mobility and deformation due to its special geometric shape and therefore can block the void in the area inaccessible to the pressure of the IDL.

This mechanism is probabilistic, since fixing it in a dynamic state is quite problematic. If charge bursts do not have sufficient mobility necessary for the development of the first mechanism of pore formation, then a mechanism for the formation of voids by mechanical overlapping of structural recesses with a sprayed charge is possible. Since the deposition mechanism is layered in nature, in the process of SL compaction in the depth of the layer, pore formation can continue along the path of decreasing pore size, elongation, spheroidization, etc. In view of the complexity of these processes that occur in a dynamic state and in a closed system, the described mechanisms are probabilistic character.

IOP Conf. Series: Materials Science and Engineering **866** (2020) 012006 doi:10.1088/1757-899X/866/1/012006

4. Conclusion

As a result of the studies, a probable mechanism of pore formation in the structure of iron ore pellets was formulated and methods for influencing the nature of the porosity of the embryos using gas-jet spraying of a wet charge on the surface of a charge skull of a plate pelletizer were shown. It was established that the porosity of the germinal centers can be controlled by the selection of moisture, fractional characteristics of the charge and the dynamic parameters of the jet apparatus.

References

[1] Pavlovets V M 2014 *Pellets in the Technology of Extraction of Metals from Ores* (Novokuznetsk: SibSIU) p 345
[2] Pavlovets V M M 2016 Expansion of the Functional Capabilities of Units for the Preparation of Iron Ore Raw Materials for Metallurgical Smelting (Novokuznetsk: SibGIU) p 373
[3] Pavlovets V M 2015 *Refractory and Thermal Insulation Materials* (Novokuznetsk: SibSIU) p 334
[4] Gnezdovets A G and Kalita V I 2007 Physics and Chemistry of Materials Processing **1** 30–39
[5] Kalita V I and Komlev D I 2003 Metals **6** 30–37
[6] Pavlovets V M 2009 *Applied Mechanics of Liquids and Gas* (Novokuznetsk: SibSIU) p 214

IOP Conf. Series: Materials Science and Engineering **866** (2020) 012007 doi:10.1088/1757-899X/866/1/012007

The use of titanium diboride to protect the cathodes of aluminum electrolysis cells

G V Galevskiy[1], V V Rudneva[1], S G Galevskiy[2], M Komroni[3] and O I Gordiyevsky[4]

[1]Siberian State Industrial University, 42 Kirova street, Novokuznetsk, 654007, Russia
[2]St. Petersburg Mining University, 2, 21st Line, St Petersburg, 199106, Russia
[3]Ministry of Industry and New Technologies
[4]JSC "RUSAL – Novokuznetsk", Novokuznetsk, 654007, Russia

E-mail: kafcmet@sibsiu.ru

Abstract. The analysis of modern technical solutions for the protection of the cathodes of electrolysis cell was carried out. The technological variant of obtaining protective coatings based on nanocrystalline titanium diboride is substantiated and proposed. A comparison of the basic and expected indicators of aluminum production is made using cathodes with protective coatings. A technical solution has been developed to organize the production of nanocrystalline titanium diboride. Its main technical and economic indicators are substantiated.

1. Introduction

Modern metallurgy provides the world economic system with a variety of metal products of both mass and special purposes. According to the World Steel Association, in the structure of metal products consumption 94% is accounted for iron alloys and non-ferrous metals – 5%, 1% is made up of a diverse group of metal-containing materials with a special set of properties.

An important place in this group is occupied by borides of metals of subgroups of titanium, vanadium and chromium, materials and alloys based on them, which, thanks to a unique combination of practically significant properties, are used in machine, aircraft and rocket science to solve applied engineering, technical and production problems requiring high temperature, superhard, heat-resistant, heat-resistant, wear-resistant structural, refractory, surfacing materials and protective coatings capable of working in extreme conditions.

This group includes titanium diboride TiB_2, investigated and put into circulation by the scientific school of the famous Russian material scientist G.V. Samsonov more than 50 years ago and still in demand. At the same time, applied interest in the production and use of titanium diboride is constantly growing and reflects the trend of the transition from the use of coarse-grained titanium diboride to micro- and nanocrystalline. This is due to the desire of scientists and technologists-practitioners to achieve a qualitatively new level of operational properties of materials and coatings based on it.

In this regard, the study and technological implementation of boride formation processes during the plasma metallurgical processing of titanium-boron-containing raw materials is an important scientific and practical task of great importance for the development of domestic titanium metallurgy and its multifunctional compounds, as well as the effective solution of innovative problems of applied materials science.

2. Wetted coating materials for cathodes of electrolysis cells

Aluminum alloys are widely used in many engineering industries due to their low density and relatively high strength. The world production of primary aluminum is constantly growing, currently reaching a volume of about 47 million tonnes/year [1]. Commercial aluminum is obtained by electrolysis of a fluoride cryolite-alumina melt containing dissolved alumina Al_2O_3 at a temperature of 1213-1243 K.

The process is implemented in horizontal cells with carbon graphite anodes and cathodes. At the same time, molten aluminum is the real cathode in the electrolytic baths, under whose layer there is a carbon-graphite lining with a life of 5-8 years. The main disadvantage of such a hearth is the non-wettability of molten aluminum. Therefore, between the bottom and liquid aluminum, a thin layer of electrolyte accumulates, which facilitates the penetration of sodium into the crystal lattice of the carbon-containing materials of the bottom and its destruction.

In this regard, over the past 20 years, the volume of research and technological proposals has been expanding in world practice aimed at selection of materials for cathode lining, formation of wettable coatings on it, or manufacturing bulk products of its components [1-6].

The cathode material must satisfy a number of requirements, namely, it must be resistant to molten aluminum and electrolyte, highly conductive, strong enough, wetted by aluminum melt well (i.e., have high adhesion to it), and its linear wear rate should not exceed 3-5 mm per year. Such a set of properties can be realized only in a composite material having a functional basis and phase additives that perform various purposes. According to [2], borides and carbides of refractory metals can be used as part of a functional framework.

Currently, titanium diboride TiB_2 is recognized as the most effective functional material for wettable cathodes of cells. This is confirmed by the results of large-scale industrial experiments conducted at different times by the Great Lakes companies, Reynolds Metal (USA), Komalko (Australia), Shenyang Sheng IED (China) [2]. Various options for cathodic coatings were studied: hot-pressed tiles made of TiB_2, a composite coating with a binder (TiB_2 + graphite powder + resin/pitch; 30-60% TiB_2 + 40-60% anthracite + 5-20% graphite powder + 5-20% pitch; TiB_2 + resin/pitch, TiB_2 + Al_2O_3 + H_2O). Coatings were applied to the hearth with tile lining, pouring and compaction with vibration, spraying, staining.

In all cases, there is a decrease in the cathodic voltage drop (up to 15-30 mV), an increase in current efficiency (by 1-2%), an improvement in the stability of the cell energy parameters, a decrease in damage to the cathode blocks, and the possibility of reducing the interpolar distance (by about 1 cm), technological preference for the use of TiB_2 + binder. Indeed, insufficient heat resistance, the cost of manufacturing monolithic products and, most importantly, the high cost of marketable diboride powder allow it to be used only in composites. The proposed thin coatings based on TiB_2 wear out quickly and are therefore ineffective, and thick coatings or bulk products are economically disadvantageous even with a diboride content reduced to 30-40%. An even lower content does not provide reliable wetting of the composite with aluminum. Partially, these shortcomings are eliminated by the use of unsintered heterophasic powder composites based on TiB_2 with fillers, which should also be resistant to liquid aluminum (as a rule, various forms of carbon, as well as corundum). A carbon or alumina binder during heat treatment in the manufacture of cathode products fastens the phase components of the material into a monolithic solid, forming a composite, the wettability of which is achieved by aluminum due to a certain volume content of TiB_2.

In [7], one of the technological options for the preparation and use of a coating based on TiB_2 wetted by aluminum was described. To obtain such a coating, an aqueous suspension is prepared containing 68-70% solid (90% TiB_2, 10% Al_2O_3). The physical and chemical bond between colloidal Al_2O_3 particles and TiB_2 particles in suspension leads to the formation of a viscoelastic, jelly-like state of the material. Such material does not emit water and behaves as a solid after drying.

The suspension is applied by spraying or staining with intermediate drying by air after applying each layer. The total drying time is 24 hours. A coating with a thickness of 1.0-2.0 mm provides wetting of the cathode with aluminum, has a high resistance to sodium penetration, at the same time combines sufficient hardness, bending strength, wear resistance, adhesion to the base, helps to reduce the cathodic voltage drop and increase the cathode current output of aluminum.

Thus, TiB_2 coating cathode protection is a powerful reserve of energy saving in modern aluminum production, estimated at 10%. This indicates the need for further development of the technological base of its production. The main methods for producing TiB_2 for wettable cathode coatings are self-propagating high-temperature and furnace synthesis. However, these methods, with the relative simplicity of the technological solution, are inefficient and make it possible to obtain TiB_2 in the form of a sufficiently large powder with particles in the size range of 5–10 μm. There is reason to believe that the introduction of TiB_2 into the suspension in the form of a finer powder with a particle size smaller or comparable with the particle size of Al_2O_3 (0.1 - 1 μm) will enhance the physicomechanical and protective properties of the coating.

3. Protective coatings of cell cathodes

In the foreign practice of aluminum production, materials for protective cathode coatings of cells are supplied by MOLTEK company and have the TINOR A, TINOR M trademarks and thickened TINOR [2]. RUSAL also shows a certain technological interest in the creation and industrial use of wettable cathodes. In 2016, its Engineering and Technology Center, together with Energoprom Group, the largest Russian producer of carbon-graphite materials, began production tests of cells with protective coatings of cathodes based on the TiB_2 + pitch composition in the conditions of RUSAL-Krasnoyarsk JSC. In this regard, an assessment of the prospects for the development and implementation of a wettable cathode technology within the company was made, some results of which are shown in table 1.

The assessment was carried out for the conditions of 2016 from the assumption of a decrease in the interpolar distance by 1 cm, an increase in the cathodic aluminum current output by 1%, a decrease in the voltage drop in the aluminum-hearth contact by 50 mV, a coating thickness of 8 mm, and a specific TiB_2 consumption of 0.26 kg/t Al, an increase in the average life of the cell from 1625 to 2555 days with a unit cost of overhaul of 2500 rubles/t Al.

The annual demand in titanium diboride at one aluminum plant, for example, Khakassky with one extra-long series of electrolysis with a voltage 1600 V, current 350 kA, 336 installed cells with capacity 280 000 tonnes of aluminum per year, is 72 tonnes.

A technical proposal has been developed for aluminum-producing enterprises to organize their own production of nanocrystalline titanium diboride as the main component of the cathode coatings of aluminum electrolysis cells, including the technological process for producing titanium diboride, a complex of basic and auxiliary equipment, and substantiation of the main technical and economic indicators of production. The hardware-technological scheme for producing nanocrystalline titanium diboride is shown in figure 1.

Table 1. Basic and expected indicators of aluminum production by RUSAL (Russia) with the use of cathodes UG and UG - TiB_2.

Indicators of aluminum production	UG cathodes	UG - TiB_2 cathodes
Production Al, t/year	3724000	3724000
Specific power consumption, kW·h/t Al	14000	12500
Energy loss reduction, kW·h/t Al	-	1500
Annual energy consumption, kW·h/t Al	52176000000	46550000000
Energy saving, kW·h/t	-	5586000000
Monetary savings, $/year	-	123203000
Equivalent production Al, t/year	-	44700
Average cathode life , day	1625	2555
Unit repair costs, rub/t Al	2500	2500
Monetary savings, $/year	-	99306000
Total cost in terms of money, $/year	-	225120000

Specific consumption TiB_2, kg/t Al	-	0.26
Need in TiB_2, t/year	-	968
Reasonable price TiB_2, \$/kg	-	230

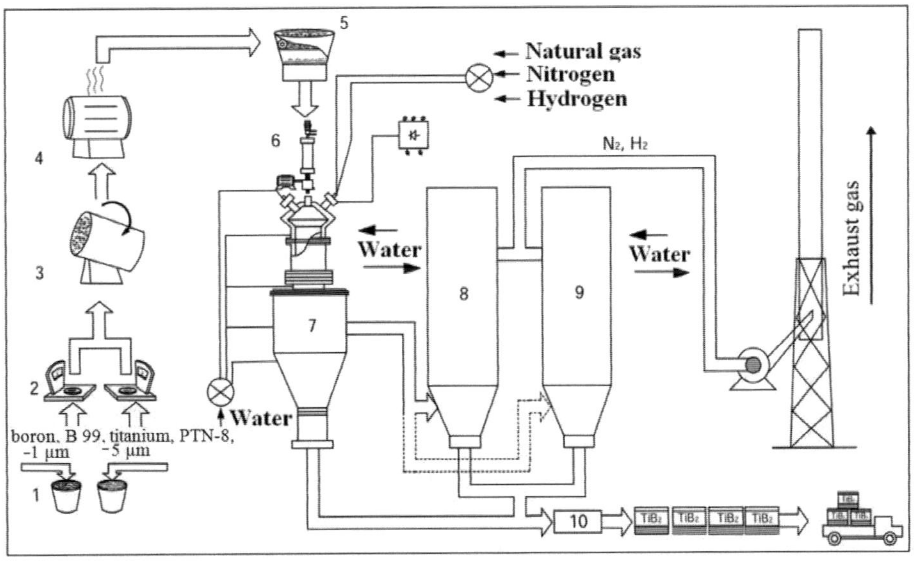

Figure 1. Process flow diagram for producing nanocrystalline titanium diboride: 1 – storage of charge materials and charge preparation; 2 – dosing; 3 – mixing; 4 – drying; 5 – rubbing; 6 – loading the charge into the dispenser; 7 – plasma treatment; 8-9 – cooling of the exhaust dust and gas flow and separation of the target product; 10 – collection, quality control and packaging.

The required investment for organizing the production of nanocrystalline titanium diboride in 3 plasma-metallurgical reactors with a total capacity of 450 kW is 93.3 million rubles. The annual demand for raw materials is PTN-8 grade titanium powder 37.3 t/year, B-99 grade boron powder 17.1 t/year, energy consumption – 2.06 million kWh/year. At the same time, annual production of 52 tonnes/year is forecasted at a selling price of 34,670 rubles/kg. The payback period for capital investments is 0.5 years.

4. Conclusion

The technical proposal for aluminum-producing enterprises concerning the organization of nanocrystalline titanium diboride production was developed, which is component of the cathode protective coatings of aluminum electrolysis cells, including the technological process, basic and auxiliary equipment, and substantiation of the main technical and economic indicators of production. The required investment with an installed capacity of 450 kW is 93.3 million rubles. It is projected to achieve annual output of 52 tonnes/year, selling price of 34,670 rubles/kg, a payback period of investment of half a year.

References

[1] Galevsky G V, Mintsis M Ya and Sirazutdinov G A 2009 *Aluminum Metallurgy: a Guide to Technology and Equipment* (Novokuznetsk: SibSIU) p 251
[2] Serlire M and Oye H A 2010 *Cathodes in Aluminum Electrolysis* (Dusseldorf : Aluminum – Verlag) p 698
[3] Grotheim K and Kvande H 1993 *Introduction to Aluminium Electrolysis* (Dusseldorf: Aluminium – Verlag) p 260
[4] Li J et al 2008 *The Journal of The Minerals* **60(8)** 32–37

[5] Gorlanov E S, Bazhin V Yu and Vlasov A A 2016 *Electrometallurgy* **6** 19–24
[6] Gorlanov E S, Bazhin V Yu and Vlasov A A 2016 *New Refractories* **3** 77–78
[7] Ivanov V V, Vasiliev S Yu, Laurinavichyute V K, Chernousov A A and Blokhin I A 2013 Pat. 2498880 RF, IPC C04B35 / 58 decl. 08/13/2012, publ. 11/20/2013

IOP Conf. Series: Materials Science and Engineering **866** (2020) 012008 doi:10.1088/1757-899X/866/1/012008

Production of prebaked anodes for aluminum electrolysis cells: dominant trends, technology, quality

G V Galevskiy[1], V V Rudneva[1] and S G Galevskiy[2]

[1]Siberian State Industrial University, 42 Kirova street, Novokuznetsk, 654007, Russia
[2]St. Petersburg Mining University, 2, 21st Line, St Petersburg, 199106, Russia

E-mail: kafcmet@sibsiu.ru

Abstract. The analysis of the current state of world and domestic production of calcined anodes for aluminum electrolysis cells is carried out. The dominant trends are identified, the basic technology is described, and directions for its improvement are indicated.

1. Introduction

For a long time, the technological strategy of leading aluminum-producing companies is based on the development of electrolysis production using prebaked anodes. This ensures the development and subsequent operation of high-ampere cells working with high technical and economic indicators, improving the sanitary and hygienic working conditions of personnel in electrolysis buildings, and generally improving the environmental safety of aluminum smelters. Currently, this technology produces more than 80% of the world primary aluminum. The implemented technologies of high-ampere electrolysis at electrolytic cells with prebaked anodes with a power of 300-500 kA have proved that a further increase in the efficiency of the process is possible by improving the design and advanced technical solutions using new materials and high-quality raw materials.

The indicators achieved at the leading aluminum enterprises confirm the effectiveness of the chosen direction: the current efficiency reaches 95%, the technological energy consumption is 12500-13000 kW·h/t Al, the anode consumption is 500-510 kg/t Al, aluminum fluoride 15-17 kg/t Al. The use of prebaked anodes significantly reduces emissions of tarry substances, including benzapyrene, and other harmful components from the cell. The possibility of using more airtight shelters with high efficiency, automated alumina feeding systems, which reduce the operating time of cells with open shelters, contributes to the reduction of emissions of harmful substances on high-ampere electrolysis cells.

The main distinguishing features of high-ampere technologies in comparison with medium-sized cells are: use of sand type alumina, fluorine salts, fluorinated alumina and reverse cryolite for feeding the cell through point feeding system; work in a given interval of low alumina concentrations of 2-5% according to technology without anode effects (0.05 pcs/day); low metal level (18-20 cm); high electrolyte level (20-22 cm); the use in the process of only acidic electrolytes with a given interval KO 2.2-2.4; KO corrections according to given algorithms; the ability to work at current densities close to critical values of 0.9-0.95 A/cm^2; high speeds (up to 18-20 cm/s) of the movement of the melt in the electrolysis bath; a new composition of covering materials, providing stabilization of the thermal and material balance; "plant-automatic" technology using multi-function cranes, mechanisms and robotics when servicing with constant scanning monitoring of the technological state of the cell; frame-type cathode device with additional cooling fins using highly graphitized lock-type cathode blocks and lateral silicon carbide lining with artificial profile overlay.

RUSAL in its medium-term and long-term projects also provides for a continuous increase in aluminum production using prebaked anodes. Sayanogorsk (530 thousand tonnes of Al/year), Khakassky (300 thousand tonnes of Al/year), partially Krasnoyarsk (1024 thousand tonnes of Al/year), Irkutsk (415 thousand tonnes of Al/year) are fully working on this technology. Novokuznetsk (213 thousand tonnes of Al/year) aluminum smelters. Boguchansky aluminum smelter (design capacity of 600 thousand tonnes of Al/year) is under commissioning, Taishet aluminum smelter (design capacity of 750 thousand tonnes of Al/year) is under construction.

2. Condition assessment and forecasts in the production of prebaked anodes

According to [1], in 2019 the global capacity of the global aluminum market may increase to 70 million tonnes with an expected deficit of ~ 1.2 million tonnes. With prevailing in 2015-2017 ratios in the volumes of primary and secondary aluminum production, it makes possible to predict the volume of primary aluminum production at the level of 43.0 million tonnes, and the need for annealed anodes – at the level of 19 million tonnes. The largest plants for the production of prebaked anodes with a production volume of more than 1 million t per year are in China. The leading aluminum-producing companies RIO TINTO ALCAN, ALCOA, RUSAL, EMAL, ALBA are making serious efforts to expand their own production of prebaked anodes.

RUSAL annually produces 3.5 - 3.7 million tonnes of primary aluminum at Russian plants, including using prebaked anodes 1.05 - 1.10 million tonnes. With the development of production capacities of Boguchansky aluminum plant, these figures can reach 1.5 million tonnes, which will require about 810 thousand tonnes per year of prebaked anodes. RUSAL prebaked anode production capacities in Russia include the existing anode production facilities at Sayanogorsk Aluminum Smelter (330 thousand tonnes per year). About 140 thousand tonnes per year are produced by the enterprises of the company ENERGOPROM [2]. The total deficit of prebaked anodes is about 440 thousand tonnes and is covered by imports from China. In this regard, the problem of import substitution is quite acute.

To solve it, RUSAL implements the following projects to increase the production of prebaked anodes:

- at the turn of 2018 – 2019 launch of the production of prebaked anodes at Volgograd aluminum smelter [3] with a production volume of 104 thousand tonnes per year. The project cost is 8 billion rubles with a payback period of 8 years.
- phased construction of the anode production of Taishet aluminum smelter with an output of 870 thousand tonnes per year [4]. Currently, the first stage is being implemented with an output of 217.5 thousand tonnes per year. The commissioning of the production facilities of the first stage is planned in the third quarter of 2019.

3. Multi-chamber furnaces for burning anodes

Anode burning is carried out in multi-chamber furnaces. Moreover, both in domestic and foreign practice, mainly open (anhydrous) furnaces are used, although at a number of enterprises one can find closed (arched) furnaces. Both types of furnaces are undergoing continuous modernization. During the construction of each new production of anodes, a technical and economic assessment of the choice of the first or second type of furnaces is carried out. Nevertheless, over the past 20 years, open furnaces have been given greater preference. A model of an open furnace in a three-dimensional image is shown in figure 1.

The main element of the furnace is firing chambers, directly adjacent to each other through the separation walls. Each chamber consists of 6-8 cassettes located in parallel along the axis of the furnace, and, therefore, is a block of unit cells – cassettes, which are completely identical both in size and in design. The cassette has the shape of a rectangular box, bounded on the sides by the walls of the fire or heating walls, at the ends – by the walls of the inter-chamber partitions, from below – by the surface of the hearth.

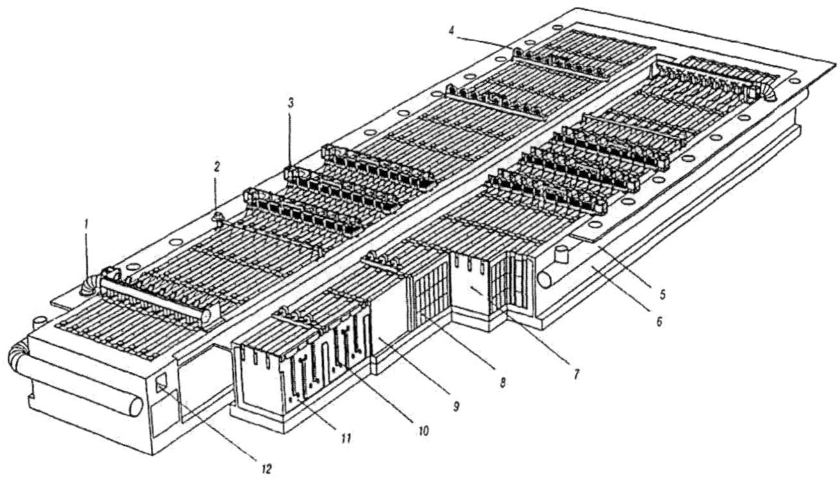

Figure 1. Open anode furnace: 1 – exhauster, 2 – rarefaction control ramp, 3 – fuel bridge, 4 – air cooler, 5 – central smoke bore, 6 – concrete frame, 7 – dividing wall, 8 – anodes, 9 – cassette, 10 – partitions, 11 – heating wall, 12 – flip channel.

To heat the kilns, both gaseous and liquid fuels are used. Fuel is supplied from the central highway to the internal network located around the perimeter of the furnace. The fuel line of the internal workshop network is equipped with individual nozzles opposite each chamber. The nozzles have transitional couplings for connecting to the fuel network of a mobile burner ramp (or bridge), on which the fuel burners are located.

4. The technology of prebaked anodes production

In almost all production of prebaked anodes, the traditional technological scheme as shown in figure 2 is implemented [5-6].

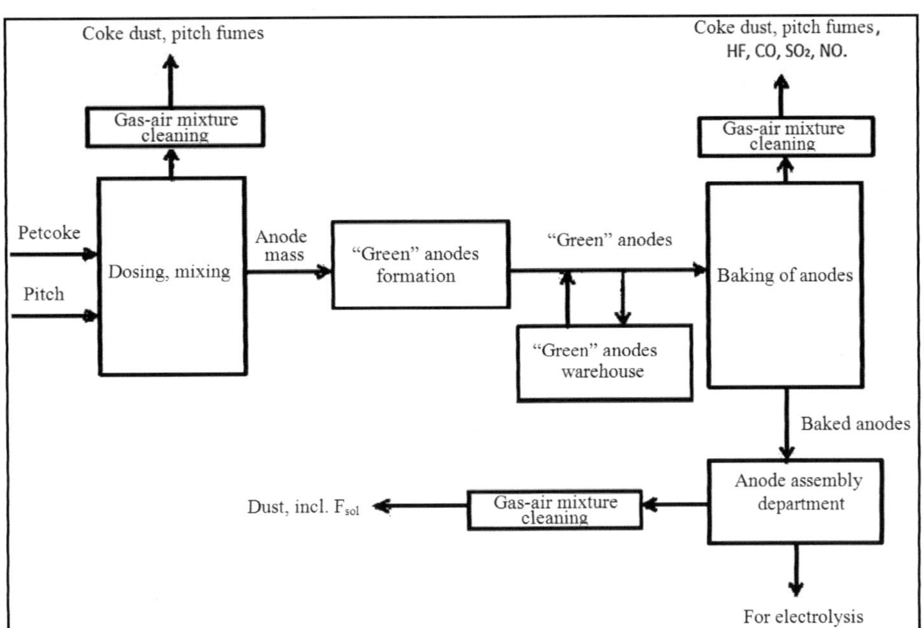

Figure 2. Technological scheme of prebaked anodes production.

In multi-chamber furnaces, prebaked anodes are at rest, i.e. motionless, and the zones of preheating, firing and cooling periodically move from chamber to chamber of the furnace. Due to this, the anodes of each chamber within one cycle go through all stages of firing and cooling in accordance with the technically sound schedule. The entire heat engineering cycle, including the preheating, forced heating and cooling zones, as well as the preparatory cycle chambers (loading, unloading) is called “fire”. With 72 cameras, the furnace operates on 4 “fires”. “Fire” moves through the furnace “at a speed” of 24-28 hours per camera.

After firing and cooling, the anodes are unloaded from the chamber using a universal bridge crane onto a conveyor with a pushing rod. Unloading one cartridge lasts in average 60 minutes. After unloading, the anodes are transported to the treatment station, where they are cleaned of coked dust with scraper knives and blown with air.

After the cleaning station, the anodes are fed to the transmission device, which lowers the anodes from the furnace site to the roller conveyor, divided into three sections. The middle section of the roller conveyor serves to stop the anodes for the purpose of visual inspection and, if necessary, manual purification. Defective prebaked anodes through the rotary table are sent to the conveyor and then to the crushing site. Useful prebaked anodes are transported by conveyors with a pushing rod to the loading area on an overhead conveyor with pallets and are transported to a warehouse. Prebaked anodes are shown in figure 3.

Figure 3. Prebaked anodes as a part of the anode device in the aluminum cell.

RUSAL enterprises produce anode blocks of various sizes and purposes. For low-power cells, type A blocks are produced in size 550x400, height 510 mm. On modern cells, anode blocks of types B and C are widely used (figure 4). The technical requirements for prebaked anode blocks were developed by the Swiss company R&D Carbon Ltd. The most significant of them are given in table 1 [7, 8].

Table 1. Quality indicators of the anode blocks.

Quality indicators	Typical value
Density:	
- volumetric, kg/dm^3,	1.53-1.58
- true, kg/dm^3,	2.05-2.08
Specific electric resistance, μOhm·m	52-60
Strength:	
- bending, MPa,	8.0-12.0
- compressive, MPa.	32.0-48.0
Thermal conductivity, W/m·K	3.0-4.5

CO_2 reactivity:	
- scaling, %,	4.0-8.0
- oxidation,%.	10.0-20.0
Ash content, %.	0.2-0.5
Impurities content, ppm:	
- S	1.2-2.4
-Si	100-300
-Fe	100-500
-V	80-260
-Na	200-600

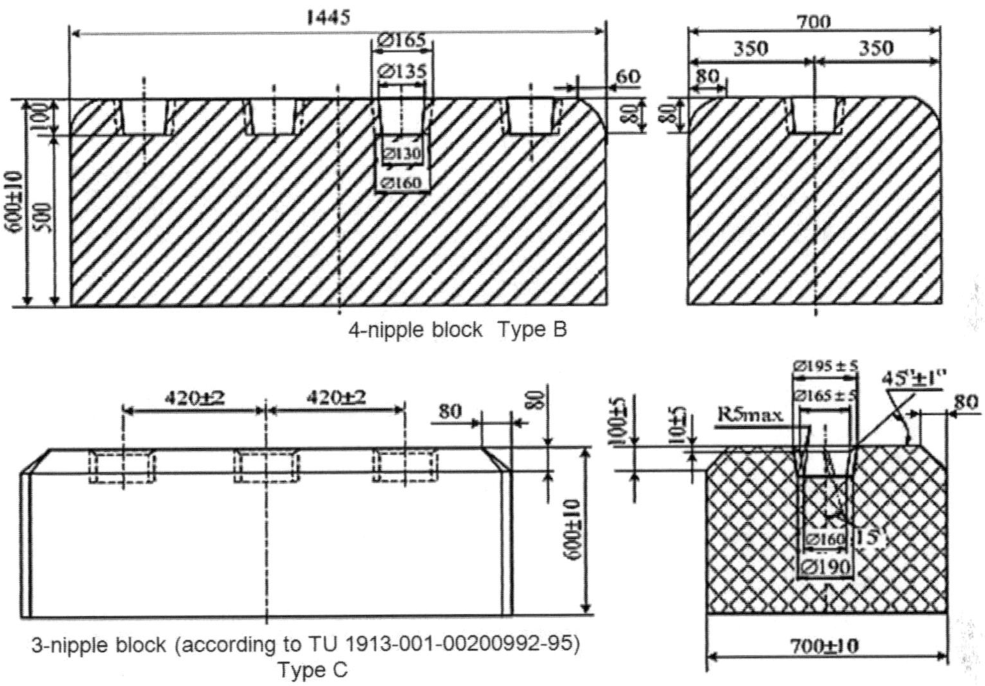

Figure 4. Anode blocks of type B and C.

5. Conclusion

The state of world and domestic production of prebaked anodes for aluminum cells was assessed. Based on the analysis of scientific and technical literature, the main trend of modern electrolysis production was confirmed – the development and operation of high-ampere electrolytic cells with prebaked anodes, which have increased environmental safety. The global demand for prebaked anodes is estimated at 19 million tonnes, RUSAL – 0.8 million tonnes. Moreover, the total shortage of prebaked anodes in UC RUSAL is 0.44 million tonnes and is covered by imports from China. In 2019, RUSAL plans to commission new capacities – 0.1 million tonnes at Volgograd and 0.22 million tonnes at Taishet aluminum smelters. The phased development of the anode production in the conditions of Taishet aluminum plant will ensure the production of 1.3 million tonnes of prebaked anodes for RUSAL.

It was confirmed that the leading aluminum-producing companies – RIO TINTO ALCAN, ALCOA, RUSAL, EMAL, ALBA are making serious efforts to expand their own production of prebaked anodes using mainly open-type multi-chamber furnaces, providing 4-stage firing with high-quality anodes production at minimal energy costs. The expansion of the production of prebaked anodes is accompanied by the improvement of furnace designs, firing technology, replacement of fuel

(fuel oil with natural gas), with the formation of protective coatings on the anodes, a change in their geometry (the presence of gas channels), etc. The use of new technological schemes for the treatment of exhaust gases significantly increases the environmental safety of anode production.

References

[1] Chernavina D A, Chernavin E A, Faller A V and Zdanovich M U 2018 *Young Scientist* **17** 206-210
[2] Prebaked anodes ENERGOPROM https://www.epmgroup.ru/ru/
[3] Pleshakov V *Volgograd Aluminum Plant is Ready to Launch the Production of Anodes* https://vpravda.ru/ekonomika/alyuminievyy-zavod-volgograda-gotov-zapuskat-proizvodstvo-anodov-58439/
[4] *Taishet Aluminum Smelter. Construction* http://visinform.ru/?action=view&id=4345&module=articles&rubric=20
[5] Nikiforov V P, Palchikov A I and Matveev Yu A 1982 *Production of Prebaked Anodes vol III book 3* (Leningrad) p 782
[6] Yanko E A 2001 *Anodes of Alumina Reduction Cells* (Moscow: Rydy I Metally) p 672
[7] *Aluminum Market Review. Aluminum Association* http://www.aluminas.ru /upload/iblock/4ab/2016 – aluminium-market-reviev.pdf-2016/
[8] Galevsky G V, Mintsis M I and Sirazutdinov G A 2009 *Aluminum Metallurgy. Reference Book on Technology and Equipment* (Novokuznetsk) p 251

IOP Conf. Series: Materials Science and Engineering **866** (2020) 012009 doi:10.1088/1757-899X/866/1/012009

Technology for the reduction of iron oxides in the processing of metal slag scrap in the oxygen converter

A V Amelin[1], E V Protopopov[2], S N Kuznetsov[2] and A N Kalinogorskiy[2]

[1]EVRAZ ZSMK JSC, 16/1 Kosmicheskoye shosse, Novokuznetsk, 654043, Russia
[2] Siberian State Industrial University, 42 Kirova Street, Novokuznetsk, 654007, Russia

E-mail: protopopov@sibsiu.ru

Abstract. The characteristic features of slag formation and oxidative metal refining in 160-tonne converters of EVRAZ ZSMK JSC using metal slag scrap in the solid metal charge are considered. Technological advantages of the process are established for the reduction of iron oxides from the slag component of the iron-containing product in the processing of waste slags.

1. Introduction

Currently, in steelmaking the issues of meeting the needs for scrap metal are of particular importance in connection with the deterioration of the quality properties of collected scrap metal and a decrease in the amount of recycled scrap due to the improvement of continuous casting technology [1, 2]. Under the current conditions, it is necessary to improve the traditional composition and technological mode for the formation of solid metal charge using, in particular, iron-containing products for processing dump converter slags, including metal slag scrap. The latter is a conglomerate of metal and slag components, which remains in the slag pots after the liquid converter slag draining into the pit of the slag compartment.

2. Technology

In BOF-1 of EVRAZ ZSMK JSC, the slag cakes are removed from the slag pots and loaded with scrap metal into the converter in the amount of 10-50% from the weight of the solid metal charge with the corresponding adjustment of the heat part of the process. The weight of the material charge (cast iron and scrap metal) is 153 – 155 tonnes, including 24 - 27% of the solid metal charge. The chemical composition of cast iron varies in the range: 0.35 – 0.65% Si; 0.35 – 0.65% Mn; 0.08 – 0.12% P and 0.020 – 0.028% S. The temperature of cast iron is 1290 – 1320 °C.

Before casting, as a rule, preheating of scrap metal takes place in the converter. Coal from the Kuznetsk basin of the TOM brand is used as the main heat carrier. Oxygen for coal combustion is supplied with a flow rate of 150 – 250 m^3/min. Duration of heating is at least 6 minutes.

The metal blowing with oxygen in the converters is performed through a 5-nozzle lance with Laval nozzles of a critical diameter 0.035 m and an angle of their inclination to the vertical axis of 15°. The operation is started at the tuyere position relative to the dead metal level 2.0 – 2.6 m with slag formation for 3 to 4 minutes, then the tuyere is lowered to 1.0 - 1.5 m. The blowing time is 20 – 21 minutes at an oxygen flow rate 350 – 450 m^3/min. The total consumption of lime for melting is 6 – 8 tonnes, high-magnesia flux FOMI 0.9 – 1.1 tonnes, coal 2.7 – 3.1 tonnes.

It should be noted that a characteristic feature of the operation of converters at EVRAZ ZSMK JSC is the blowing of cast iron with a low manganese content, which creates certain difficulties in

IOP Conf. Series: Materials Science and Engineering **866** (2020) 012009 doi:10.1088/1757-899X/866/1/012009

organization of the slag melting mode and oxidative refining of metal. The use of high-magnesia fluxes for steel smelting and increase in the lining stability of oxygen converters determines an increase in the viscosity of converter slags and a decrease in their refining ability [3]. An increase in the temperature of the bath during the preheating of scrap metal and slag-forming materials improves the kinetic parameters of the process. However, slag formation often proceeds at an insufficient speed and is characterized by incomplete absorption of lime by slag and prolongation of the processes for harmful impurities removal.

It is known that the composition of slag and its physical properties change during converter smelting according to almost the same scheme [4, 5]. Slag in the converter is formed from oxides generated during oxidation of oxides impurities and slag-forming materials introduced into the bath: lime and high-magnesia flux. Primary slags, which are formed as a product of iron and iron impurities oxidation (primarily silicon and manganese), are characterized by a high assimilative capability with respect to calcium and magnesium oxides. Depending on the temperature conditions of the process and the properties of the slag-forming materials, the dissolution of lime and high-magnesian flux proceeds with a greater or lesser rate: the basicity of the slag and, simultaneously, the content of magnesium oxide in it increase.

3. Results and discussion

In the slag of the metal slag scrap the part of CaO is in a state bound to SiO_2 and P_2O_5, while the other part is in the free state or in the composition of ferrites and calcium aluminoferrites is involved in the slag formation. So with the characteristic composition of the slag component: 35 – 40% CaO; 13 – 15% SiO_2; 1 – 1.5% P_2O_5, 24.31 – 28.05% of calcium oxide in the composition of silicates will be bound; in the composition of phosphates – from 0.79 to 1.19%. Thus, technological value is 5.69 – 14.83% CaO.

Moreover, the eutectic of the slag component of the loaded slag metal scrap is enriched with CaO-containing mineral components, which is due to the relatively low melting temperature of the latter [3]. As a result, their rapid transition to the liquid state and participation in the process of slag formation from the first minutes of the operation can be assumed. In accordance with the calculated data, when the consumption of the loaded slag scrap is 10-50% from weight of the solid metal charge in the initial period of metal blowing, it is possible to increase the content of technologically valuable calcium oxide in the slag by 4 – 9%.

The slag component of the loaded slag metal scrap also contains about 10% MgO, mainly in the form of magnesiuustite and magnesioferrite, and provides an increase in the content of magnesium oxide in the converter slag at the beginning of the blowing. In this case, at the end of the blowing its excessive supersaturation with MgO is prevented. The aforementioned contributes to a decrease in the viscosity of the primary slag and increases the degree of homogenization of the final slag, which generally increases the kinetic possibilities of metal refining from harmful impurities.

Thus, the dissolution of the slag component of the metal slag scrap allows acceleration of slag formation at the beginning of the operation, and the total amount of slag increases. At the maximum consumption of metal slag scrap – 50% from the mass of solid metal charge, the calculated increase in the amount of slag is 40 – 45%, which, of course, changes the hydrodynamic picture in the unit and requires the introduction of appropriate control actions to correct the smelting charge to prevent slag overrunning from the unit.

The analysis of the calculated data showed that for the conditions and characteristic temperatures under consideration, iron oxides can be reduced from the slag component of the loaded metal slag scrap. An industrially realized process has been developed for introduction of iron-containing products from waste slags processing using carbon-containing materials into the metal charge of oxygen converter and agitating the melt by the mixture of neutral gas and oxygen to recover iron oxides entering the bath [6].

In this case, the development of a solid-phase reduction is possible, when the reduction of iron oxides of the slag component occurs in the solid phase, followed by melting and reduction of iron oxides from the melt, as well as a liquid-phase reduction, with the reduction of iron oxides of the slag component from the melt. In case of solid-phase reduction, the development of reactions in the direction of iron oxides reduction is possible only if the actual process temperature is higher than the temperature of the reaction onset, which is also provided by the technology under consideration. During liquid-phase reduction, the kinetic parameters of the process are favorably affected by the elevated temperature and intensive agitation of the melt.

At a relatively low bath temperature at the beginning of converter smelting, the oxidation of phosphorus is carried out by iron oxides, including those coming from the slag component of the dissolving metal slag scrap. This is due to the relatively low concentration of silicon in cast iron (0.4 – 0.5%) and, accordingly, silica in the primary slag, as well as to a slow increase in the bath temperature due to the greater cooling effect of metal slag scrap and the presence of calcium oxide binding P_2O_5 into complexes that firmly hold phosphorus in the slag and lower its activity.

The phosphorus content in the metal before its tapping using slag metal slag scrap is on average 0.017%. The basic slag basicity is 2.66 units, the iron oxide content is 24.45%, the metal temperature is 1629 °C, the phosphorus concentration in the cast iron is 0.08 – 0.12%. As for the phosphorus content in the metal before being tapped according to the traditional technology, its average value is at the same level. However, this ensures a higher distribution coefficient of phosphorus between the slag and the metal, which is apparently due to its increased entry into the bath.

In the normal converter smelting conditions, as a rule, the slag is formed with a high content of iron oxides, which is generally unfavorable for sulfur removal. The sulfur content in the metal before being tapped using metal slag scrap is on average 0.029%, and in the melting according to the traditional technology – 0.028%. It should be noted that in melts with metal slag scrap the slag oxidation is slightly higher, and its basicity and temperature conditions of the process are almost identical.

The analysis of the chemical composition of the metal at the tapping using metal slag scrap showed a marked decrease in the content of undesirable impurities of non-ferrous metals, in particular Cu, Ni, Cr, due to a change in the qualitative composition of the solid charge. The usage of the metal slag scrap and a corresponding reduction in the consumption of scrap metal, which is the main source of non-ferrous metal impurities in the converter bath with the traditional technological scheme for steel production, made it possible to reduce the concentration, in particular, of copper in the metal before its tapping from the unit by an average of 16%, and nickel – by 12.5%.

4. Conclusion

The proposed technology for the iron oxides reduction from the slag component of metal slag scrap allows the iron in the oxide form to be returned into production. Processing of metal slag scrap without efficient use of the slag component reduces the economic opportunities of the process.

References

[1] Grigorovich K V 2018 *Proceedings of the XV International Congress of Steelmakers and Metal Producers* pp 42–59
[2] Parshin V M 2015 *Steel* **2** 13–17
[3] Amelin A V, Protopopov E V, Kalinogorskiy A N and Feyler S V 2014 *Steel in Translation* **44(7)** 513–516
[4] Protopopov E V, Kalinogorskiy A N and Yakushevich N F 2016 *Bulletin of the Russian Academy of Natural Sciences. West Siberian Branch* **18** 64–69
[5] Kuznetsov S N et al 2017 *Bulletin of the Russian Academy of Natural Sciences. West Siberian Branch* **20** 95–101
[6] Kuznetsov S N, Protopopov E V, Kalinogorskiy A N and Ganzer L A 2017 *RF Patent 2641587 Ru*

IOP Conf. Series: Materials Science and Engineering **866** (2020) 012010 doi:10.1088/1757-899X/866/1/012010

Technology for producing high-quality manganese concentrates from ferromanganese concretions and polymetallic manganese-containing ores

O I Nokhrina[1], I D Rozhikhina[1] and I E Proshunin[2]

[1]Siberian State Industrial University, 42 Kirova Street, Novokuznetsk, 654007, Russia
[2]EVRAZ Consolidated West Siberian Metallurgical Plant, 16 Kosmicheskoye shosse, 654043, Russia

E-mail: rogihina@mail.ru

Abstract. The results of studies of ferromanganese concretions enrichment using various solvents are presented. Universal technologies were proposed for processing manganese raw materials, including its autoclave leaching with aqueous solutions of chlorides, which allow manganese to be extracted into solution from compounds of various oxide and carbonate forms. Manganese recovery is over 90%. The introduction of ferrous manganese concretions into the chlorine-calcium enrichment process allows the extraction of manganese to be increased by 2.3, cobalt by 3.2, and nickel by 2.5 times. The leaching of ferromanganese concretions by aqueous solution of hydrochloric acid with the neutralization of its excessive concentration by the introduction of carbonate manganese ore makes it possible to extract non-ferrous metals in the solution of 99.80 – 99.88% and up to 95.4% Mn. As a result of enrichment, manganese, copper-nickel-cobalt and iron concentrates can be obtained.

1. Introduction

Ferromanganese concretions (FMC) are promising sources of raw materials. The most valuable components of FMCs are manganese, cobalt, nickel and copper. Concretions are aggregates of extremely finely fused hydroxides of iron, manganese and clastic-clay material, which occur at the bottom of the Pacific Ocean and the Gulf of Finland. Enrichment of ferromanganese concretions is a necessary stage in its preparation for metallurgical processing.

The analysis of literature data and a physicochemical evaluation of the extraction of manganese and other impurities from manganese-containing raw materials showed that the calcium chloride method is the most promising [1, 2], with which it is possible to obtain a manganese concentrate that meets the highest industry requirements [3,4]. In terms of hardware design, this method is similar to autoclave leaching in alumina production, which has been working smoothly for many years and allows one plant to achieve a productivity of 1 million tonnes or more per year.

A significant part of manganese in ferromanganese concretions is presented in oxide form and only an insignificant part in the form of carbonates, and the calcium chloride process is effective in treating only carbonate manganese ores. An analysis of the literature data, as well as the works of the authors in [3], and preliminary experiments with pure oxide materials, showed that it is most expedient to develop a method for enriching concretions using aqueous solutions of calcium chloride and the addition of solutions of iron chlorides or sulfates, or usage of a salt solution as a solvent acids. In

IOP Conf. Series: Materials Science and Engineering **866** (2020) 012010 doi:10.1088/1757-899X/866/1/012010

addition to FMC, polymetallic manganese-containing ores deposits of which are located in Kemerovo Region are of particular interest as raw materials. Along with manganese (up to 54%), these ores contain non-ferrous metal impurities (Ni – 0.5%, Co - 3%, Cu – 0.2%) and a rather high phosphorus content (0.2%).

2. Materials and methods of research

Studies of the chemical enrichment of manganese-containing raw materials were carried out by the method of sampling a solution, precipitation, slag, metal.

The chemical and phase composition of the starting materials and products of their processing was determined by chemical, spectral, and X-ray microanalysis. Extraction of manganese, cobalt, nickel and copper from concretions was evaluated by analyzing the dry solids after leaching and filtrate.

For optimal leaching conditions, we studied the effect of processing time, process temperature, coarseness of concretions crushing, the concentration of calcium chloride in the initial solution, the ratio of iron in the form of sulfates to the consumption of manganese in the charge (kg/kg), as well as the effect of preliminary heat treatment of concretions on the extraction of valuable components.

For this, the concretions of the above composition were crushed to a certain size, mixed with iron sulfate (in the form of $FeSO_4 \cdot 7H_2O$) at different ratios between them and autoclaved at temperatures of 180 – 220 °C with an aqueous solution of calcium chloride (600 – 900 g/l). Each experiment was repeated three to four times.

The experiments were carried out in autoclaves with volumes of 75, 100, and 10,000 cm^3. During the experiments, exactly the same temperature conditions and pulp mixing were provided. The temperature was automatically adjusted with an accuracy of ± 5 °C.

For the experiments were used:

- concretions - Mn_{tot} – 15.17 %; MnO_2 – 21.2 %; Co – .275 %; Ni – 0.46 %; Cu – 0.22 %; Fe_2O_3 – 23.1 %; SiO_2 – 13.74 %; CaO – 2.6 %; MgO – 2.01 %; Al_2O_3 – 3.67 %; P – 0.255 %; moisture – 16.68 %;
- polymetallic manganese ore - Mn_{tot} – 48.8 %; Co – 3,0 %; Ni – 0.49 %; Cu – 0.23 %; Fe_{tot} – 3.1 %; SiO_2 – 17.5 %; CaO – 2.6 %; MgO – 0.68 %; Al_2O_3 – 3.67 %; P – 0.255 %.
- carbonate manganese ore – 26 – 31 % manganese, 8 – 11 % calcium oxide and 1 – 3 % MgO, 2 – 7 % Fe_2O_3, 8 – 17 % SiO_2.
- solutions of iron chloride - $FeCl_2$ – 220 – 230 *g/l*, HCl – 60 – 70 *g/l*, $FeCl_3$ – 15 – 25 *g/l*, H SO_4 – 30 – 50 *g/l*, $FeSO_4$ – 150 – 200 *g/l*.

3. Results and discussion

When using calcium chloride as a solvent with the addition of iron sulfate solution, manganese oxide passes into the solution according to the following reaction:

$$MnO_2+MnO+2FeSO_4+2CaCl_2+3H_2O=2MnCl_2+2CaSO_4+2Fe(OH)_3$$

From the data of table 1 it can be seen that the introduction of iron sulfates into the calcium chloride process can increase the extraction of manganese by 2.3; cobalt – 3.2; nickel – 2.5 times. The results of experiments with concretions of varying degrees of grinding showed that with a decrease in their size, the extraction of manganese, cobalt, nickel and copper increases markedly and when using concretions of the fraction 0.125, the extraction of manganese, cobalt, nickel and copper is 73.4%, 75.5%, 51.2% and 18.6%, respectively.

The study of the effect of the calcium chloride concentration on process indicators allowed us to conclude that, within the studied limits (600 - 900 g/l), it does not lead to significant changes in the extraction of valuable components. Therefore, it is preferable to work with calcium chloride concentrations of 600 - 700 g/l, because in this case, the density of the solution decreases and both the filtration of the pulp and the further processing of the solutions are simplified.

The study of the effect of concretion calcination at 105-500 °C on manganese extraction showed that with the increase in the calcination temperature, manganese extraction decreases: when using raw concretions, manganese extraction is 87.8%, and when using calcined concretions, it is 58.3%, which is associated with a decrease in raw materials porosity at the increase of calcination firing temperature.

The study of the effect of temperature and leaching duration on process indicators showed that temperature has a significant effect on the extraction of manganese, cobalt, nickel and copper from concretions. At a temperature of 180 °C, with an increase in the duration of the experiment from 1 to 3 hours, the extraction of manganese and cobalt increases. At 200 and 220 °C, a change in the duration of the experiment from 1 to 4 hours does not significantly affect the extraction rates of manganese and cobalt. With the increase in the temperature from 180 to 200 and 220 °C, the leaching rate of manganese and cobalt increases significantly. As a result, at 200 - 220 °C for 1 hour the same recovery is achieved as at 180 °C for 4 hours. The temperature and duration of leaching are more difficult to influence on the dissolution of nickel and, especially, copper. This is due to the fact that iron sulfates for binding leaching products are consumed faster than non-ferrous metals dissolve. As a result, especially at elevated temperatures, hydrolysis of a part of the dissolution products, especially copper, is possible.

Table 1. The effect of temperature and leaching duration on the extraction of manganese, cobalt, nickel and copper from concretions.

$Fe_{FeSO4.7H2O}$ / $Mn_{concr.}$	Experiment temperature, °C	Leaching time, h	Precipitation content, %				Precipitation analysis recovery, %			
			Mn	Co	Ni	Cu	Mn	Co	Ni	Cu
1.17	180	1	2.64	0.03	0.11	0.06	59.0	74.3	46.1	35.7
1.16	180	2	2.26	0.03	0.11	0.09	67.4	76.1	50.0	10.5
1.17	180	3	2.17	0.03	0.10	0.10	74.0	80.2	60.4	17.0
1.17	180	4	2.48	0.03	0.09	0.08	79.5	79.5	68.4	40.6
1.17	200	1	2.72	0.03	0.11	0.06	68.9	81.1	62.3	52.7
1.17	200	3	2.26	0.03	0.10	0.06	73.0	80.2	60.7	50.7
1.17	200	4	2.64	0.03	0.10	0.04	69.3	80.7	61.6	67.9
1.16	220	1	1.91	0.02	0.12	0.07	73.4	84.7	45.0	32.8

The most significant effect on the extraction of manganese, cobalt, nickel and copper is exerted by the mass ratio of iron in the form of sulfates to the manganese consumption in the charge. From the results shown in figure 1, it can be seen that the optimal ratio is in the range 1.25 –1.35.

Thus, when using calcium chloride leaching with the sulfate iron (II) introduction into the solvent and the optimal parameters: temperature 180 - 220 °C;. duration 1 to 3 hours; the concentration of calcium chloride in the initial solution 600 – 700 g/l; fineness of concretions less than 0.125 mm; $FeFeSO_4 \cdot 7H_2O/Mn_{conc.}$ 1.25 – 1.35, 81.0 – 88.0% Mn, 78.0 – 85.0% Co, 51.6 – 68.4% Ni, 40.6 – 51.0% Cu.

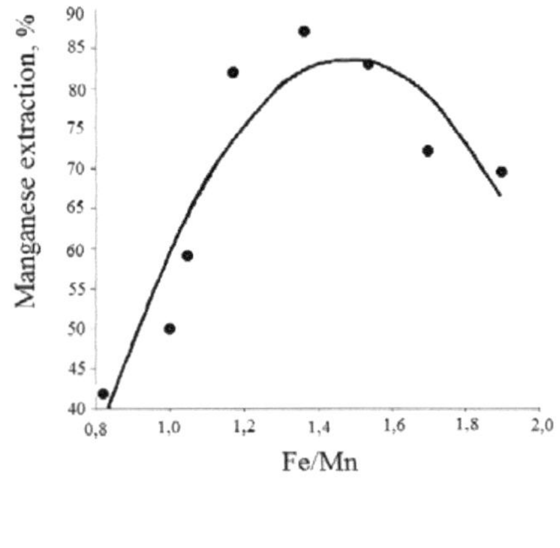

Figure 1. Effect on the manganese extraction of the mass ratio of iron (II) in sulfates to manganese in concretions.

The use of aqueous solutions of iron chloride (table 2) allows up to 99.2% of manganese to be extracted, with the extraction of cobalt up to 99.1%, nickel up to 82.7%, copper up to 73.8%.

Table 2. The effect of the mass ratio of iron (II) in chlorides to manganese in concretions on the extraction of manganese, cobalt, nickel and copper.

$Fe_{FeSO4.7H2O}$ / $Mn_{concr.}$	Extraction, %			
	Mn	Co	Ni	Cu
0.56	18.80	-	-	-
1.11	71.08	51.34	54.57	61.04
1.65	98.83	93.75	65.75	71.72
2.19	99.19	99.15	82.77	73.80
2.74	99.25	99.28	71.38	58.73

For experiments on the leaching of valuable components from polymetallic ores, the ratio of iron (II) in chlorides to manganese in the ore equal to 1.65was used. The experimental results in table 3 show that as for the extraction of manganese, nickel, cobalt and copper, they are in the same range as in the FMC leaching.

The resulting concentrate is characterized by high quality, especially in its content of phosphorus, sulfur and silica and contains 55 – 64 % Mn; 0.1 – 2.31 % SiO_2; < 0.55 % Fe; < 0.073 % S and < 0.005 % P. When using hydrochloric acid as a solvent to neutralize its excessive amount, carbonate manganese ore was used.

Table 3. Indicators of extraction of manganese, cobalt, nickel and copper from polymetallic manganese ore.

Test No.	Precipitation analysis recovery,%			
	Mn	Co	Ni	Cu
1	95.35	78.50	55.22	62.42
2	98.83	93.75	65.75	71.72
3	99.17	97.76	74.79	68.97
4	99.19	99.15	82.77	73.80

It is advisable to maintain the ratio of HCl_{pr}/HCl_{theor} in the range of 1.9 – 2.80, and the process temperature in the range of 85 – 95 °C. At a concentration of HCl in the solution of 200-300 g/dm^3, metals leaching is carried out in 70-120 minutes. The neutralization of the excess amount of hydrochloric acid was carried out by carbonate ore with different manganese contents (12.95 and 29.8%). From the solution of carbonate ore passes from 48.2 to 95.4 % Mn.

Residues from carbonate ore during neutralization, containing about 7% Mn, can be used in the initial stage of neutralization.

Precipitation of non-ferrous metals, Co and Ni from solution using sodium sulfide or hydrogen sulfide allows high extraction of metals from raw materials to be achieved. Manganese hydroxide is precipitated from the solution by hydrolysis at pH 2 – 3. From the purified solution – at pH 8.5 – 9.0.

Extraction of non-ferrous metals in the solution is 99.80 – 99.88%. Manganese concentrate obtained from manganese hydroxide by calcination contains up to 60% manganese and less than 0.005% phosphorus.

All products of FMC processing according to these methods are suitable for use. Manganese concentrate is a raw material for the production of manganese alloys, iron ore material is used for the production of agglomerate and metallized iron, and non-ferrous metal concentrate is used for the production of cobalt and nickel [5].

4. Conclusion

The introduction of ferrous manganese concretions into the chlorine-calcium enrichment process allows the extraction of manganese to be increased by 2.3, cobalt by 3.2, and nickel by 2.5 times. During processing of raw ferromanganese concretions and polymetallic ores, non-ferrous metals are also well recovered: 98-99 % Co; 83-98 % Ni; 74 % Cu at 95-99 % Mn. As a result, it is possible to obtain manganese, copper-nickel-cobalt and iron concentrates. Leaching of FMC with an aqueous solution of hydrochloric acid with neutralization of its excessive concentration by the introduction of carbonate manganese ore makes it possible to extract non-ferrous metals in a solution of 99.80 – 99.88% and up to 95.4% Mn.

Acknowledgments

The study was carried out with the financial support of the Russian Foundation for Basic Research and Kemerovo Region – Kuzbass within the framework of scientific project No. 20-48-420001/20".

References

[1] Litvintsev E G et al 2001 *Coll. of scientific papers "The State of the Russian Manganese Ore Base and Issues of Providing Industry with Manganese* (Krasnoyarsk) pp 138–142

[2] Malov E I et al 2001 *Proc. of the Second All-Russian Scientific and Technical Conf. on the state of the Manganese Ore Base in Russia and Issues of Providing Industry with Manganese* (Krasnoyarsk) pp 188–193

[3] Nokhrina O I et al 2009 *Izv. Vuzov. Ferrous Metallurgy* **10** 16–17

[4] Tolstoguzov N V et al 1986 *Manganese Metallurgy* pp 46–47

[5] Nokhrina O I et al 2015 *Izv. Vuzov. Ferrous Metallurgy* **58(5)** 309–31

IOP Conf. Series: Materials Science and Engineering **866** (2020) 012011 doi:10.1088/1757-899X/866/1/012011

Development of a method for calculating time and solidification coefficient of castings in sand-clay molds

V B Deev[1,2,3], O G Prikhodko[3], E S Prusov[4], E V Protopopov[3], M V Temlyantsev[3], A I Kutsenko[3], Mei Shunqi[1], E Kh Ri[5], S V Smetanyuk[6], K V Ponomareva[3] and G N Gavrilov[7]

[1]Wuhan Textile University, Wuhan, China
[2]National Research Technological University "MISiS", Moscow, Russia
[3]Siberian State Industrial University, Novokuznetsk, Russia
[4]Vladimir State University n.a. Alexander and Nikolay Stoletovs, Vladimir, Russia
[5]Pacific State University, Khabarovsk, Russia
[6]LLC Litmash, Shuya, Russia
[7]Nizhny Novgorod State Technical University n.a. R.E. Alekseev, Nizhny Novgorod, Russia

E-mail: deev.vb@mail.ru

Abstract. The mathematical model for calculation of the coefficient and solidification time of castings in sandy-clay form was developed. It takes into account the basic thermophysical parameters of the casting metal and mold material, heat transfer conditions at the crystallization front, at the 'casting-mold' interface and on the mold surface.

1. Introduction

In traditional foundry practice, it is customary to describe the solidification process of castings according to the square root law, which determines the thickness of the hardened rim:

$$\varepsilon = k\sqrt{\tau}, \tag{1}$$

where k is a constant value that received the name of the solidification constant in relation to foundry processes.

However, the square root law cannot be considered as a universal formula for calculating the solidification of castings and ingots. The assumptions of the authors who investigated this problem are reduced to the possibility to neglect the inhomogeneity of the temperature field in the casting and to take the form by the semi-bounded body having tight contact with the casting.

Thus, the use of the square root law in the analysis of real hardening processes is possible by adjusting the hardening coefficient, the value of which for different authors, even for the same mold materials and castings, varies widely (for example, [1-3], etc.). Therefore, the calculations performed according to the square root law often do not coincide with the actual experimental data.

During crystallization, the thermophysical parameters of the casting materials and molds are constantly changing and they must be taken into account when studying the process of solidification and crystallization of the casting [4-6]. A number of thermophysical parameters for calculation according to existing models are either taken as constant values or given by a piecewise constant function, which is not true. The regression models proposed in the literature give true results about the process under study only for a certain type of castings and under certain technological limitations. In this regard, the search for directions for improving methods of calculating various indicators of castings solidification process is an urgent task in the field of foundry.

The purpose of the work is to develop a methodology for calculating the time and solidification coefficient of castings in sandy-clay form, taking into account the main thermophysical parameters of the casting metal and mold material, heat transfer conditions at the crystallization front, at the 'casting-mold' interface and on the mold surface. This work is the continuation of a many years of research in the field of alloys crystallization and castings solidification modeling, performed under the guidance of Prof. I.F. Selyanin [7, 8].

2. Methods

Figure 1 presents a diagram of the temperature distribution in the mold and hardened casting. We describe the temperature distribution curve in the casting section in the form of an nth-order parabola [9]:

$$\frac{T_{CR}-T_M}{T_{CR}-T_C} = \left(\frac{y}{\varepsilon}\right)^n = \left(1-\frac{x}{\varepsilon}\right)^n, \tag{2}$$

where T_{CR} is the crystallization temperature of the alloy (for alloys crystallizing in the temperature range – liquidus temperature), K; T_C — contact temperature at the "cast-mold" interface, K; T_M is the temperature of the hardened metal at time τ at the point with coordinate y, K; y is the coordinate measured from the solidification front in the direction of the surface of the casting, m; ε is the thickness of the hardened rim, m; x is the coordinate, measured from the surface of the casting in the direction of the solidification front, m

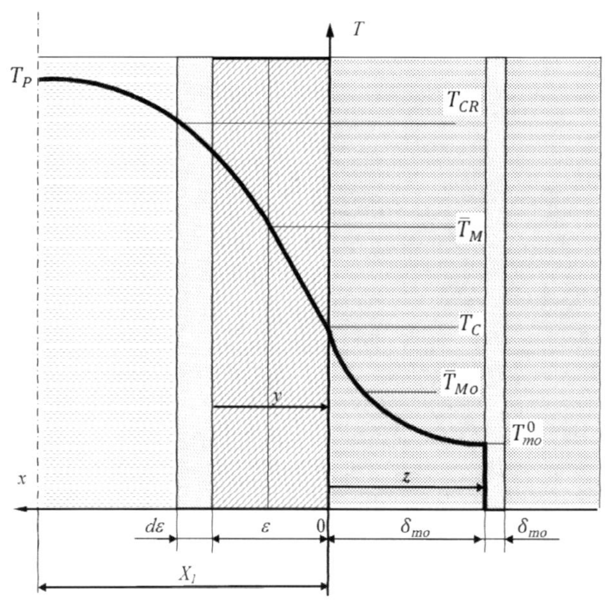

Figure 1. Scheme of temperature distribution in the mold and hardened casting.

In the formulation of the Stefan-Schwartz problem T_C at the 'casting-mold' boundary remains constant during the entire crystallization period, therefore, a quasistationary approach can be applied to the problem.

The contact temperature T_K at the casting-mold interface is determined by the formula proposed by N.I. Khvorinov [10]:

$$T_C = T_{mo}^0 + \frac{T_{CR} - T_f^0}{1 + \frac{b_f}{b_m}}, \tag{3}$$

where T_f^0 – the initial temperature of the form, K; b_{mo}, b_m – heat storage coefficient of the mold and metal, W·s.$^{0.5}$/(m^2·K).

$$b_{mo} = \sqrt{c_{mo} \cdot \rho_{mo} \cdot \lambda_\phi}\,;\; b_m = \sqrt{c_m \cdot \rho_m \cdot \lambda_m}\,. \tag{4}$$

The temperature distribution curve in the cross section of the form will be described in the form of a parabola of n'-order

$$\frac{T_{mo}^0 - T_{mo}}{T_{mo}^0 - T_C} = \left(\frac{z}{\delta_{mo}}\right)^{n'}, \tag{5}$$

where T_{mo} is the temperature of the mold at time τ, at the point with coordinate z, K; z is the coordinate measured from the surface of the casting in the mold, m; δ_f – thickness of the heated layer of the mold, m. According to [9], the average temperature of the hardened rim $\bar{T}_M$ for an infinite flat casting

$$\bar{T}_M = \frac{T_C - T_{CR}}{n+1} + T_{CR}\,, \tag{6}$$

for infinite cylindrical casting

$$\bar{T}_M = \frac{T_C - T_{RP}}{n+1} \cdot \frac{1 - \frac{\varepsilon/R}{n+2}}{1 - \frac{\varepsilon}{2R}} + T_{CR}, \tag{7}$$

for ball casting

$$\bar{T}_M = \frac{T_C - T_{CR}}{n+1} \cdot \frac{1 - \frac{2\varepsilon/R}{n+2} + \frac{2(\varepsilon/R)^2}{(n+2)\cdot(n+3)}}{1 - \frac{\varepsilon}{R} + \frac{1\varepsilon^2}{3R^2}} + T_{CR}, \tag{8}$$

where R is the reduced size of the casting, $R = V_0/F_0$ (where V_0 and F_0 are the volume and surface area of the casting), the introduction of the reduced size of the casting leads the solidification of the casting to a one-dimensional version.

From (5) follows that the average values of the temperatures of the heated layer of the mold for an infinite flat casting can be expressed as

$$\bar{T}_F = \frac{T_C - T_{mo}^0}{n'+1} + T_{mo}^0, \tag{9}$$

For casting of infinite cylinder

$$\bar{T}_F = \frac{T_C - T_{mo}^0}{n'+1} \cdot \frac{1 + \frac{\delta_{mo}/R}{n'+2}}{1 + \frac{\delta_{mo}}{2R}} + T_{mo}^0, \tag{10}$$

for ball casting

$$\bar{T}_{Mo} = \frac{T_C - T_{mo}^0}{n'+1} \cdot \frac{1 + \frac{\frac{2\delta_{mo}}{R}}{n'+2} + \frac{2\left(\frac{\delta_{mo}}{R}\right)^2}{(n'+2)(n'+3)\cdot}}{1 + \frac{\delta_{mo}}{R} + \frac{\delta_{mo}^2}{3R^2}} + T_{mo}^0\,. \tag{11}$$

The thickness of the growing rim and the thickness of the heated layer of the form change in the scale of the reduced size; in real scale, their values will be determined as follows:

$$\varepsilon_l = \frac{\varepsilon l}{R};\; \delta_{fl} = \frac{\delta_{mo}\, l}{R}, \tag{12}$$

where l is the actual size of the casting in the direction of the heat sink, m

The heat balance equation during casting crystallization in the sand-clay mold has the following form:

$$L_{CR}\rho_M F_{CR} d\varepsilon + c_M \rho_M F_{CR} d\varepsilon (T_P - T_{CR}) + c_M \rho_M F_{CR} d\varepsilon (T_{CR} - \bar{T}_M) = c_{mo}\rho_{mo} F_F d\delta_{mo} (\bar{T}_{Mo} - T^0_{mo}), \quad (13)$$

where L_{CR} – the heat of metal crystallization, J/kg; ρ_m, ρ_{mo} – density of the alloy and the molding mixture, kg/m^3; c_m, c_{mo} is the heat capacity of the alloy and the molding mixture, J/(kg·K); T_P – pouring temperature of the alloy, K; F_{CR}, F_{Mo} – surface area of the crystallization front and the heated layer of the mold, m^2.

On the left side of equation (13), the heat input is recorded, which is the sum of the heat of crystallization and the heat of superheat, the heat accumulated by the hardened metal rim. On the right side of the equation, heat consumption is the heat spent on heating the mold layer.

To take into account the multidimensionality of the problem, the parameter $F_{CR} \div F_O \div F_{Mo}$ was introduced, which takes into account the spreading of the heat flux from the crystallization front (F_O is the surface area of the casting). For infinite flat casting

$$F_{CR} = F_O = F_{Mo}. \quad (14)$$

For the infinite cylinder casting

$$F_{CR} \div F_O \div F_{Mo} = \left(1 - \frac{\varepsilon}{R}\right) \div 1 \div \left(1 + \frac{\delta_{mo}}{R}\right). \quad (15)$$

For ball casting and casting of finite sizes of arbitrary shape (plate, cylinder, ball)

$$F_{CR} \div F_O \div F_{Mo} = \left(1 - \frac{\varepsilon}{R}\right)^2 \div 1 \div \left(1 + \frac{\delta_{mo}}{R}\right)^2. \quad (16)$$

From (2) it follows

$$d\delta_{mo} = b\frac{F_{CR}}{F_{Mo}} d\varepsilon, \quad (17)$$

where b is a dimensionless coefficient showing how many times for an infinite plate the thickness of the heated layer of the mold δ_{mo} is greater than the thickness of the hardened metal rim ε, and

$$b = \frac{L_{CR}\rho_m + c_m \rho_m (T_P - \bar{T}_M)}{c_{Mo}\rho_{mo}(\bar{T}_F - T^0_{mo})}. \quad (18)$$

The heat flow from the crystallization front is transferred to form heating through the hardened metal rim and a heated layer of the molding sand. In accordance with this, we can write the following equality

$$K_H (T_P - T^0_{mo}) F_O d\tau = c_{Mo}\rho_{mo} F_{Mo} d\delta_{mo} (\bar{T}_{Mo} - T^0_{mo}), \quad (19)$$

where

$$\delta_{mo} = b \cdot \varepsilon,$$

K_H is heat transfer coefficient for a flat multilayer wall, W/(m^2·K):

$$K_H = \left(\frac{1}{\alpha_1} + \frac{\varepsilon}{\lambda_M} + \frac{1}{\alpha_2} + \frac{\delta_{mo}}{\lambda_{mo}}\right)^{-1},$$

α_1, α_2 – heat transfer coefficients at the crystallization front, at the boundary 'casting-mold', W/(m^{2}K); λ_f, λ_m – the thermal conductivity of the mold and metal, W/(m·K).

The heat transfer coefficient at the crystallization front is determined as follows [1]

$$\alpha_1 = Nu \cdot \lambda_{lm}/d, \tag{20}$$

where Nu – the Nusselt criterion; λ_{lm} – thermal conductivity coefficient of liquid metal, W/(m·K); d – the reduced characteristic size of the liquid part of the casting, m

Thus, the differential equation for determining the solidification time of castings in sand-clay molds is

$$d\tau = B \cdot b \cdot \left(\frac{1}{\alpha_1} + \frac{\varepsilon}{\lambda_m} + \frac{1}{\alpha_2} + \frac{b \cdot \varepsilon}{\lambda_f}\right) \frac{F_{CR}}{F_O} d\varepsilon, \tag{21}$$

where B is the coefficient, J/(m^{3}K);

$$B = c_{Mo} \rho_{mo} \frac{(\bar{T}_{Mo} - T^0_{mo})}{(T_P - T^0_{mo})}. \tag{22}$$

3. Results and discussion

Based on the mathematical transformations carried out, the solidification time of castings in the sand-clay molds can be determined by the formula:

$$\tau = A_1 \varepsilon + A_2 \varepsilon^2 + A_3 \varepsilon^3 + A_4 \varepsilon^4, \tag{23}$$

where A_1, A_2; A_3; A_4 – coefficients determined by the expressions below.

For casting of "infinite plate" type

$$A_1 = B \cdot b \cdot \left(\frac{1}{\alpha_1} + \frac{1}{\alpha_2}\right), \text{ s/m; } A_2 = \frac{B \cdot b}{2} \cdot \left(\frac{1}{\lambda_m} + \frac{1}{\lambda_{mo}}\right), \text{ s/m}^2\text{; } A_3 = 0; \ \ A_4 = 0. \tag{24}$$

For casting of "infinite cylinder" type

$$A_1 = B \cdot b \cdot \left(\frac{1}{\alpha_1} + \frac{1}{\alpha_2}\right), \text{ s/m; } A_2 = \frac{B \cdot b}{2} \cdot \left(\frac{1}{\lambda_m} + \frac{1}{\lambda_f}\right) - \frac{B \cdot b}{2R} \cdot \left(\frac{1}{\alpha_1} + \frac{1}{\alpha_2}\right), \text{ s/m}^2\text{;}$$
$$A_3 = \frac{B \cdot b}{3R} \cdot \left(\frac{1}{\lambda_m} + \frac{b}{\lambda_f}\right), \text{ s/m}^3\text{; } A_4 = 0. \tag{25}$$

For ball casting

$$A_1 = B \cdot b \cdot \left(\frac{1}{\alpha_1} + \frac{1}{\alpha_2}\right), \text{ s/m; } A_2 = \frac{B \cdot b}{2} \cdot \left(\frac{1}{\lambda_m} + \frac{1}{\lambda_{mo}}\right) - \frac{B \cdot b}{R} \cdot \left(\frac{1}{\alpha_1} + \frac{1}{\alpha_2}\right), \text{ s/m}^2\text{;}$$
$$A_3 = \frac{B \cdot b}{3R} \cdot \left[\frac{1}{R} \cdot \left(\frac{1}{\alpha_1} + \frac{1}{\alpha_2}\right) - 2\left(\frac{1}{\lambda_m} + \frac{b}{\lambda_{mo}}\right)\right], \text{ s/m}^3\text{; } A_4 = \frac{B \cdot b}{4R^2} \cdot \left(\frac{1}{\lambda_m} + \frac{b}{\lambda_{mo}}\right), \text{ s/m}^4. \tag{26}$$

Transforming formula (23), we can obtain an expression for determining thickness of the hardened rim during crystallization of castings:

$$\varepsilon = \sqrt{\left(\frac{A_1}{\varepsilon} + A_2 + A_3 \varepsilon + A_4 \varepsilon^2\right)^{-1}} \cdot \sqrt{\tau}. \tag{27}$$

Formula (27) is equivalent to the square root law (1). From it follows that the solidification constant

$$K = \frac{1}{\sqrt{\frac{A_1}{\varepsilon} + A_2 + A_3 \varepsilon + A_4 \varepsilon^2}}. \tag{28}$$

The full solidification time (τ_S) with perfect contact at the crystallization front ($\alpha_1 \to \infty$) and at the casting-mold boundary ($\alpha_2 \to \infty$) for casting of any configuration with a given size R can be calculated by the formula

$$\tau_S = \frac{Bb}{2a} \cdot \left(\frac{1}{\lambda_m} + \frac{b}{\lambda_{mo}}\right) R^2, \tag{29}$$

where a is a coefficient depending on the configuration of the casting: for an infinite plate $a = 1$; for an infinite cylinder $a = 3$; for the ball $a = 6$.

The coefficient of solidification (K) in the square root law, respectively, is determined from the expression

$$K = \sqrt{\frac{2a\lambda_m\lambda_{mo}}{Bb(b\lambda_m + \lambda_{mo})}}. \tag{30}$$

When determining the solidification time of a casting with a complex configuration, it is divided into elements, each of which approaches a plate, cylinder and ball. Analysis of the design of castings of the industrial nomenclature shows that flat walls are the predominant of the forming elements, cylinders or prisms are less common, balls or cubes are even less common.

In the initial period of solidification, the thickness of the growing rim is relatively small; therefore, the process can be considered as plate solidification (the rim is considered as a thin plate). Based on the above formulas, the dependence of the thickness of the hardened rim at the surface of the 'casting-mold' section (the beginning of the process of solidification of the casting) is linear, since at $\varepsilon \to 0$ the most significant member of equation (23) will be $A_1\varepsilon$ will be. That is, at $\varepsilon \to 0$, $\tau = A_1\varepsilon$. At the final stage of solidification, the castings of final dimensions harden as cylinders if their width (or height) is less than their length, or like balls if their width (or height) does not differ significantly from their length. This is due to the dissipation of the heat flow, while the solidification front in the middle part of the hardened casting is a surface of revolution.

The developed mathematical model takes into account the basic thermophysical parameters of the casting metal and the mold material, heat transfer conditions at the crystallization front, at the 'casting-mold' interface and on the mold surface, and can be used to determine theoretical values of the coefficient and solidification time of castings in sand-clay mold.

4. Conclusion

A method for calculating the time and solidification coefficient of castings in sand-clay mold was developed. A distinctive feature of the technique is the possibility to take into account both the thermophysical parameters of the casting material and the mold, as well as the heat transfer conditions at the crystallization front, at the 'casting-mold' contact boundary, and on the mold surface. The proposed methodology can be applied in production and research laboratories and centers involved in obtaining high-quality castings for engineering purposes.

Acknowledgments

The work was performed within state task of the Ministry of Science and Higher Education of Russia in the field of scientific activity for 2017-2019 "Organization of Scientific Research" (task No.11.5684.2017/6.7).

References

[1] Shrewe H 1989 *Continuous Casting of Steel* (Germany: Stahleisen)
[2] Mehrara H, Santillana B et al 2011 *IOP Conf. Series: MSE* **27** 012046
[3] Xiao C, Zhang J M et al 2012 *Journal of University of Science and Technology Beijing* **34(9)** 1011-16
[4] Campbell J 2003 *Casting* (Butterworth-Heinemann, Elsevier)
[5] Dantzig J A and Rappaz M 2009 *Solidification* (Taylor & Francis Group, CRS Press)
[6] Stefanescu D M 2015 *Science and Engineering of Casting Solidification* (Springer International Publishing AG Switzerland)
[7] Selyanin I F, Deev V B et al 2015 *Russian J. of Non-Ferrous Metals* **56(4)** 434–436

[8] Deev V B, Selyanin I F et al 2015 *Metallurgist* **58(11-12)** 1123–27
[9] Veinik A I 1960 *Theory of Solidification of a Casting* (Moscow) 1960
[10] Chvorinov N I 1954 *Crystallization and Non-Homogeneity of Steel* (Prague: Publishing Czechoslovak Academy of Sciences)

IOP Conf. Series: Materials Science and Engineering **866** (2020) 012012 doi:10.1088/1757-899X/866/1/012012

Development of methods for calculating the solidification time of castings and ingots in metal mold

V B Deev[1,2,3], O G Prikhodko[3], E S Prusov[4], E V Protopopov[3], M V Temlyantsev[3], A I Kutsenko[3], Mei Shunqi[1], E Kh Ri[5], T A Bazlova[2], S V Smetanyuk[6] and A A Sokorev[2]

[1]Wuhan Textile University, Wuhan, China
[2]National Research Technological University "MISiS", Moscow, Russia
[3]Siberian State Industrial University, Novokuznetsk, Russia
[4]Vladimir State University n.a. Alexander and Nikolai Stoletovs, Vladimir, Russia
[5]Pacific State University, Khabarovsk, Russia
[6]Litmash LLC, Shuya, Russia

E-mail: deev.vb@mail.ru

Abstract. A technique has been developed for calculating the solidification time of castings and ingots in metal form, taking into account not only the thermophysical properties of the casting materials and molds, but also the heat transfer conditions at the "casting – mold" interface, the dissipation of the heat flux from the crystallization front to the heating of the chill mold and into the environment.

1. Introduction

The manufacture of castings and ingots by gravity casting in metal molds (gravity die casting) is an important industrial method for the production of high-quality engineering blanks [1, 2]. The widespread use of this method in foundry practice is due to a number of obvious advantages, which include a reduction in material consumption and mass of workpieces, an increase in the utilization of metal, an improvement in the mechanical and operational properties of cast products, an increase in efficiency, and improvement in working conditions.

In order to purposefully control the structure and achieve the specified operational characteristics of castings and ingots hardening in metal molds, it is necessary to understand the kinetics of the processes that occur during solidification [3-5]. In this regard, the choice of rational technological parameters of die casting should, to a large extent, be based on the use of reliable calculation and analytical methods for determining all significant physical and thermotechnical parameters characterizing the solidification process.

In this case, one of the most important parameters is the total solidification time of the casting or ingot. In particular, the relationship between the solidification time and the formation of various casting defects, for example, shrinkage porosity in castings was confirmed [6]. Many research activities are devoted to modeling the parameters of alloy crystallization and solidification of castings, including [7, 8], the development of which is the present work.

The purpose of the work is to develop a methodology for calculating the solidification time of castings and ingots in metal form.

2. Methods

It was experimentally recorded [9, 10] that during the solidification of castings in metal mold, the temperature of all points of the system tends to a certain level, which can be defined as the average calorimetric temperature of the system – $\bar{T}_C$ [9]. The average calorimetric temperature of the system remains constant provided that the loss of heat to the environment from the surface of the chill mold does not significantly affect the solidification process. This is possible when the mass of the chill mold m_{mo} is significantly higher than the mass of the casting m_O, that is, it is true for a massive chill mold.

In this case, the heat balance equation for the 'casting – mold' system has the form

$$L_{CR}\rho_M V_O + c_m\rho_m V_O(T_P - T_{CR}) + c_m\rho_m V_O(T_{CR} - \bar{T}_C) = c_{mo}\rho_{mo}V_{Mo}(\bar{T}_C - T^0_{mo}), \quad (1)$$

where L_{CR} – the heat of metal crystallization, J/kg; ρ_m, ρ_{mo} – density of the alloy and the material of the mold, kg/m^3; V_O, V_{Mo} – volume of casting and chill mold, m^3; c_m , c_{mo} – heat capacity of the alloy and the mold material, J/(kg·K); T_P – pouring temperature of the alloy, K; T_{CR} – crystallization temperature of the alloy, T^0_{mo} – the initial temperature of the form, K.

In the left side of the equation, the heat input is written, which consists of the heat of superheat, the heat of alloy crystallization and the heat given off by the hard rim during cooling from T_{CR} to $\bar{T}_C$; in the right side there is a heat heating the mold from the initial temperature to the average calorimetric. From equation (1) it follows that the average calorimetric temperature

$$\bar{T}_C = \frac{L_{CR}\rho_m V_O + c_m\rho_m V_O T_P + c_{mo}\rho_{mo}V_{Mo}T^0_{mo}}{c_m\rho_m V_O + c_{mo}\rho_{mo}V_{mo}}, \quad (2)$$

For real castings hardening in metal molds, the condition $m_O \ll m_{mo}$, is often not fulfilled, since the chill mold is cooled due to heat transfer to the environment. The amount of heat removed from the outer surface of the chill mold varies with time and can be calculated by the formula

$$Q_{env} = \alpha_3 F_{sur}(T_{sur} - T_{env}) \cdot \tau, \quad (3)$$

where α_3 – the heat transfer coefficient from the outer surface of the chill mold to the environment, W/(m^{2}K); the coefficient α_3 affects when the thickness of the mold heated layer is greater than or equal to the thickness of the mold wall ($\delta_{Mo} \geq x_{mo}$); F_{sur} – the surface area of the chill mold, m^2; T_{sur}, T_{env} – the temperature of the outer surface of the mold and the environment, K.

For a casting in the form of an infinite plate

$$F_{sur} = F_O, \quad (4)$$

for an infinite cylinder

$$F_{sur} = F_O \cdot \left(1 + \frac{x_{mo}}{R}\right), \quad (5)$$

for a ball

$$F_{sur} = F_O \cdot \left(1 + \frac{x_{mo}}{R}\right)^2, \quad (6)$$

where x_{mo} – is the thickness of the mold wall, m.

Thus, the average calorimetric temperature of the 'casting – mold' system will decrease with time and at time τ will be equal to

$$\bar{T}^{\tau}_C = \frac{L_{CR}\rho_m V_O + c_m\rho_m V_O T_P + c_{mo}\rho_{mo}V_{Mo}T^0_m - Q_{env}}{c_m\rho_m V_O + c_{mo}\rho_{mo}V_{Mo}}. \quad (7)$$

The heat flux from the crystallization front is spent on heating the chill mold and released into the environment through the hardened metal rim. Therefore, we can write the equality

$$K_H F_O(T_P - T_{env})d\tau = c_{mo}\rho_{mo}F_{Mo}(\bar{T}_C - T_{mo}^0)d\delta_{Mo} + \alpha_3 F_{sur}(T_{sur} - T_{env})d\tau, \qquad (8)$$

where K_H –the heat transfer coefficient for a flat multilayer wall, W/(m^{2}K):

$$K_H = \left(\frac{1}{\alpha_1} + \frac{\varepsilon}{\lambda_m} + \frac{1}{\alpha_2} + \frac{\delta_{mo}}{\lambda_{mo}} + \frac{1}{\alpha_3}\right)^{-1}, \qquad (9)$$

where α_1 – the heat transfer coefficient at the crystallization front; $\delta_{mo} = b \cdot \varepsilon$ – the thickness of the heated mold layer for a flat casting, m. Here b is a dimensionless coefficient showing how many times the thickness of the heated mold layer δ_{mo} is greater than the thickness of the hardened metal rim ε.

The heat transfer coefficient at the border 'casting – mold' α_2 can be determined from the conditional heat resistances of the gas gap and the paint layer:

$$\alpha_2 = \left(\frac{x_{gas}}{\lambda_{gas}} + \frac{x_{pa}}{\lambda_{pa}}\right)^{-1}, \qquad (10)$$

where λ_{gas}, λ_{pa} – the heat conductivity of the gas layer and paint, W/(mK); x_{gas}, x_{pa} – the thickness of the gas layer and the thickness of the paint layer, m.

$$x_{gas} = \alpha_m^{Ca}(T_{CR} - \bar{T}_M)X_O + \alpha_m^{Mo}(\bar{T}_{Mo} - T_{mo}^0)X_O, \qquad (11)$$

where α_m^{Ca}, α_m^{Mo} – the linear expansion coefficients of the casting solid metal and the mold material, 1/K; X_O – is the wall thickness of the casting in the direction of the heat sink, m.

$$\alpha_3 = Nu \cdot \frac{\lambda_m}{l},$$

where $l = 2r + 2x_{mo}$ is the characteristic size of the system; Nu is the Nusselt criterion characterizing the intensity of heat transfer by contact.

$$Nu = C_1(Gr \cdot Pr)^{C_2},$$

where C_1 and C_2 – tabular coefficients [10], Gr – the Grashof criterion characterizing the intensity of free motion of the medium (natural convection), Pr – the Prandtl criterion (similarity criterion for temperature and velocity fields);

$$Gr = \frac{gl^3\rho}{\mu} \cdot \frac{\Delta T}{T_{env}},$$

where g – the acceleration of gravity, g = 9.81 m/s^2; ρ – the gas density, kg/m^3; μ is the dynamic viscosity of the gas, N s/m^2; $\Delta T = T_{sur} - T_{env}$ – temperature difference between the surface of the chill mold and the environment, K.

$$Pr = \frac{\nu}{a_{env}},$$

ν is the kinematic viscosity coefficient of the medium, m^2/s; a_{env} – coefficient of thermal diffusivity of the medium, m^2/s.

Since the heat is absorbed by the mold only until it is completely heated, then to determine the thickness of the heated layer of the mold, we write the heat balance equation

$$L_{CR}\rho_m F_{CR} d\varepsilon + c_m \rho_m F_{CR} d\varepsilon (T_P - T_{CR}) + c_m \rho_m F_{CR} d\varepsilon (T_{CR} - \bar{T}_C^{\tau}) = c_{mo}\rho_{mo} F_{MO} d\delta_{mo} (\bar{T}_C - T_{mo}^0), \quad (12)$$

On the left side of equation (12), the heat input is recorded, which is the sum of the heat of crystallization and the heat of overheating, as well as the heat accumulated by the hardened metal rim. On the right side of the equation is the heat consumption, which is the heat spent on heating the mold layer. To take into account the multidimensionality of the problem, the parameter $F_{CR} \div F_O \div F_{Mo}$, is introduced, which takes into account the scattering of the heat flow from the crystallization front (F_{CR}, F_{Mo} – surface area of the crystallization front and the heated layer of the mold, m^2, F_O – casting surface area, m^2).

For an infinite flat casting

$$F_{CR} = F_O = F_{Mo}. \quad (13)$$

For an infinite cylindrical casting

$$F_{CR} \div F_O \div F_{Mo} = \left(1 - \frac{\varepsilon}{R}\right) \div 1 \div \left(1 + \frac{\delta_{mo}}{R}\right), \quad (14)$$

here R is the reduced casting size, $R = V_0/F_0$ (where V_0 and F_0 – are the volume and surface area of the casting). For ball casting and casting of finite sizes with an arbitrary shape (plate, cylinder, ball)

$$F_{CR} \div F_O \div F_{Mo} = \left(1 - \frac{\varepsilon}{R}\right)^2 \div 1 \div \left(1 + \frac{\delta_{mo}}{R}\right)^2. \quad (15)$$

Based on the introduced criterion, the thickness of the heated layer of the mold can be expressed as

$$d\delta_{mo} = b\frac{F_{CR}}{F_{Mo}} d\varepsilon, \quad (16)$$

Thus, the thickness of the heated layer of the mold is determined from equation (16), while

$$b = \frac{L_{CR}\rho_m + c_m \rho_m (T_P - \bar{T}_C^{\tau})}{c_{Mo}\rho_{mo}(\bar{T}_C^{\tau} - T_{mo}^0)}. \quad (17)$$

3. Results and discussion

Based on the described transformations, a differential equation is obtained for determining the solidification time of castings in metal molds, which for the general case has the following form:

$$d\tau = \frac{b \cdot c_{Mo}\rho_{mo}(\bar{T}_C^{\tau} - T_{mo}^0)}{K_C(T_P - T_{env}) - \alpha_3 \frac{F_{sur}}{F_O}(T_{sur} - T_{env})} \cdot \frac{F_{CR}}{F_O} d\varepsilon, \quad (18)$$

For a casting of an infinite plate type

$$d\tau = \frac{b \cdot c_{Mo}\rho_{mo}(\bar{T}_C^{\tau} - T_{mo}^0)}{K_C(T_P - T_{env}) - \alpha_3(T_{sur} - T_{env})} \cdot d\varepsilon, \quad (19)$$

for a casting of an infinite cylinder type

$$d\tau = \frac{b \cdot c_{Mo}\rho_{mo}(\bar{T}_C^{\tau} - T_{mo}^0)}{K_C(T_P - T_{env}) - \alpha_3\left(1 + \frac{x_{mo}}{R}\right)(T_{sur} - T_{env})} \cdot \left(1 - \frac{\varepsilon}{R}\right) \cdot d\varepsilon, \quad (20)$$

IOP Conf. Series: Materials Science and Engineering **866** (2020) 012012 doi:10.1088/1757-899X/866/1/012012

for a casting of a ball type

$$d\tau = \frac{b \cdot c_{Mo} \rho_{mo} (\bar{T}_C^{\tau} - T_{mo}^0)}{K_H (T_P - T_{env}) - \alpha_3 \left(1 + \frac{x_{mo}}{R}\right)^2 (T_{sur} - T_{env})} \cdot \left(1 - \frac{\varepsilon}{R}\right)^2 \cdot d\varepsilon . \quad (21)$$

The developed mathematical models make it possible to take into account while calculating the solidification time, not only the thermophysical properties of the casting materials and molds, but also the heat transfer conditions at the 'casting – mold" boundary, the heat-flux dissipation from the crystallization front for the mold heating and into the environment.

4. Conclusion

The methodology for calculating the solidification time of castings and ingots in metal mold was developed. The method takes into account the thermophysical properties of casting materials and molds, the heat transfer conditions at the 'casting – mold' interface, the heat-flux dissipation from the crystallization front for the mold heating and into the environment.

Acknowledgments

The work was performed within state task of the Ministry of Science and Higher Education of Russia in the field of scientific activity for 2017-2019 "Organization of Scientific Research" (task No.11.5684.2017/6.7).

References

[1] Major J and Hartlieb M 2009 *International Journal of Metalcasting* **3** 43–53
[2] Wang Y, Neff D, Schwam D, Zhu X and Chen C 2013 *International Journal of Metalcasting* 2013 **7** 25-38
[3] Shepel S V and Paolucci S 2002 *Applied Thermal Engineering* **22(2)** 229–248
[4] Vijayaram T R, Sulaiman S, Hamouda A M S and Ahmad M H M 2006 *Journal of Materials Processing Technology* **178(1-3)** 29–33
[5] Skrzypczak T, Węgrzyn-Skrzypczak E and Sowa L 2018 *Applied Mathematics and Computation* **321** 768–779
[6] Stefanescu D M 2015 *Science and Engineering of Casting Solidification* (Switzerland: Springer International Publishing AG) 3rd ed
[7] Selyanin I F, Deev V B, Belov N A, Prikhodko O G and Ponomareva K V 2015 *Russian Journal of Non-Ferrous Metals* **56(4)** 434–436
[8] Deev V B, Selyanin I F, Kutsenko A I, Belov N A and Ponomareva K V 2015 *Metallurgist* **58(11-12)** 1123–1127
[9] Veinik A I 1960 *Theory of Solidification of a Casting* (Moscow)
[10] Balandin G F 1979 *Fundamentals of the Theory of Casting Formation. Part 1* (Moscow)

IOP Conf. Series: Materials Science and Engineering **866** (2020) 012013 doi:10.1088/1757-899X/866/1/012013

Small-scale production of cast porous and composite materials

S V Knyazev, A A Usoltsev, A I Kutsenko, A A Kutsenko and S B Sokolov

Siberian State Industrial University, 42 Kirova Street, Novokuznetsk, 654007, Russia

E-mail: krookia@mail.ru

Abstract. Modern foundry production should be equipped with advanced technologies, equipment and materials, have a significant reserve for its development in the future, which is unthinkable without a transition to digital production, which includes integrated automation of technological processes, digital control of technology parameters, materials and quality indicators of the finished products by the non-destructive control methods, as well as the use of new foundry materials obtained with the help of nanotechnologies.

1. Introduction

The creation of a new enterprise should be aimed at meeting the existing significant needs in the society, equipped with advanced technologies and equipment, lay the significant groundwork for its development in the future. Russia's choice to develop a digital economy is unthinkable without the industry switching to "digital" production, which includes comprehensive automation of technological processes, digital control of technology parameters, materials and quality indicators of finished products using non-destructive control methods.

The foundry production can be referred to customized small-scale production.

The foundry industry is dominated by obsolete technologies created by the middle of the 20th century; there is a low level of technological and innovative development in this area.

Creation of a foundry enterprise requires:

- significant capital costs for the construction or reconstruction of production facilities;
- expensive equipment;
- very high energy intensity of the cast products manufacturing;
- low coefficient of metal utilization, high allowances for machining, lack of flexible production (quick remounting for new products manufacturing), severe restrictions on the range of manufactured products depending on the selected casting technology;
- high price competition;
- stringent requirements from Rostekhnadzor and environmental specialists;
- lack of qualified personnel in the region;
- the need for a large amount of working capital;
- long payback period of financial investments (8-12 years).

2. Results and discussion

The foundry production is an integral part of the production cycle in mechanical engineering. More than half of the enterprises in machine building and military-industrial complex of Russia have their own foundry shops. Foundry production as a sector of the national economy, of course, requires innovations. Most enterprises do not have a single information space, that would help all enterprise automated control systems, as well as industrial equipment and production personnel quickly and timely exchange information; automation and robotization of technological processes is not sufficient; at a basic level, technologies of virtual modeling and engineering analysis have been mastered, which does not allow the digital mockup of the product to be controlled; there is no interdisciplinary modeling at different levels of abstraction (from a conceptual design to a detailed single model); only in single enterprises a PDM system (product data management system) is used [1, 8, 9, 12-15].

Currently, the most promising direction in foundry is the introduction of additive and digital technologies at all stages of the production of cast products. The rapid growth of the market for additive and digital technologies – approximately by a quarter annually, is directed towards greater automation in the framework of "Industry 4.0".

These technologies successfully work at enterprises of various industries, such as:

- automotive industry;
- aerospace industry;
- tool engineering;
- military industry;
- medicine (including dentistry);
- gold work;
- art and architecture.

Within the framework of the created enterprise, it is supposed to form a team of specialists and scientists, which in practice will develop the content of the "fundamental transformations" of an ordinary enterprise into a digital one, which uses information technology as a competitive advantage in all areas of its activity: production, business processes, marketing and interaction with customers.

Moreover, the newly created enterprise will operate in the real sector of the economy, produce final products, provide services to the population, teach and train personnel for various areas of the digital economy [2,10].

One of the problems of modern society is the repair, namely the restoration of good condition and the extension of the equipment life:

- restoration of worn parts, the repair and replacement of which is difficult or expensive;
- restoration of original products discontinued or manufactured in other countries.

In both cases, the most difficult process is the restoration of parts obtained by casting.

The second problem is the development of new and innovative conceptual approaches for transferring an industrial enterprise to a higher technological level – the level of a digital enterprise.

A small foundry enterprise is the most versatile testing ground for this kind of fundamental and applied research, because it combines a whole range of related technologies of other industries: melting and casting, heat treatment, shaping, chemical technologies, powder metallurgy, automation, product quality control, etc. The first step can and should combine all of the above into a single information infrastructure.

For this, the key condition is the use of the MDC system (MachineDataCollection – collection of machine data), which allows data on the work of all production facilities (equipment, workplaces of the main workers, service and control services, etc.) to be collected in order to manage a small enterprise and to make more optimal technological decisions in real time [1, 3, 11].

At SibSIU it is proposed to create a research and production self-supporting innovative center for foundry additive and digital technologies and manufactures – customized small-scale production. The main activities of the center should be as follows:

- production of single cast products from non-ferrous metals and cast iron (investment casting);
- production of single molded products from plastics and plastic (cold casting);
- production of composite cast billets based on aluminum alloys, including porous, by vacuum suction;
- production of unique single products;

- teaching and training of personnel for the implementation of digital and additive foundry technologies in industry;
- development and promotion of digital, additive and foundry technologies for consumers;
- development and testing of "transformations" of operating models into new digital models for the concept of digital production.

To expand the reach of customers, an online store of customized products will be used that performs the functions of accepting the initial order, its digital processing, coordination and sending of the finished product to the customer – the "client-server" technology.

The base for the functioning of the center has already been created at the SibSIU, the prototyping center "Form" is operating; the technologies of "cold casting", casting of composite and porous blanks based on aluminum alloys by vacuum absorption have been developed.

3. Materials

The porous cast aluminum was first obtained in the United States by G. Kuchek in 1961, it was used as a structural material for helicopter construction, however, production ceased due to competition with foamed aluminum. The second birth of porous cast aluminum occurred simultaneously in Japan (S. Nagata, Kyushu University), Switzerland (A. Mortensen, Federal Institute of Technology in Lausanne) and the USSR (E.L. Furman, Ural Polytechnic Institute). These were the people who realized that porous cast aluminum is not a competitor to foamed aluminum, but a competitor to sintered metal powders. In addition, an important incentive for the development of technology in the USSR was the use of porous cast aluminum for the production of weapons (silencers of small arms, active protection of armored vehicles). The technology for porous aluminum production by impregnation at first glance impressed with its simplicity. However, like any new technology, it was replete with hidden problems: defects of unknown origin, extremely low metal utilization, marketing. These problems were too difficult for Japanese and Swiss science. In 1990 the industrial production of porous cast aluminum, which is currently the only one in the world, was organized in the city of Kirovgrad. Since then, the director of Composite Materials LLC has been L.E. Cherny, the scientific consultant is the student of E.L. Furman, Doctor of Technical Sciences, A.B. Finkelshtein, Marketing Director –M.L. Cherny, Ph.D. [4, 5].

Currently, among the organizations using porous aluminum products for various applications are Sibneft OJSC, Kurgankhimmash OJSC, Transpnevmatika OJSC, RAAZ AMO ZIL OJSC, Salavathydromash OJSC, Pnevmatika OJSC and many others – in total more than 300 enterprises in Russia and abroad.

Since 2014 China has actively begun to offer blanks and products from porous aluminum on the world market. For example, Alibaba.com offers anodized porous aluminum plate at a price of $ 2400-4600 per tonne and porous aluminum plate at a price of $ 1.3-7.1 per kilogram.

At present, porous aluminum is used to produce gas and liquid filters, air silencers, sound absorbing panels, dampers, bubblers, heat exchangers, flame arresters, heat pipes, wicks, gas and liquid permeable protective housings, liquid evaporators, liquid aeration, vacuum tables, current collectors in new generation rechargeable batteries, structural engineering products with lower weight and thermal expansion, cooling radiators for electronic components, design of high-tech objects, etc. (figure 1).

Filters of various designs

Pneumatic silencer

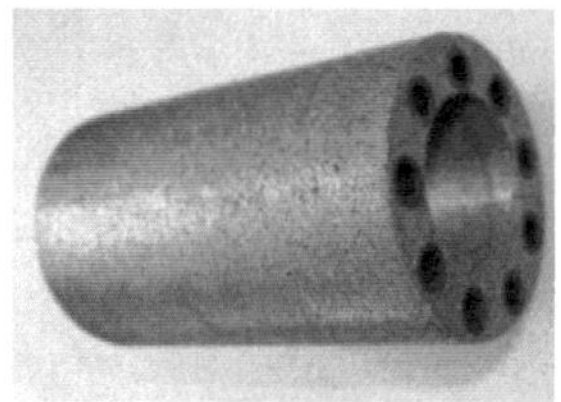
Heat exchanger

Sound absorbing panel

Ready product

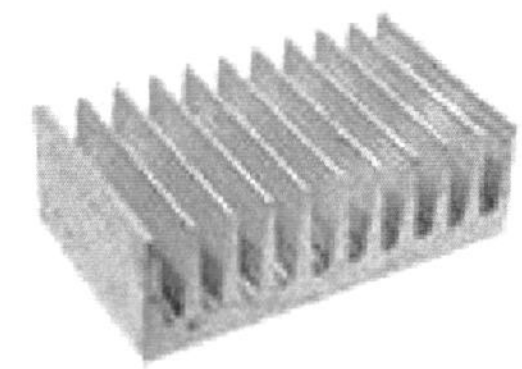
Radiator

Permeable sensor housing

Hi-tech table lamp

Figure 1. Products made of porous aluminum.

The unique properties of the products are as follows:

- Products made of porous aluminum have high permeability, heat resistance, dirt capacity and mechanical strength.
- Stronger, more durable and more efficient than sintered, mesh and polymer silencers.

Porous cast aluminum is demanded by the industry due to its high specific strength and corrosion resistance. Currently, the technology of melt impregnation of a water-soluble filler is the only one that allows a wide range of porous aluminum products to be obtained – from dampers to filter elements, of any size and configuration, combining the porous and monolithic part in one product [4-7].

The schematic diagram of the production of porous aluminum products is presented in figure 2.

Основной проблемой при получении отливок из пористого алюминия является высокий уровень дефектности, не подлежащие заделке. Коэффициент выхода годного находится на низком уровне. Обратное использование пористого алюминия в шихте дает угар порядка 50%, при этом расплав насыщается пленами.

The main problem in obtaining porous aluminum castings is a high level of defectiveness that cannot be sealed. The yield coefficient is low. The reverse use of porous aluminum in the charge gives a burn of about 50%, while the melt is saturated with films.

The traditional introduction of nanomaterials into the melt is fraught with a number of difficulties:

- the formation of a conglomerate of nanoparticles during their mechanical introduction into the melt;
- different densities of nanomaterial and melt and, as a consequence, the emergence of nanopowders upon their mechanical introduction into the melt;
- complexity in the technological provision of uniform distribution of nanomaterial particles in the entire melt volume;
- low reproducibility of introduction of a given of nanomaterials concentration into a liquid melt;

- oxidation and decrease in the wettability of the surface of nanoparticles at high temperatures;
- the presence of oxide films and slags on the surface of the melt;
- low adhesion of the metal matrix with nanoparticles;
- need for cladding nanoparticles with certain metals or substances.

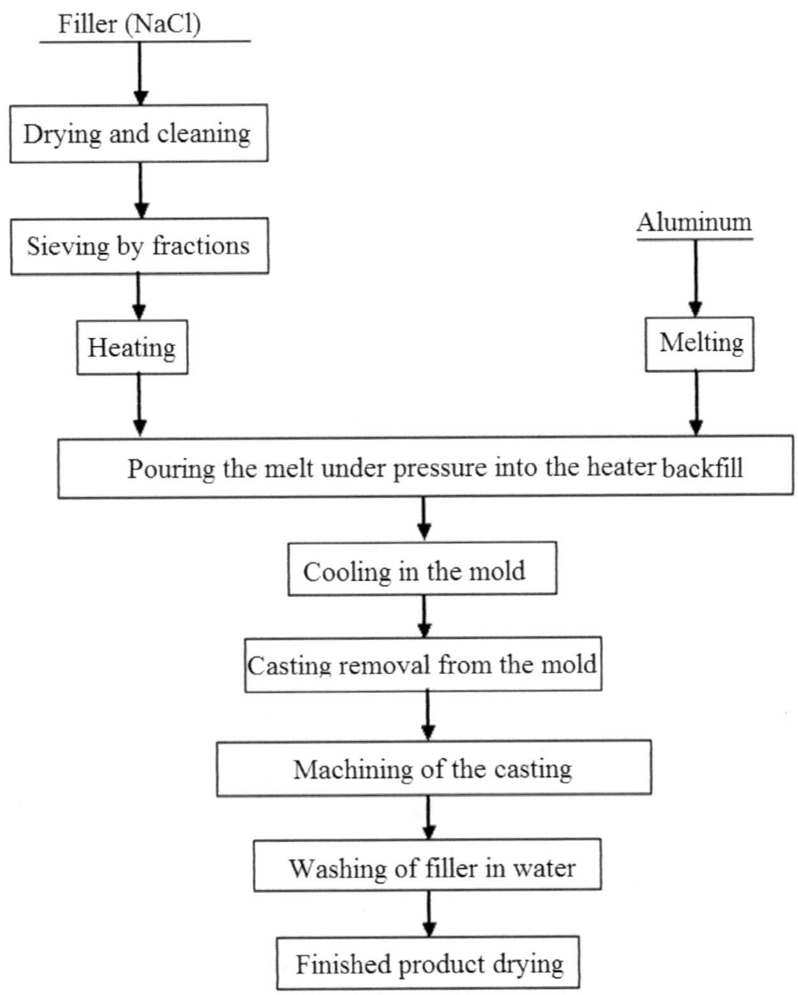

Figure 2. Schematic diagram of porous aluminum products manufacturing.

The technology for the production of porous cast billets from various metals and alloys has been improved at SibSIU in order to use it for the subsequent introduction of nanoparticles into cast alloys and the subsequent production of cast billets with a unique set of properties:

1. Obtaining a porous preform from a mother alloy with predetermined porosity parameters (figure 3).
2. Filling the pores of the porous preform with nanoparticles.
3. The introduction of a porous preform with nanoparticles into a liquid melt in a melting furnace to obtain a lump modifier (ligature) and subsequent modification (alloying) of the required cast alloys with its help.

Figure 3. A porous preform from mother alloy for introduction into ultra- and nanomaterials pores.

The developed technology provides:

- uniform distribution of nanomaterial particles in the entire melt volume;
- controlled concentration of nanomaterials in liquid melt;
- low amount of oxide films and slags in the body of the cast billet;
- does not require additional technological methods for introducing a modifier (ligature) directly into the melt.

4. Conclusion

The technology for the modifier introduction into the furnace and into the ladle does not require significant capital expenditures for its implementation. The technology is also applicable for introduction of ultrafine and finely dispersed or other specific materials into the melt. The performed studies provide the basis for the development of a new group of fundamentally new types of modifiers and ligatures. From the point of view of fundamentals of foundry, issues such as the hereditary effect of a mixture subjected to nanomodification on the properties of cast alloys, the creation of highly effective compositions of modifiers and ligatures for cast alloys should be studied.

Acknowledgments

The study was carried out with the financial support of the Russian Foundation for Basic Research and the Government of the Kemerovo Region – Kuzbass in the framework of scientific project No. 20-43-420013 r_a.

References

[1] Volostnov B I 2018 *Problems of Mechanical Engineering and Automation* **1** 4–37
[2] Finkelstein A B *Theory and Practice of Obtaining Porous Castings from Aluminum Alloys by Impregnation* https://dlib.rsl.ru/01004600562
[3] Davydov S V *Nanomodifier as a Tool for Genetic Engineering of the Structural State of Cast Iron Melt* http://rudocs.exdat.com/docs/index-70430.html
[4] Gostev A S, Gosteva E G and Gulevsky V A 2010 *Young Scientist* **11(1)** 53–55
[5] Knyazev S V, Usoltsev A A and Kutsenko A I 2019 *Litejnoe Proizvodstvo* **9** 5–7
[6] Knyazev S V, Usoltsev A A, Sokolov B M and Dolgopolov A E 2018 *Problems and Prospects for the Development of Innovative Technologies* (Orenburg) pp 33–35
[7] Knyazev S V, Usoltsev A A, Dmitrienko A V and Mamedov R O 2018 *Problems and Prospects of the Development of Science in Russia and the World* (Chelyabinsk) pp 14–17
[8] Knyazev S V, Usoltsev A A et al 2017 *Metallurgy: Technology, Innovation, Quality* (SibSIU) pp 205–208
[9] Knyazev S V and Khazan G L 1992 *Litejnoe Proizvodstvo* **12** 24–25
[10] Shiryaeva L S, Kutsenko A A et al 2017 *Young Russia* (KuzSTU) 36005
[11] Knyazev S V, Usoltsev A A et al 2017 *New Technologies in Industry, Science and Education* (Orenburg) pp 75–79
[12] Knyazev S V, Skopich D V et al 2019 *Izvestiya Vuzov. Ferrous Metallurgy* **62(2)** 134–140
[13] Antipenko V I and Knyazev S V 1987 *Soviet Castings Technology* **7** 34–37
[14] Knyazev S V, Usoltsev A A et al 2016 *IOP Conf. Series: MSE* **150** 012039
[15] Cheprasov A I, Knyazev S V et al 2016 *IOP Conf. Series: MSE* **150** 012026

Research of non-metallic inclusions of E90KhAF rail steel

A S Simachev, T N Oskolkova and A A Umansky

Siberian State Industrial University, 42 Kirova Street, Novokuznetsk, 654007, Russia

E-mail: simachev_as@mail.ru

Abstract: This article is devoted to the research of the non-metallic inclusions of the E90KhAF rail steel in different zones of continuous cast billets. It is established that in the rim zone the main non-metallic inclusions are silicates of iron and alumosilicates. Alumosilicates and silicates of iron are identified in the zone of columnar crystals. Sulfides of manganese (MnS), silicates of manganese, aluminum and iron are found in the central zone of the billet.

1. Introduction

The quantity, composition and distribution of non-metallic inclusions are important indicators of the steels quality that determine their mechanical properties and service characteristics. Among all possible ways of classifying inclusions, their separation by origin and time of formation is generally recognized. Endogenous inclusions are formed during phase-structural transformations in liquid and solid steel, and exogenous inclusions are introduced into the liquid from the outside as a result of the destruction of steelmaking and casting equipment lining, capture of charge particles, exothermic mixtures and slags [1-3].

2. Results and discussion

The authors of [4-6] previously studied the role of nonmetallic inclusions in rail steels of grades E76 and E76F in the process of hot plastic deformation. The purpose of these studies was to obtain information on non-metallic inclusions in the areas of continuous cast billet (CCB) of E90KhAF rail steel, which will be further used in the search for new technologies and improving the existing technology for the production of rail products.

When studying the CCB rim zone of E90KhAF steel, the presence of iron silicates ($FeO\cdot SiO_2$), as well as the inclusion of aluminosilicates, were revealed by X-ray diffraction after high-temperature torsion (figures 1-3).

In the zone of CCB columnar crystals of E90KhAF steel, aluminosilicates were identified by X-ray diffraction after high-temperature torsion (figure 4). Metallographic studies, in addition to aluminosilicates (figure 5), revealed the presence of iron silicates (figure 6).

The manganese sulfides (MnS), manganese, aluminum and iron silicates (figures 7-9) were revealed by X-ray and metallography in the CCB central zone of E90KhAF steel after high-temperature torsion.

IOP Conf. Series: Materials Science and Engineering **866** (2020) 012014 doi:10.1088/1757-899X/866/1/012014

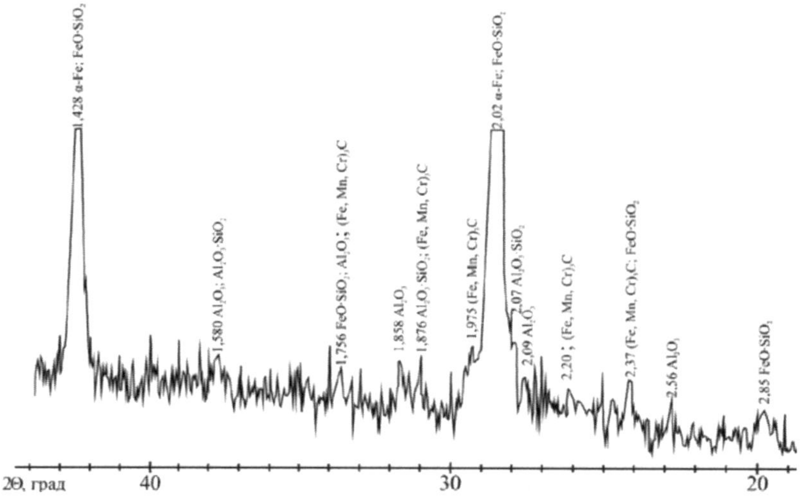

Figure 1. The fragment of diffraction pattern of CCB rim zone of E90KhAF steel.

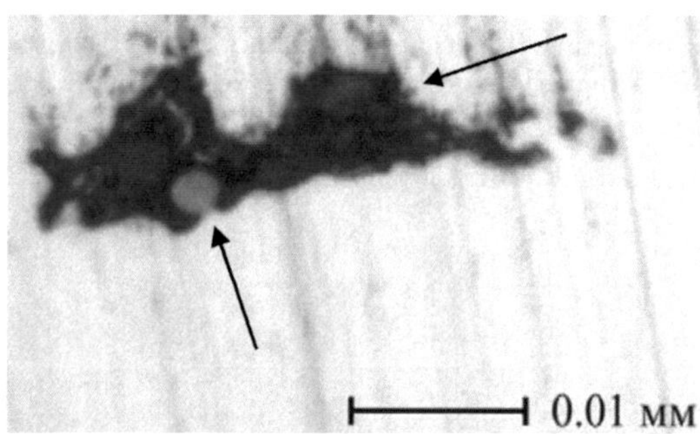

Figure 2. Iron silicates of CCB rim zone of E90KhAF steel after high temperature torsion.

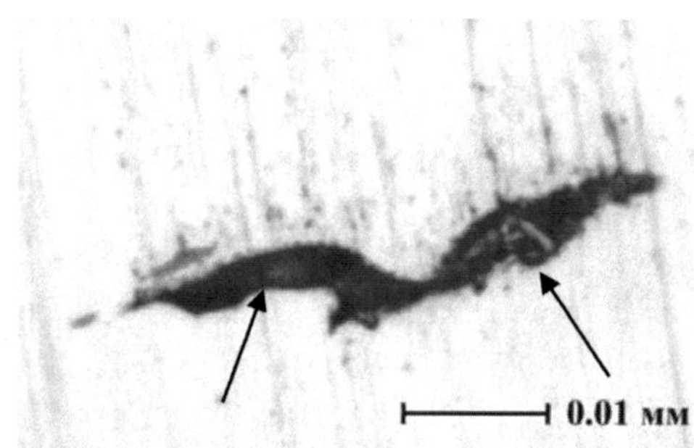

Figure 3. Aluminosilicates of the CCB rim zone of E90KhAF steel after high temperature torsion.

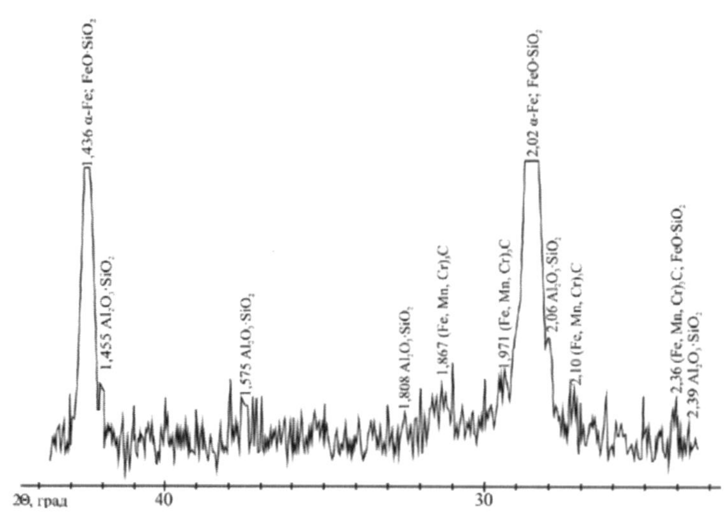

Figure 4. The fragment of diffraction pattern of the zone of CCB columnar crystals of E90KhAF steel.

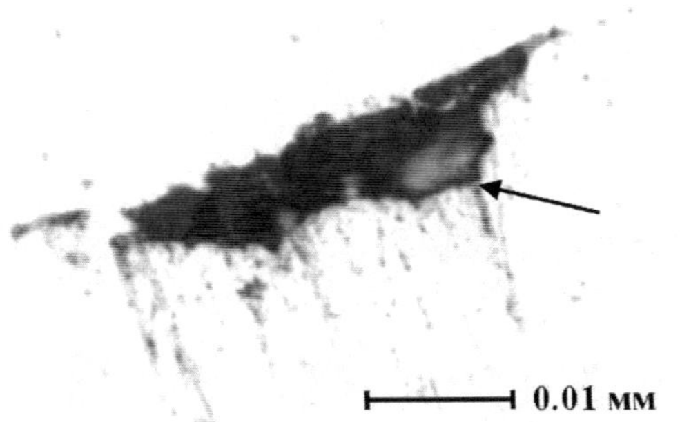

Figure 5. Aluminosilicates in the CCB zone of columnar crystals of E90KhAF steel after high temperature torsion.

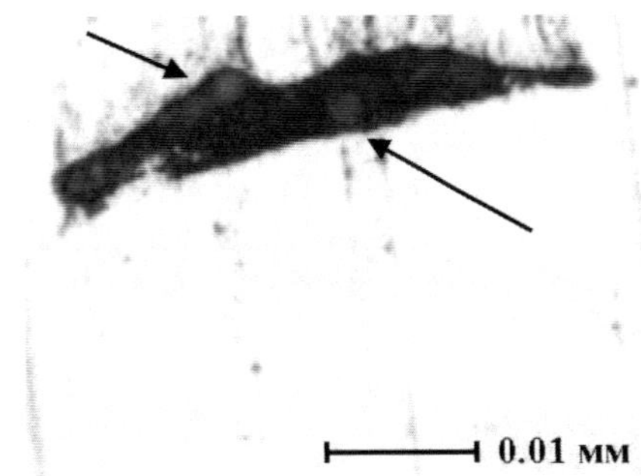

Figure 6. Iron silicates in the CCB zone of columnar crystals of E90KhAF after high temperature torsion.

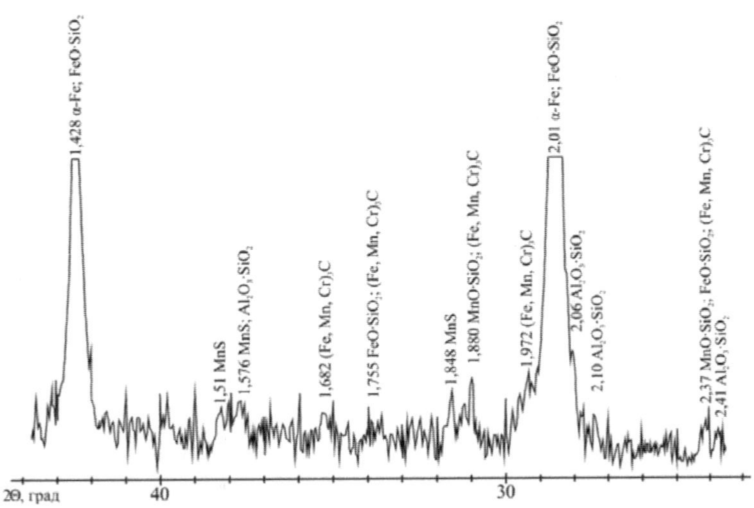

Figure 7. The fragment of diffraction pattern of CCB central zone of E90KhAF steel.

The presence of non-deformable aluminum silicates ($Al_2O_3{\cdot}SiO_2$), iron silicates ($FeO{\cdot}SiO_2$) and manganese ($MnO{\cdot}SiO_2$), in the locations of which the integrity violations of metal may be formed (figures 2, 3, 5, 6, 8, 9), will contribute to the localization of deformation and reduction of the shear deformation degree.

Non-metallic inclusions contained in steel are a set of stress concentrators, the value of which depends on the type and size of the inclusion, temperature and deformation conditions, the ratio of the physico-mechanical properties of the inclusion and the steel matrix.

The heterogeneity of the mechanical properties of steel can be caused not only by a change in the number of inclusions, but also by a change in their shape. During pressure treatment, plastic silicate and sulfide inclusions are stretched in the direction of deformation, and non-deforming inclusions of oxides; silicates and complex spinels are redistributed and form stitching clusters, which contributes to anisotropy of mechanical properties, which manifests itself mainly in ductility indicators.

Two-phase inclusions consisting of silicate and oxide or spinel exhibit inhomogeneous deformability. The silicate phase is well deformed, elongating in the direction of deformation (figures 3, 5), and corundum or spinel, being in the silicate matrix, is not deformed.

Inclusions of most oxides and spinels during high-temperature torsion are brittle. During deformation, the resulting fragments of inclusions rotate along the torsion axis, while being smoothed as a result of friction forces at the interface.

IOP Conf. Series: Materials Science and Engineering **866** (2020) 012014 doi:10.1088/1757-899X/866/1/012014

Multiphase inclusions exhibit inhomogeneous deformability depending on the nature of the phases. They are particles of oxides or spinels enclosed in a silicate matrix (figures 3, 5). The latter is plastically deformed with the metal matrix of steel, and the oxide particles are not deformed (particles of destroyed refractories that got into the steel during smelting), rotate in the silicate matrix in the direction of its flow, being destroyed during high-temperature deformation.

The observation data on the distribution and behavior of non-metallic inclusions during high-temperature plastic deformation do not contradict the studies of non-metallic inclusions in steels performed by Gubenko S.I., Starov R.V., Parusov V.V., Derevyanchenko I.V. [1, 6-9].

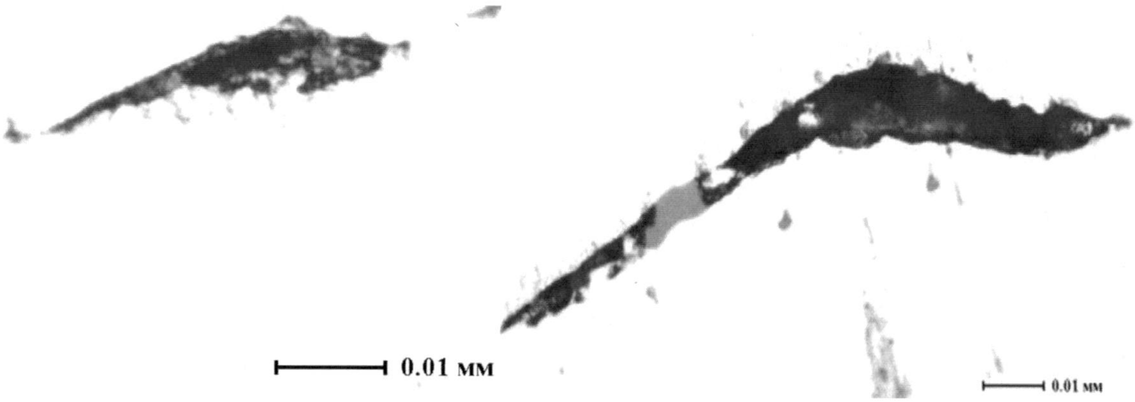

Figure 8. Manganese sulfides in the CCB central zone of E90KhAF steel after high temperature torsion.

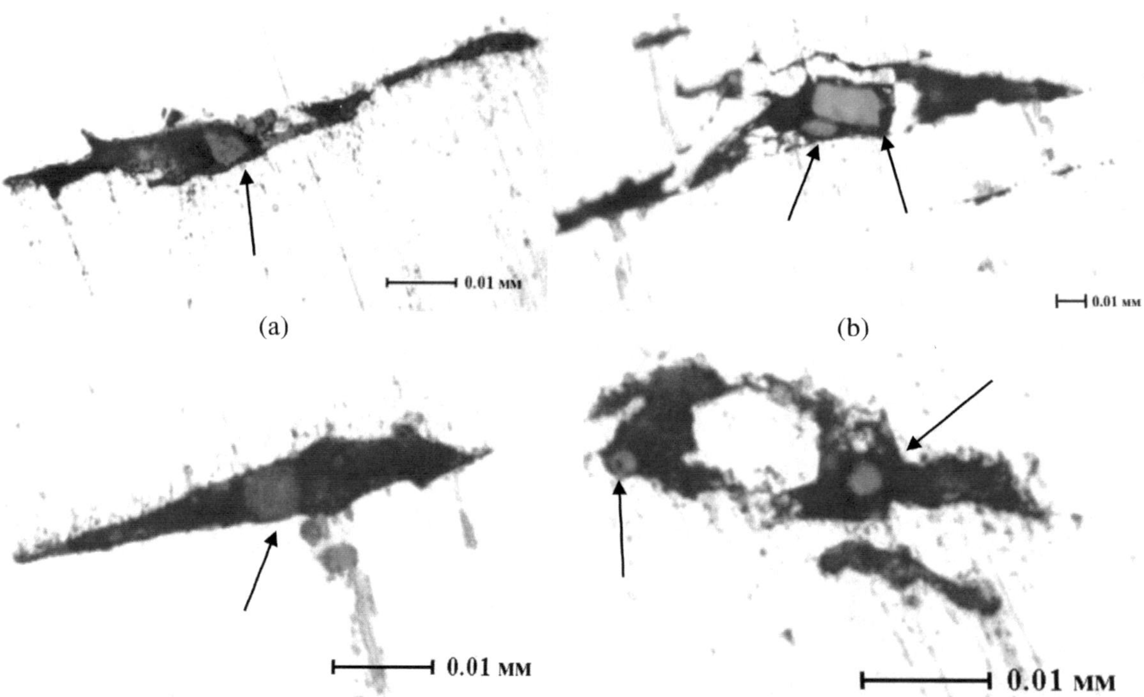

(a) (b)

(c) (d)

Figure 9. Non-metallic inclusions of the CCB central zone of E90KhAF steel after high temperature torsion: a, b – silicates of aluminum and manganese; c, d – iron silicates;

IOP Conf. Series: Materials Science and Engineering **866** (2020) 012014 doi:10.1088/1757-899X/866/1/012014

It should be noted that during plastic deformation at the "inclusion-matrix" interfaces, deformation and contact stresses arise due to the different deformability of the inclusions and the steel matrix. The inclusion and matrix constitute a system of a stressed (inclusion) and plastic (matrix) layer with dislocations at the interface [9].

3. Conclusion

Thus, the presence of various types and forms of non-metallic inclusions in steels allows the technological plasticity and mechanical properties of steels to be predicted as a whole.

Acknowledgments
The study was carried out with the financial support of the Russian Foundation for Basic Research and Kemerovo Region within the framework of scientific project No. 20-48-420011.

References

[1] Gubenko S I 1991 *Transformation of Non-Metallic Inclusions* (Moscow: Metallurgy) 225
[2] Starov R V, Parusov V V and Nesterenko A M 2000 *Int. Scientific-Technical Conference on Steel Production in the 21st Century. Forecast, Processes, Technology, Ecology* (Kiev) pp 167–168
[3] Starov R V, Litvinov L F and Derevyanchenko I V 2003 *Metallurgical and Mining Industry* **8** 25–27
[4] Simachev A S, Temlyantsev M V and Oskolkova T N 2016 *Steel in Translation* **2** 112–14
[5] Simachev A S 2014 *IV Intern. Scientific-Practical Conf. on Modern Innovations in Science and Technology* (Kursk) vol 4 pp 101–103
[6] Simachev A S 2014 *Int. Research Journal* **3-2** 63–64
[7] Starov R V, Derevyanchenko I V, Parusov V V and Sychkov A B 2005 *Steel* **1** 79–82
[8] Dobuzhskaya A B, Smirnov L A et al 2015 *Steel* **7** 82–6
[9] Gubenko S I, Parusov V V and Derevyanchenko I V 2005 *Non-Metallic Inclusions in Steel* (ART-PRESS) p 536

IOP Conf. Series: Materials Science and Engineering **866** (2020) 012015 doi:10.1088/1757-899X/866/1/012015

Patterns of changes in the plastic and deformation characteristics of rail steels alloyed with chromium in the temperature range of hot rolling

A A Umansky[1], A V Golovatenko[2], A S Simachev[1] and V V Dorofeev[2]

[1]Siberian State Industrial University, 42 Kirova Street, Novokuznetsk, 654007, Russia
[2]EVRAZ Consolidated West Siberian Metallurgical Plant, 16 Kosmicheskoye Shosse, 654043, Russia

E-mail: umanskii@bk.ru

Abstract. Based on the set of experimental studies for chromium rail steels of grades E76KhF, E76KhSF and E90KhAF, the basic laws governing the influence of hot rolling temperature on ductility and the regularities of temperature and strain rate influence in the temperature range of hot rolling on plastic deformation resistance are determined and scientifically substantiated. It was shown that for E76KhF steel at deformation temperatures of 1050 °C, a sharp decrease in ductility occurs due to the presence of cementite-type carbides $(Fe, Mn, Cr)_3C$ at the grain boundaries. For rail steels with a lower chromium content, the indicated ductility drop is not observed. It was established that an increase in the deformation temperature of rail steel E76KhSF causes a decrease in deformation resistance, and an increase in the strain rate significantly reduces this indicator. Also the fact that the deformation resistance decreases with distance from the surface to the central zone of continuously cast billets is established, which is due to an increase in grain size. Based on the obtained data, a new mode of rail rolling was developed in the conditions of the rail and beam workshop of EVRAZ ZSMK JSC, the introduction of which allowed the quality of the rails to be significantly improved as well as the technical and economic indicators of their production with an economic effect of approximately 120 million rubles/year.

1. Introduction

One of the trends in the development of rail rolling in Russia and the world is to increase the degree and complexity of alloying rail steels. The indicated trend was also reflected in the latest version of the state standard for the production of railway rails, which is currently in force (GOST R 51685-2013). So in this standard there are four types of rail steels alloyed with chrome in addition to vanadium (E76KhF), silicon (E76KhSF) and nitrogen (E76KhAF, E90EAAF), and in its previous version (GOST R 51685-2000) there was only one steel grade, at the same time doped with vanadium and chromium (E78KhSF).

A significant (up to 2 times) increase in the maximum permissible chromium content in such steels should also be noted. At EVRAZ ZSMK JSC, which is today the leading manufacturer of railway rails in Russia, currently E76KhF and E76KhSF steels are used as the main steels for the production of railway rails instead of previously used E76F steel.

The analysis of the literature data shows [1-5] that the theoretical justification of the optimal temperature ranges of deformation (at which the ductility has the maximum and the deformation

IOP Conf. Series: Materials Science and Engineering **866** (2020) 012015 doi:10.1088/1757-899X/866/1/012015

resistance is minimal) for complex alloyed steels is significantly complicated due to the complexity and multivariance of structural changes. Moreover, this problem is exacerbated in relation to the production of rail profiles, since their complex cross-sectional shape causes a significant non-uniformity of deformations during rolling.

Thus, we can conclude that, today, the determination of the patterns of change in ductility and deformation resistance of complex alloyed rail steels of a newly mastered range of products depending on the rolling temperature is an urgent task.

2. Methods of research

In the framework of the presented work, continuously cast billets with a cross-section 300×360 mm of current meltings in the electric steel-smelting shop of EVRAZ ZSMK JSC were used as an object of research to determine the ductility and deformation resistance. Samples were taken directly from the surface of billets, as well as at a depth of 50 mm, 75 mm and 110 mm from the surface of the billets in order to obtain maximum information about the plastic and deformation characteristics of steel in various zones of continuously cast billet. It is known that continuously cast billets are characterized by a three-zone structure, which includes a near-surface cortical zone, a columnar crystal zone and a central zone. According to the actual data of the structure analysis of the studied billets, the depth of the cortical zone is about 15–25 mm, the zones of columnar crystals are 50–60 mm, the rest is the central zone. That is, samples from all three zones of non-cast-in-place blanks were used for analysis.

Plasticity studies were carried out for samples of E76KhF steel, and studies of deformation resistance were carried out for steel E76KhSF, which is conditioned by the almost similar chemical composition of these steel grades according to the requirements of GOST R 51685-2013 (differences exist only in the ranges of variation of the allowable silicon and chromium content, while the indicated ranges overlap) and minor differences in the actual chemical composition of the analyzed bottoms of the current production.

To study the ductility, the hot torsion method was used in the temperature range 900–1200 °C, and the degree of shear strain was taken as a criterion for ductility. The studies of plastic deformation resistance were carried out on the unit Gleeble System 3800 using hot settling in the temperature range 900–1150 °C at strain rates of 1 s^{-1} and 10 s^{-1}.

The metallographic studies of rail steel samples were carried out using OLYMPUS GX-51 optical microscope and TESCAN MIRA 3 LMH scanning electron microscope with a Schottky field emission cathode. Shimadzu XRD-6000 X-ray diffractometer was used to determine the phase composition.

3. Results and discussion

According to the data obtained, the ductility of the near-surface and 50-mm-thick layers of continuously cast billets of rail steel E76KhF due to the deformation temperature has a complex behavior (figure 1): at the beginning of the indicated temperature range of deformation (with an increase in temperature from 900 to 1000 °C), plasticity intensively increases, then with a further increase in temperature, a noticeable decrease occurs with a minimum value at temperatures of 1025-1050 °C, after which the ductility sharply increases, reaching maximum at 1100 °C, and at the end of the considered temperature range (1100-1200 °C) is a smooth reduction in ductility.

The obtained experimental data indicate the absence of sharp changes in the plastic properties of continuously cast billets from E90KhAF steel and the absence of plasticity drop (figure 2) recorded for E76KhF steel, which is due to a significantly lower chromium content. It can also be noted that the absolute values of ductility for steel E90KhAF are lower compared to steel E76KhF, which is primarily associated with a higher carbon content.

It should be noted that according to the results of earlier studies of the ductility of continuously cast billets of E76F rail steel [6], there was also no recorded decrease in ductility at temperatures of 1050 °C. The effect of chromium on ductility is confirmed by the results of metallographic and X-ray diffraction studies of steel samples E76KhF: cementite carbides $(Fe, Mn, Cr)_3C$ were found at grain boundaries – figure 3. These carbides do not completely dissolve in austenite at temperatures up to

IOP Conf. Series: Materials Science and Engineering **866** (2020) 012015 doi:10.1088/1757-899X/866/1/012015

1050 °C and prevent the degree of growth shear deformations, that is, in fact, reduce the ductility of steel.

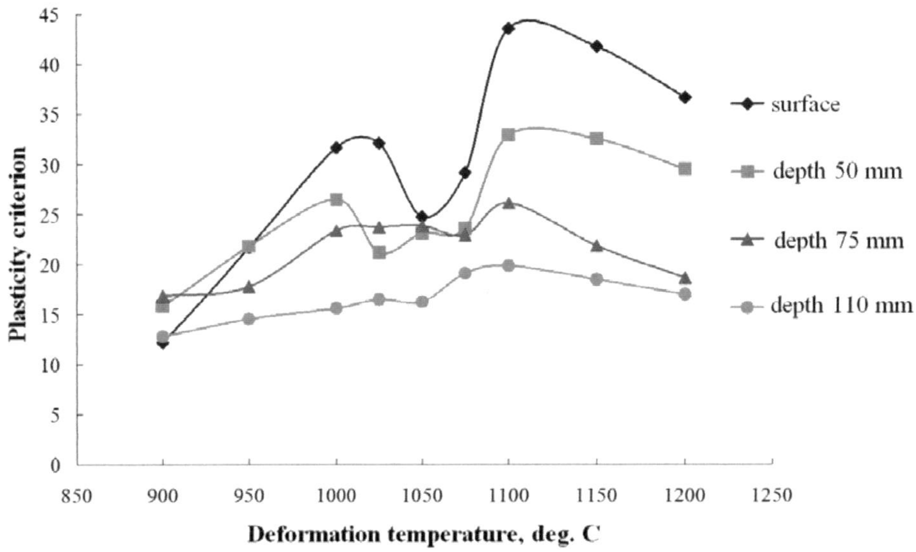

Figure 1. The ductility dependence continuously cast billets of rail steel E76KhF on the deformation temperature.

According to the research results, one can also note a general tendency for both considered steels to decrease ductility with distance from the surface to the central zone (figures 1, 2), which is due to the difference in grain size. Thus, according to the data obtained in the near-surface layer, the average grain diameter is 1.3–2.1 times smaller compared to the central zone of the billets, and there is a tendency for the increase in the difference in grain sizes and in strain temperature.

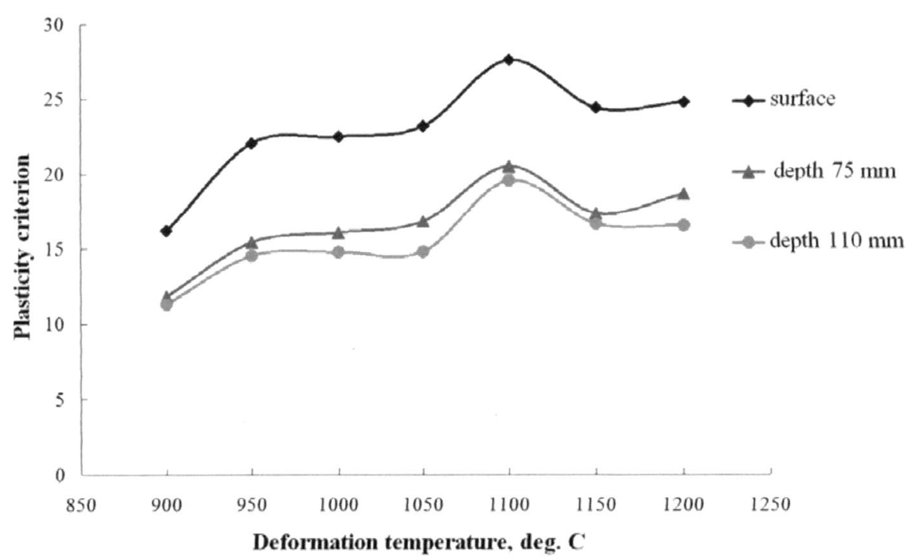

Figure 2. The ductility dependence of continuously cast billets of rail steel E90KhAF on the deformation temperature.

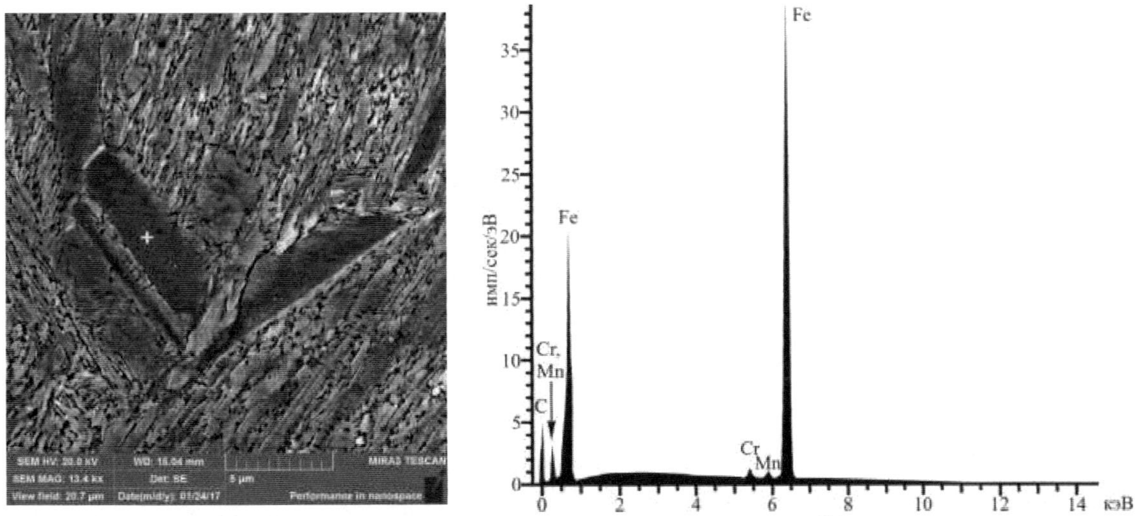

(a) (b)

Figure 3. Cementite carbides in steel grade E76KhF:
a – steel microstructure; b – distribution of elements in the carbide phase.

Based on studies of plastic deformation resistance, it was found that an increase in the deformation temperature of E76XSF rail steel causes a decrease in this indicator (figure 4). There is also a decrease in deformation resistance as the distance from the surface to the central zone of continuously cast billets increases, which, as shown above, is due to an increase in grain size. The significant effect of the increase in the strain rate on the increase in deformation resistance was established (figure 4), which is due to the acceleration of hardening processes, which are realized due to an increase in the density of dislocations and point defects, with an almost constant speed of the opposing softening processes [7–9].

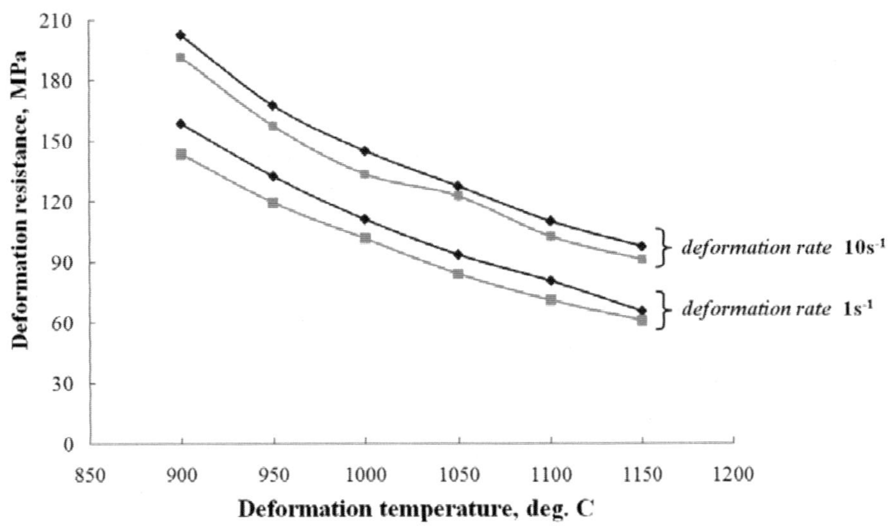

Figure 4. The deformation resistance dependence of continuously cast billets of rail steel E76KhSF on the deformation temperature.

Using the obtained experimental data on the ductility of E76HF rail steel, a new mode of rail production at the universal rail and beam mill of EVRAZ ZSMK JSC was developed, which made it possible to reduce the rejection of finished rails according to surface defects [10]. Improvement of the rails quality is achieved by redistributing the reductions between the passes in such a way that rolling in the split gauge (where the unevenness of the deformation has the greatest values) is carried out at temperatures close to the temperature of maximum ductility (1100 °C), and compression at temperatures of ductility drop (1025-1050 °C) have minimum values. To obtain the specified distribution of compression, the heating temperature of the billets for rolling was reduced from 1200 °C to 1170-1180 °C and the number of passes in the crimping stands was reduced. The limit of increase in reductions in individual passes from the point of allowable rolling force and moments on the shaft of the load engine is justified using the obtained experimental dependences of the strain resistance on temperature and strain rate.

The introduction of the developed rail production mode at the rail and beam mill of EVRAZ ZSMK JSC allowed the quality of the finished rails to be improved, and the specific energy consumption for rolling, fuel for heating billets, the specific consumption of rolling rolls to be reduced and the productivity of the mill to be improved. The total economic effect is estimated at 1250 million rubles/year.

4. Conclusion

The experimental studies established the dependences of the ductility and deformation resistance of rail steel alloyed with chromium on temperature-strain conditions. The data obtained made the theoretical basis for the development of a new mode of rail rolling, providing a significant increase in their quality and technical and economic indicators of production.

Acknowledgments

The reported study was funded by RFBR and Kemerovo region, project number 20-48-420011.

References

[1] Dimatteo A, Lovicu G, DeSanctis M and Valentini R. 2012 *Proceedings of Crack Paths* 131–139

[2] Lopez-Chipres E, Mejıa I, Maldonado C, Bedolla-Jacuinde A and Cabrera J M 2013 *Materials Science and Engineering: A* **460-461** 464–470

[3] Gladkovskii S V, Potapov A I and Lepikhin S V 2015 *Diagnostics, Resource and Mechanics of materials and structures* **4** 18–28

[4] Banks K M, Tuling A, Klinkenberg C, and Mintz B 2011 *Materials Science and Technology* **89** 537–545

[5] Konovalov A V, Smirnov A S et al 2016 *Metallurgist* **59** (11) 1118-1121

[6] Simachev A S, Temlyantsev M V and Oskolkova T N 2016 *Steel in Translation* **46** (2) 112–114

[7] Manonukul A and Dunne N 1999 *Acta Mater* **47(7)** 4339–4354

[8] Umanskii A A, Golovatenko A V and Kadykov V N 2017 *Izvestiya Ferrous Metallurgy* **60** (10) 804–810

[9] Umansky A A, Golovatenko A V, Kadykov V N and Dumova L V 2016 *IOP Conference Series: Materials Science and Engineering* **150** *012029*

[10] Umanskiy A A, Dorofeev V V, Golovatenko A V and Dobryanskiy A V 2018 *Cherniye Metally* **10** 38–42

IOP Conf. Series: Materials Science and Engineering **866** (2020) 012016 doi:10.1088/1757-899X/866/1/012016

New production technology of screw sections by drawing

A R Fastykovskii[1], E V Chinokalov[2], A N Prudnikov[1] and T N Oskolkova[1]

[1]Siberian State Industrial University, 42 Kirova Street, Novokuznetsk, 654007, Russia
[2]EVRAZ United West Siberian Metallurgical Plant, 16 Kosmicheskoe shosse, 654043, Novokuznetsk, Russia

E-mail: omd@sibsiu.ru

Abstract. The development of engineering and technology necessitates the search for new solutions related to the release of innovative products with high added value, which include long-length cold-worked screw sections. An analysis of the methods for producing long screw profiles is made, as a result of which a new method is proposed that differs from the known methods by the absence of a twisting operation for forming a helical surface. This feature of the new method for producing long screw profiles simplifies the design of the device and reduces energy consumption. The influence of the workpiece diameter and the number of passes of the helical surface on energy consumption was studied, and a comparative schedule was constructed that proves the advantages of the proposed method for producing screw sections by drawing. For the strength analysis of equipment, a dependence is proposed by which the force that occurs when drawing long screw profiles can be found, taking into account the deformation conditions and design features of the device forming the screw surface. Using the fourth theory of strength, a dependence is obtained to determine the condition that guarantees the absence of breaks in the process of drawing a screw profile.

1. Introduction

The development of engineering and technology necessitates the search for breakthrough solutions in various fields, especially in basic ones, such as metallurgy. Thanks to the development of new types of products, the greatest effect can be obtained, which, in turn, contributes to the development of machine building and construction industries. With this interaction, the development efficiency becomes more significant.

New perspective sections rightfully include screw sections, from which various products used in construction are produced. In Japan, the USA, and China, extensive work is being carried out on the industrial production of screw sections on a large scale. From the point of view of the development prospects of this area, special attention is paid to screw sections for construction purposes: reinforcing bars, nails. Constantly growing requirements for buildings and structures dictate the need to improve reinforcing sections in terms of finding optimal surface shapes, while guaranteeing high strength and reliable adhesion to concrete.

The evolutionary change in the cross-sectional shape of reinforcing sections led consumers to the conclusion that reinforcement with a helical surface obtained by cold working has optimal technological characteristics. Compared to conventional, such reinforcement with a lower mass and similar strength has a greater adhesion area to concrete. However, despite all the advantages, multi-way screw reinforcement is mastered with great difficulty by enterprises. The reason for this state of affairs is a lack of theoretical knowledge of the process of helical surface formation, the structural complexity of the equipment, the lack of reasonable recommendations for choosing rational modes of forming. The same situation is with screw nails.

As follows from the literature, at present, cold-deformed screw profiles are obtained by drawing in a forcefully rotary die [1], by drawing, followed by twisting to obtain a given pitch of the screw surface [2]. In both cases, when forming a screw section, it is necessary to additionally apply torque. Such feature, on the one hand, increases energy costs for production, and on the other hand, complicates equipment, making it necessary to use, in addition to the drawing mill, additional devices that provide twisting at a certain angle during the drawing process. To eliminate the disadvantages of the known solutions, a new drawing method [3] and a tool [4] have been developed, the design of which, due to the application of longitudinal force, provides drawing with simultaneous rotation of the tool around an axis with a certain adjustable frequency, and a screw profile is formed on the surface of the workpiece. This process is stable and guarantees a constant pitch of the helical surface when the drawing speed changes. The geometrical parameters of the tool for obtaining screw profiles are similar to the parameters of a standard die. This feature greatly simplifies the development of screw sections on existing equipment, without requiring additional costs for reconstruction.

2. Theoretical aspects of screw sections production by drawing

It was noted above that the proposed method and equipment that implements it do not provide for the presence of an additional drive, thereby reducing energy costs for obtaining screw profiles. Let us evaluate the effectiveness of the proposed method by comparing the energy consumption for its implementation with the energy consumption of the combined drawing with twisting. Reduction in energy costs when obtaining cold worked screw sections by drawing with the help of the proposed method in comparison with combined drawing with twisting can be found using the dependence:

$$\Delta N = N_{w1} + N_{tw} - N_{w2} - N_{rot} , \tag{1}$$

where ΔN is the additional power required when forming a helical surface by drawing with subsequent twisting; N_{w1} – power to obtain longitudinal grooves for subsequent twisting; N_{w2} – power for forming a helical surface by rolling; N_{tw} – power when twisting a screw profile; N_{rot} is the power for rotating the die in bearings.

After substitution and transformations, we obtain the calculated dependence for determining ΔN in the form:

$$\Delta N = \frac{6,28v}{h \cdot k}\left(0,12\sigma_s D^3 - Pd_{bear} f'\right), \tag{2}$$

where v – the speed of drawing; h – the pitch of the screw profile; k – the number of runs of the helical surface; σs – the deformation resistance of the workpiece material; D – the diameter of the workpiece; P – the drag force; d_{bear} – diameter of the thrust bearing; f′ – the coefficient of friction in the bearing.

The absolute value of ΔN shows how much more power is required to obtain screw sections by drawing with twisting. According to formula (2), technological factors and the design of the deforming device influence ΔN. Usage of the absolute value ΔN gives an idea of the energy efficiency of the proposed method, however, it does not allow the benefit in rubles to be evaluated. To determine the material benefits of the proposed solutions, energy consumption must be presented in specific values kW·h/t. Using dependence (2), we obtain a formula for estimating the specific energy consumption (Δe) in the form:

$$\Delta e = \frac{v \cdot t}{593,2G \cdot h \cdot k}\left(0,12\sigma_s D^3 - Pd_{bear} f'\right), \tag{3}$$

where t – the drawing time of one workpiece; G – the mass of the workpiece.

A more obvious possible change in energy consumption depending on the geometrical parameters of the screw section is shown in figure 1.

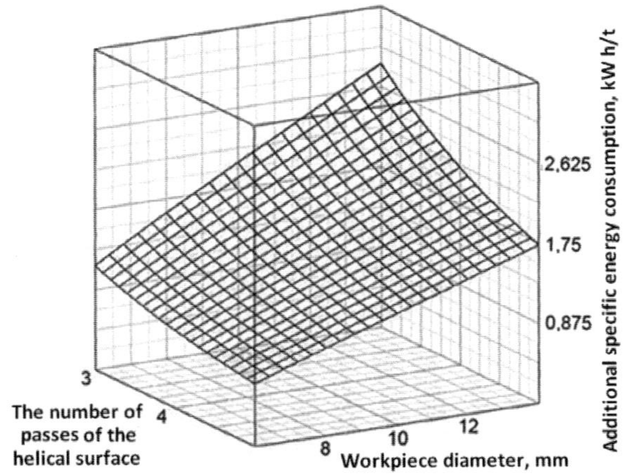

Figure 1. The diagram of the dependence of the additional specific energy costs required for drawing with subsequent twisting on the number of passes of the helical surface and the diameter of the workpiece.

The economic effect (ΔE) from the use of the developed method and tool when receiving helical cold worked sections can be calculated by the formula:

$$\Delta E = \Delta e \cdot P_E \, , \qquad (4)$$

where P_E is the price of kW·h of electricity for enterprises.

An analysis of the effectiveness of using the proposed method for producing multi-helical screw sections in relation to the conditions of JSC EVRAZ ZSMK was made using the above formulas to determine energy consumption. According to the data obtained, an increase in the diameter of the drawn wire from 6 to 14 mm during the formation of a three-way surface allows production costs to 9-10 rub/t to be reduced.

To verify the operability of technological equipment, it is necessary to carry out strength calculations based on knowledge of the force required during deformation. The drawing force that occurs when obtaining long multi-pass screw sections can be found by the formula;

$$r = 1,6k\sigma_s\sqrt{\Delta h^3 R_p}\left(1 + f \cdot ctg\varphi\right), \qquad (5)$$

where Δh – the amount of compression during the formation of the helical groove; φ – the angle depending on the pitch of the helical surface; f – the coefficient of friction in the deformation zone; R_r is the radius of the deforming roller.

$$\varphi = \arccos\left(\frac{3hS_1 \ln\dfrac{S_0}{S_1}}{4\pi\Delta h\sqrt{R_r\left(D-\Delta h\right)^3}}\right), \qquad (6)$$

where S_0, S_1 are the cross-sectional areas of the workpiece and the screw section.

The permissible resulting stress at the end located in the grips during drawing can be determined using the fourth strength theorem:

$$[\sigma] \geq \sqrt{\left(\frac{1,6k\sigma_s\sqrt{\Delta h^3 R_r}\left(1+f\cdot ctg\varphi\right)}{S_1}\right)^2+3\left(\frac{P\left(D-\Delta h\right)\sin 2\varphi}{0,8D_g^3}\right)^2}, \quad (7)$$

where D_g is the diameter of the end of the profile located in the grips.

The given dependences for determining the drawing force of long-length screw sections, the condition for the strength of the cross-section to which the load is applied, are necessary for strength calculations of equipment, choosing the drive of the drawing mill, and determining the probability of breaks in the drawing process. Successful experience in obtaining long screw sections in the current production environment is given in works [5 - 9].

3. Conclusions

1. The possibility of obtaining innovative products – cold-rolled screw sections in the conditions of drawing production of EVRAZ ZSMK JSC was shown.
2. The energy efficiency of the proposed solution for the production of cold-rolled screw sections in comparison with known methods is proved.
3. Formulas have been obtained, the use of which allows energy-efficient modes for producing cold-rolled screw sections to be determined, check the equipment for strength, and assess the probability of breaks during drawing.

References

[1] Parshin S V 2007 *Production of Rolled Products* **12** 27–30
[2] Kargin V R 1994 *Processes for Producing Screw Profiles and Pipes* (M: Metallurgiya) p 96
[3] Fastikovskii A R et al 2015 *Patent of the Russian Federation* No 2553728 publ. 06.20.2015
[4] Fastikovskii A R, Chinokalov E V et al 2014 *Utility Model Patent of the Russian Federation* No 143099 publ. 07.10.2014
[5] Fastikovskii A R, Chinikalov E V and Milovanov A G 2014 *Metallurg* **6** 128–131
[6] Fastikovskii A R, Chinikalov E V et al 2014 *Steel* **10** 48–50
[7] Fastikovskii A R, Chinokalov E V et al 2013 *Production of Rolled Metal* **10** 40–42
[8] Fastikovskii A R and Chinokalov E V 2019 Steel 7 50–51
[9] Fastikovskii A R and Chinokalov E V 2019 *Ferrous Metals* **6** 35–37

IOP Conf. Series: Materials Science and Engineering **866** (2020) 012017 doi:10.1088/1757-899X/866/1/012017

The influence of technological and labor processes on the efficiency of the rolling complex

V N Kadykov and A I Musatova

Siberian State Industrial University, 42 Kirova Street, Novokuznetsk, 654007, Russia

E-mail: musatova-ai@yandex.ru

Abstract. Studies on the organization of technological and labor processes in the existing rolling complex (long product rolling mill) were carried out. Using normative clock models of equipment operation in the areas of the complex, a "bottleneck" was identified that constrains the work of the human-technical system, which is the packaging section for finished products, where baling is performed by workers manually. In order to eliminate the "bottleneck", it is proposed to modernize the baling area by introducing units for mechanizing and automating the processing of finished products, to eliminate manual labor, improve the formation of current information. The possibility of improving technical and economic indicators (reducing the number of workers, increasing production, labor productivity, etc.) is shown.

1. Introduction

The production process at a modern metallurgical enterprise is a complex production flow, starting from the receipt of raw materials, materials, semi-finished products and ending with the receipt and shipment of finished products. However, the resulting quantitative and qualitative indicators of this complex production flow is not decided by one unit or person, but by the general totality and optimal coordination in the actions of personnel, the operation of units and the organization of material flows.

2. Analysis of the organization of production and labor by sections in the long product rolling mill

The rolling production is a part a complex production process, which occupies 75% of the total metallurgical cycle, and which is the final stage in the production of finished products [1 – 4].

For a more detailed organizational analysis of the production process, the long product rolling mill of a metallurgical enterprise was selected, which is intended for the manufacture of products of a sectional iron assortment. The conducted comprehensive research and analysis of the organization of production and labor by sections of the mill allowed the following conclusions to be drawn. Highly automated and mechanized equipment operates in the areas of heating, rolling, heat hardening, cooling and cutting. The technological processes proceed continuously, the sections are interconnected by a single hot metal stream; any delays, stops in its operation reduce productivity of the mill, labor productivity, output and profitability of products (figure 1).

In the study of the organization of technological and labor processes, it turned out that the least mechanized and most labor-intensive area using manual labor is the metal baling area, which employs 32 packaging workers and 16 ready-made stampers. Currently, at the bailing site, metal baling is carried out in stationary bins by workers manually. Such work is associated with high physical efforts and performed at elevated temperatures.

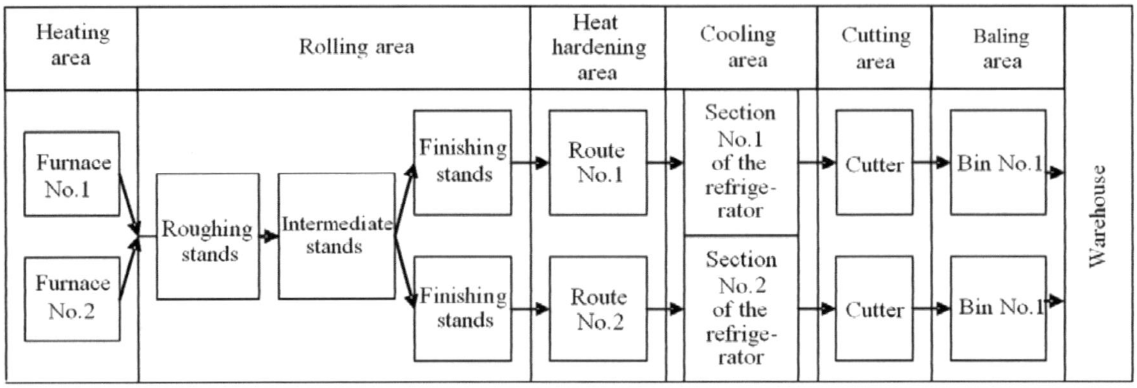

Figure 1. Layout of production areas of a long product rolling mill.

The main disadvantages of the existing organization of labor at the baling site are: a significant number of workers due to the large number of manual operations associated with baling, stamping, marking and cleaning of metal; low labor productivity, especially during baling of small-tonnage packs of finished products; a tight connection between the operation of the baling and rolling sections leads to a restraint of the production process, since the time for baling, labeling and cleaning of the steel exceeds the duration of rolling; low quality of rolled metal baling does not always meet the requirements of consumers; the impossibility of tight baling of rolled metal bundles and a durable knot in knitting; lack of automated metal-roll metering in bundles.

3. Clock approach to justification of the bottleneck in the system (mill)

In order to justify the degree of influence of the baling area on the coordinated work of the mill, functioning as a continuous-flow human-technical system, the technically possible and normative clocks time at production sites were determined. The clocks of the system (mill), subsystem (areas), elements (equipment), microelements (mechanisms and actions of workers) have their own internal structure, which corresponds to the structure of technological and labor operations. Clock should be considered as an important temporal characteristic affecting the performance of the system, which is one of the main design elements in planning the volume of production, which means labor productivity and many other technical and economic indicators [5].

The clock period is the time interval between the processing moments of the same name (beginning or end) of two adjacent units of production (blanks, bags or bundles) on the same equipment.

In the structure of a complex production process, each unit of production (equipment, site) works with its own clock period, depending on the type of manufactured product (assortment) and production situations. Situational clock cycles of the system are associated with the corresponding clock cycles of its subsystems and elements. Moreover, each clock cycle has boundaries ("fixing points"), which must be determined during the study of the operation of equipment, process conditions, material flows, and the work of maintenance personnel.

In general, the equipment operation cycle consists of the following elements:

$$T_{eq} = t_1 + t_2 + t_3 + t_4, \tag{1}$$

where t_1 is the time of workpiece feeding (loading) into the unit (equipment); t_2 – workpiece processing time in the unit; t_3 – time of unloading of the workpiece from the unit; t_4 – preparation time of the processed workpiece for the next operation.

The authors developed the clock models to assess the degree of rationality of the functioning production processes in rolling mills [6 – 9]. Their use makes it possible to solve the problems of improving multivariate standardization, planning, forecasting and operational management of technological and labor processes. When rationing, a combination of methods was used: simulation modeling, nominal optimum, mathematical analysis, expert estimates, statistical data processing.

The single technically possible (normative) clock period of work of a long product rolling mill was selected from the clock cycles of the areas according to the criterion of the maximum time costs:

$$T_s^{tp(n)} = max\left\{T_{y1}^{tp)n)},\quad T_{y2}^{tp(n)},\dots,\ T_{y6}^{tp(n)}\right\}, \qquad (2)$$

where $T_{y1}^{tp(n)},\dots,\ T_{y6}^{tp(n)}$ – technically feasible (normative) clock period of the site operation, s.

The procedure for selecting technically possible clock cycles of the system (mill) for each section size was carried out by modeling the coordinated work of the main and auxiliary equipment in all areas of the mill. At the same time, a lot of estimated cycles were formed, options for the occurrence of production situations were considered depending on changes in the external and internal conditions of the human-technical system functioning.

According to the developed regulatory model, technically feasible clock periods, durations of operations and cycles were calculated, standard values of equipment utilization coefficients were established, then normative clock periods of areas were determined taking into account the assortment and production situations by formula (1). The results of the implementation of the clock model are shown on the example of the section production, table 1.

Table 1. The choice of the clock period of a long product rolling mill on the example of a profile (reinforcing bar No. 14).

Clock periods of the production areas in the mill												Clock period of the mill
Heating in the furnace		Workpiece rolling in the mill stand		Heat hardenings on the line		Cooling on the sections		Cutting by the cutters		Baling on the line		
one	two	one	two	one	two	one	two	one	two	one	five	
1. Technically possible ckock periods before modernization, s												
30.0	15.0	42.5	21.75	41.8	20.9	33.0	16.5	36.5	18.25	119.0	**23.8**	**23.8**
Regulatory utilization rates of equipment before modernization												
0.90		0.92		0.92		0.91		0.93		**0.85**		**0.85**
2. Standard clock periods before modernization, s												
33.3	16.7	47.2	23.6	45.4	22.7	36.3	18.7	17.7	8.9	140.0	**28.0**	**28.0**
3. Technically possible ckock periods after modernization, s										one	four	
30.0	15.0	42.5	**21.75**	41,8	20.9	33.0	16.5	16.5	8.25	56.0	14.0	**21.75**
Standard utilization rates of equipment after modernization												
0.90		**0.92**		0.92		0.91		0.93		0.90		**0.92**
4. Standard clock periods after modernization, s												
33.3	16.7	47.2	**23.6**	45.4	22.7	36.3	18.7	17.7	8.9	62.2	15.6	**23.6**

A comparative analysis of the obtained results of technically possible and standard values of the clock periods showed that the “bottleneck” in the continuous flow system is the baling area, which limits the work of the leading rolling area.

In order to improve conditions and reduce manual labor of workers at the baling site, as well as to ensure the clock operation of the mill in a continuous process flow, it is proposed to mechanize and automate the system for baling the rolled metal into packages, which will make it possible to: improve the quality of finished packages; reduce time spent on getting the finished package; reduce the cost of strapping materials; to increase the competitiveness of products on the market due to presentation and compliance of rolled products with the requirements of world standards; to reduce the number of manual operations by installing tying machines, which eliminates the human factor in the strapping process; to increase accounting, control and tracking of current package information by installing a monitoring tracking system, expand the ability to supply packages depending on the weight, length and number of rods.

Currently, there are 44 people (28 packers and 16 stampers) at the baling site according to the staff schedule and according to the norm of working personnel. As a result of the introduction of production lines, 15 people are freed (9 packers and 6 stampers). After the proposed implementation of the automated production complex, the technically feasible and normative clock period for the baling site was determined.

Due to the fact that the proposed modernization is being implemented in the baling area, the values of the technically feasible and normative clock periods in this area are changing, and in the rest they remain unchanged, this is clearly shown in Table 1.

Based on the simulation of the operation of the main and auxiliary equipment in mill areas, the system clocks (technically possible and normative) for a given assortment were determined (circles with a diameter of 12-14 mm, reinforcing bars No. 10-14, angle section). Then the indicators of the mill production program are calculated, table 2.

Table 2. Prediction of indicators of the production program of the long product rolling mill.

Indicators	Unit of measure	Calculated values for PM_1	Actual values	Standard indicators before modernization	Standard indicators after modernization	Calculated values for PM_2
1. Calendar time	day	365	365	365	365	365
	h	8760	8760	8760	8760	8760
2. General maintenance	day	–	–	–	3	–
	h	–	–	–	72	–
3. Preventive maintenance	day	15	15.83	15.83	12.83	15
	h	360	380	380	308	360
4. Rated time	h	8400	8380	8380	8380	8400
5. Ongoing downtime:						
– technical	h	140	164.3	158	158	140
– technological	h	185	210.8	192.6	192.6	185
– organizational	h	360	395.5	372	365	354
(including delays in the baling area)	(h)	(6)	(10)	(7)	(–)	(–)
– independent	h	–	40.2	–	–	–
Total current downtime	h	685	810.8	722.6	715.6	679
6. Current downtime at nominal time	%	8.2	9.7	8.6	8.5	8.1
7. Actual time	h	7715	7569.2	7657.4	7664.4	772.1
8. Mill productivity	t/h	107.3	76.4	91.1	107.8	117.1
9. Output	t	–	578286.88	697589.14	826222.32	–
10. Productive capacity PM_1 and PM_2	t	827819.5	–	–	–	904129.1

Thus, after the proposed modernization in the baling area by introducing an automated process for baling rolled products aimed at eliminating manual labor, the work cycles of this area will be reduced. As a result, this area will not restrain the operation of the main (leading) section, and the cycle of the system will correspond to the cycle of the rolling section. After the proposed modernization in the baling area (taking into account capital investments), output will increase by 18%. In addition, the number and wages of workers will be reduced, labor productivity will increase, labor intensity of production will decrease, production costs will decrease, profit and profitability of marketable products will increase.

4. Conclusion

Based on the analysis of technical and economic indicators, a detailed study of equipment, technology, organization of production and labor, comprehensive research at each area of the long product rolling mill it was proved that the efficiency of the production process is affected not only by technological, but also to a large extent by labor processes. The developed regulatory model for the functioning of the long product rolling mill allowed the regulatory framework for technical and economic indicators necessary for planning, forecasting and operational management of production to be formed.

References

[1] Musatova A I , Kulakov S M and Kadykov V N 2011 *Steel in Translation* vol 41 No 294–300
[2] Genkin B M 2013 *Organization, Regulation and Remuneration of Labor at Industrial Enterprises* (Moscow: NORM) p 480
[3] Aniskin Yu P 2018 *Production Organizer* vol 26 No 1 67–72
[4] Kulakov S M, Musatova A I, Kadykov V N and Baranov P P 2015 *Steel in Translation* vol 45 No 5 367–375
[5] Musatova A I, Kulakov S M and Kadykov V N 2010 Izv. Vuzov. Ferrous Metallurgy 2 47–51
[6] Musatova A I, Kadykov V N, Kulakov S M 2014 *Scientific Review* **5** 253–262
[7] Musatova A I, Kulakov S M 2014 *Scientific Review* **3** 139–149
[8] Kulakov S M, Musatova A I and Kadykov V N 2018 *Izv. Vuzov. Ferrous Metallurgy* vol 61 no 2 164–167

IOP Conf. Series: Materials Science and Engineering **866** (2020) 012018 doi:10.1088/1757-899X/866/1/012018

Deformation, heat treatment and properties of piston hypereutectic silumins

V K Afanasyev[1], A N Prudnikov[1], V A Prudnikov[2], M V Popova[1] and A R Fastykovskii[1]

[1]Siberian State Industrial University, 42 Kirova Street, Novokuznetsk, 654007, Russia
[2]LLC "MSB Engineering", 11-A Pavlovskogo Street, 654007, Russia

E- mail: a.prudnikov@mail.ru

Abstract. A series of piston hypereutectic silumins based on Al – (15 ÷ 20) % Si alloyed with copper, magnesium, nickel, and chromium was studied. The alloys were modified with phosphorus and hydrogen-containing reagents to increase the deformability and mechanical properties. The experimental ingots, cast in the steel mold, had dimensions: diameter 190 mm, height 500-550 mm, weight 32-40 kg. The mechanical properties and temperature coefficient of linear expansion (TCLE) of the experimental ingots and forgings made from them are determined. Temporary tensile strength and relative elongation of deformed ingots from hypereutectic silumins were 227-306 MPa and 5.7-7.5%, respectively. The optimal heat treatment of deformed silumins was determined: quenching with step heating and aging, which allows the strength of forgings to be increased up to 370-470 MPa. Moreover, the ductility indicators remain at a high level, and the average TCLE of the alloys is $(18.0 \div 19.2) \cdot 10^{-6}$ K^{-1} in the range of 50 ÷ 200 °C.

1. Introduction

The analysis of the working conditions of pistons in modern heavily loaded engines [1-3] shows that in the process of operation they perceive significant dynamic loads that change during one cycle in sign and direction. Moreover, the gas pressure on the piston can reach 800 MPa. Another feature of the piston operating conditions is the high temperature of fuel combustion products (~ up to 2000 °C). This determines the heating of the piston bottom under certain engine operating conditions to 350-400 °C. Given the piston operating conditions, hypereutectic alloyed silumins are currently one of the most promising materials for the manufacture of pistons for internal combustion engines. This is determined by the favorable combination of low specific gravity, low temperature coefficient of linear expansion (TCLE) with good mechanical and technological properties. However, the presence of coarse crystals of primary silicon in the structure of hypereutectic silumins reduces their ductility and does not allow pistons to be produced by the pressure treatment in the industrial conditions [2-7].

To ensure the necessary complex of physicomechanical characteristics of workpieces from piston alloys, the final heat treatment is applied [3, 8-10]. Therefore, the study of the tendency of experimental high-silicon deformable silumins to hardening heat treatment and to TCLE change is of practical interest and is the aim of this work.

2. Materials and methods of research

A series of experimental ingots from hypereutectic silumins based on Al – (15÷20) % Si alloyed with copper, magnesium, nickel, chromium and other elements was made. In order to ensure good deformability and a high complex of mechanical properties, the alloys were modified with phosphorus and hydrogen-containing reagents [2,3]. The diameter of the experimental ingots was 190 mm, the height was 500-550 mm, and the weight was 32-40 kg. The chemical composition of the alloys is given in table 1.

Table 1. The chemical composition of the experimental piston hypereutectic silumins.

Alloy No.	Alloy components, weight. %									
	Si	Cu	Mg	Ni	Mn	Cr	Ti	P	H	Al
1	15	3	0.2	-	-	-	-	0.01	0.00008	res.
2	15	3	0.2	1.0	-	-	-	0.01	0.0001	res.
3	15	5	1.1	1.0	-	-	-	0.01	0.0003	res.
4	18	4	0.6	-	-	-	-	0.02	0.0003	res.
5	18	4	0.6	1.0	-	-	-	0.02	0.0006	res.
6	20	5	1.1	-	-	-	-	0.03	0.0006	res.
7	20	1	0.4	-	0.8	0.3	0.1	0.01	0.00008	res.

After removing the cast surface and cutting the ingots into billets, they were hot forged. The total degree of deformation was $\varepsilon = 94\%$, and the total forging coefficient was $K_{tot} = 28$. Forging of blanks was carried out on a pneumatic forging hammer MV 412 with a hammer mass of 160 kg. Before forging, the ingots were annealed to obtain an equilibrium structure in them. The annealing temperature and the holding time were 450-470 °C and 1.5–3 h. During the forging process, intermediate annealings were performed at a temperature of 450±10 °C for 0.5–1 h [11-14].

The heat treatment of the forgings was carried out in resistance furnaces of SNOL type. For metallographic analysis of ingots and forgings, an OLYMPUS GX-51F optical microscope was used. The microstructure of the ingots was studied on microsections prepared from transverse templates cut at the same height from the bottom of the ingot. The TCLE of the alloys was determined using a high-temperature differential dilatometer. The error in determining the TCLE was $0.1\cdot10^{-6}$ K^{-1}.

3. Results and discussion

The microstructure of the experimental ingot of doped hypereutectic silumin based on Al-18% Si is shown in aigure 1. It is established that the microstructure of the alloys in the cast state is inhomogeneous over the cross section of the ingot. According to its cross section, three characteristic zones can be distinguished:

- surface zone – is a coarse-crystalline rim 3 ÷ 5 mm wide with the largest volume fraction of primary silicon crystals;
- intermediate zone – consisting of needle-type eutectics, sections of α-solid solution and primary silicon crystals. Dark desorption phase of Mg_2Si and light rounded ones of $CuAl_2$ are also clearly visible.
- central zone – the microstructure consists of double and more complex eutectics, sections of an α-solid solution of complex composition and crystals of primary silicon (CPS). Moreover, the volume fraction of CPS in the ingot central zone is much less than in the intermediate.

Table 2 shows the mechanical properties and temperature coefficient of linear expansion of alloys in the cast state. It can be seen that despite the structural differences between the edge (intermediate zone) and the central parts of the ingots, their hardness differs slightly and amounts to 840-1020 and 813-1006 MPa, respectively. Temporary tear resistance of ingots from experimental hypereutectic silumins is in the range of 80-153 MPa. Their average value of the coefficient of linear expansion, determined in the range of 50-200 °C, is $(17.2\div19.5)\cdot10^{-6}$ K^{-1}, and in the range of 200-450 °C – $(17.7\div21.4)\cdot10^{-6}$ K^{-1} respectively.

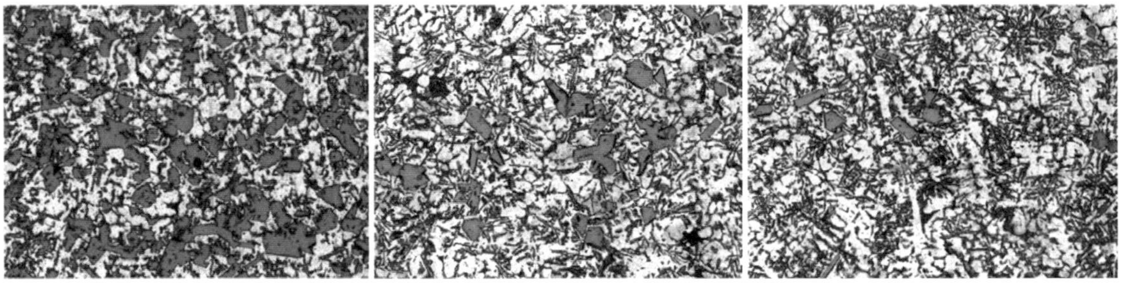

(a) (b) (c)

Figure 1. The ingot microstructure from an experimental alloyed silumin Al-18 % Si-4 % Cu-0.6 % Mg: a – surface zone; b – intermediate zone; c – central zone; ×100.

Before forging, the ingots were annealed at 450–470 °C for 1.5–3 h to eliminate the heterogeneity of the structure and composition of the alloys. The mechanical properties of the experimental forgings from silumin are σ_r = 227÷306 MPa, δ = 5.7÷7.5 %.

Heat treatment of forgings from alloyed hypereutectic silumins included hardening and aging. Hardening in this case is achieved by fixing the maximum supersaturated α-solid aluminum solution during quenching and its decomposition during subsequent aging due to the decrease in the solubility of the alloying components with decreasing temperature.

Table 2. Physico-mechanical properties of ingots from alloyed hypereutectic silumins.

Alloy No.	Hardness, HB, MPa		σ_r, MPa	Average TCLE, $\alpha \cdot 10^{-6}$ K^{-1} in the range of temperatures, °C	
	edge	centre		50-200	200-450
1	873	853	153	19.1	20.8
2	840	813	148	18.9	19.7
3	1020	1006	125	19.2	20.6
4	986	928	138	19.5	21.4
5	980	876	124	16.7	17.7
6	943	941	86	18.7	19.9
7	958	950	80	19.0	20.5

The analysis of the state diagram of the Al-Si-Cu-Mg system, to which these alloys belong, shows that they have a complex phase composition. Seven phases can be in equilibrium with the aluminum solid solution: Si, θ ($CuAl_2$), β (Al_2Mg_3) or (Al_8Mg_5), Mg_2Si, S (Al_2CuMg), T (Al_6Cu Mg), W ($Cu_2Mg_8Si_6Al_5$) [4]. The main hardeners are the phases $CuAl_2$, Mg_2Si, S (Al_2CuMg) and partially W ($Cu_2Mg_8Si_6Al_5$). The complex phase composition of the alloys makes it possible to carry out a large number of invariant reactions in a wide temperature range of 444–577 °C [15] and, consequently, the formation of complex fusible eutectics, the melting temperature of which determines the quenching heating conditions, is possible. To ensure the complete dissolution of the alloying elements and prevent the possibility of burnout, it is recommended to use step heating for hardening. The need for such heating for these alloys is confirmed by the results of experiments on the influence of the heating temperature for quenching in cold water and the exposure time on the mechanical properties of alloys (aging mode – 150 °C, 5 h), shown in table 4 for alloy composition No. 4 (see table 1).

The analysis of the data in table 3 shows that the satisfactory combination of strength and ductility is achieved after quenching in modes 2 and 3, providing for step-by-step heating under quenching at temperatures of 480 ± 10 °C (first stage) and 490 and 500 ± 10 °C (second stage) with holding at these temperatures 0.5-1 h and subsequent aging. After carrying out the indicated heat treatment regimes, the temporary resistance of the forgings is 450-460 MPa, the elongation is 4%, the relative narrowing is 5.8-8.3%.

Table 3. Mechanical properties of forgings from alloys Al-18 % Si-4 % Cu-0.6 % Mg, depending on quenching conditions (aging 150 °C, 5 h).

No.	Quenching temperature and holding time	Mechanical alloys		
		σ_r, MPa	δ, %	Ψ, %
1	Without treatment	306	7.2	13.4
2	480±10 °C, 2 h → 490±10 °C, 2 h	450	4.0	5.8
3	480±10 °C, 2 h → 500±10 °C, 0.5 h	460	3.8	8.3
4	480±10 °C, 2 h → 500±10 °C, 1 h	425	1.0	0
5	490±10 °C, 1 h	437	1.2	0
6	490±10 °C, 3 h	437	1.0	0

After quenching according to regimes 5 and 6, which do not provide for the use of step heating, burnout processes occur in the alloys, which leads to a sharp decrease in ductility to 1-1.2%.

Thus, the following was adopted as the optimal heat treatment regime for piston alloys: quenching – stepwise heating 480 ± 10 °C → (490-500)±10 °C, with a holding 2 and 1 h, respectively, at the lower and upper stages, cooling in cold water and subsequent artificial aging at 150 °C for 5 hours. Table 4 shows the physicomechanical properties of the studied deformable alloys after heat treatment according to the specified mode.

Table 4. Physico-mechanical properties of forgings after quenching and aging.

Alloy No.	Mechanical properties			Average TCLE, $\alpha \cdot 10^{-6}$ K^{-1} in the range of temperatures, °C	
	σ_r, MPa	δ, %	Ψ, %	50-200	200-450
1	425	6.4	9.9	18.6	22.4
2	372	4.6	7.4	19.0	22.1
4	450	4.0	5.8	18.0	20.6
5	400	2.8	6.1	19.2	21.5
6	470	5.0	5.2	18.8	22.0
7	400	3.5	-	19.0	22.7

From these tables it can be seen that as a result of quenching and aging, the tensile strength of the deformed hypereutectic aluminum-silicon alloys significantly increases to 370-470 MPa. Moreover, the ductility indicators remain at a high level for piston silumins, and the average TCLE of the alloys after heat treatment is $(18.0\div19.2)\cdot10^{-6}$ K^{-1} in the range 50÷200 °C and $(20.6\div22.7)$ 10^{-6} K^{-1} in the range of 200 ÷ 450 °C.

Metallographic studies of deformable doped hypereutectic silumins after quenching and aging showed that in the microstructure of heat-treated forgings, particles of primary and eutectic silicon take a more round and spheroidized shape. The phases Mg_2Si and $CuAl_2$, which did not completely dissolve upon heating and deformation of the workpieces, after quenching dissolve in the α-solid solution of aluminum

4. Conclusion

The mechanical characteristics and TCLE of ingots and forgings from experimental hypereutectic silumins based on Al – (15 ÷ 20) % Si alloyed with copper, magnesium, nickel, chromium, modified with phosphorus and hydrogen-containing reagents are studied.

Through the use of optimal heat treatment modes (hardening and aging), the possibility of improving the complex of physico-mechanical properties of deformed piston hypereutectic silumins is shown. A significant increase in strength for such alloys was obtained while maintaining high ductility in comparison with cast piston silumin and a low TCLE in relation to deformed alloys.

IOP Conf. Series: Materials Science and Engineering **866** (2020) 012018 doi:10.1088/1757-899X/866/1/012018

References

[1] Khokhlev V M 1980 *Production of Foundry Aluminum-Silicon Alloys* (Moscow: Metallurgy) p 68
[2] Stroganov G B, Rotenberg V A and Gershman G B 1977 *Alloys of Aluminum with Silicon* (Moscow: Metallurgy) p 271
[3] Prudnikov A N 2013 *Structural and Technological Basis for the Development of Precision Silumin with Regulated Hydrogen Content* (Novosibirsk: NSTU) p 40
[4] Deyev V B et al 2008 *Polzunovsky Almanac* **3** 77–81
[5] Prudnikov A N 2009 *Foundry Production* **2** 2–5
[6] Prudnikov A N 2014 *Technology of Metals* **2** 8–11
[7] Prudnikov A N and Prudnikov V A 2017 *Actual Problems in Mech. Engineering* **4(3)** 78–83
[8] Prudnikov A N 2009 *Steel in Translations* **39** 391–3
[9] Prudnikov A N 2004 *Izv. Vuzov. Ferrous Metallurgy* **4** 40–2
[10] Song D, Kang G and Kan Q 2014 *Smart Materials and Structures* **23(1)** 1–7
[11] Prudnikov A N 2014 Deformation and Destruction of Materials **2** 14–20
[12] Prudnikov A N and Prudnikov V A 2015 *Metallurgy: Technologies, Innovations, Quality* vol 2 (Novokuznetsk: SibSIU) pp 15–18
[13] Prudnikov A N 2009 *Metal Processing (Technology, Equipment, Tools)* **1** 8–11
[14] Afanasyev V K, Prudnikov A N and Prudnikov V A 2010 *Metal Processing (Technology, Equipment, Tools)* **3** 28–31
[15] Mondolfo L F 1979 *Structure and Properties of Aluminum Alloys* (Moscow: Metallurgy) p 640

IOP Conf. Series: Materials Science and Engineering **866** (2020) 012019 doi:10.1088/1757-899X/866/1/012019

The working conditions analysis of the electrodes of plasma torch EDP-104A in nitrogen plasma

G V Galevskiy[1], V V Rudneva[1], E G Orshanskaya [1], S G Galevskiy[2] and I Mishne[3]

[1]Siberian State Industrial University, 42 Kirova Street, Novokuznetsk, 654007, Russia
[2]St. Petersburg Mining University, 2, 21st Line, St Petersburg, 199106, Russia
[3]EST - Environmental System & Treatment

E-mail: kafcmet@sibsiu.ru

Abstract. The analysis of the electrode conditions and performance of plasma torches EDP-104A in nitrogen plasma is carried out. It is shown that the life of the plasma torch is determined by the erosion of the tungsten thermal cathode. It is confirmed that the erosion depends on the strength of the current and the oxygen content in the plasma-forming gas. At the current strength of 1 kA with an increase in oxygen content from 0.1 to 0.5%, erosion increases from 10-12 to 10-9 kg/C. To reduce cathode erosion, it is necessary to limit the oxygen content in technical nitrogen.

1. Introduction

At present, plasma chemical-metallurgical reactors of various designs are used for the implementation of nanotechnology, the most common of which are three-jet ones. In the reactor of this type constructed by SibSIU, Kutateladze Institute of Thermophysics of the Siberian Branch of the Russian Academy of Sciences and LLC "Polimet' three electric arc gas heaters (plasma torch) EDP-104A with capacity of 50 kW each are used. They are installed in the mixing chamber at an angle of 30° to the axis of the reactor and working on a common cylindrical channel. In a 150 kW three-jet plasma metallurgical reactor, the EDP-104A plasmatrons operate at a constant current of up to 200 A and a voltage of up to 250 V on an electric arc. The stabilization of the electric arc is gas-vortex due to the tangential introduction of plasma-forming gas through a special twist ring.

In the last decade, as a result of theoretical and experimental studies, it was possible to improve significantly the quantitative indicators of specific erosion of tungsten thermal cathodes, reducing it to a record low value of 10-13 kg/C with a current strength not exceeding 1 kA. Successes have also been noted in reduction of the erosion rate of copper cooled cylindrical anodes with a movable supporting arc spot. Typically, the average values of specific erosion of such anodes are at the level of 10–9 kg/C at a current strength in the range 0.1–4 kA, pressure 105 Pa, and for a wide range of gases (air, nitrogen, oxygen, hydrogen).

The physical nature of the near-electrode processes is extremely complex, and their characteristics significantly depend on the material, shape and method of cooling the electrodes, pressure, temperature, nature of the flow and type of gas, as well as the method of organizing the arc discharge in a gas-discharge device. In addition, at high gas pressures (of the order of atmospheric and higher), the thickness of the electrode layers turns out to be very small, which at high current densities significantly complicates the experimental study of the electrode processes. Therefore, in recent years, theoretical methods for studying near-electrode processes have been increasingly used, along with experimental ones, which significantly expands the possibilities of elucidating their nature and determining the features of the process itself.

According to the passport data, the service life of the cathodes of plasmatrons EDP-104A in non-oxidizing environments is at least 100 hours. When using nitrogen of technical purity as a plasma-forming gas with an oxygen content of 0.5 – 2.0%, this figure is reduced by almost 2 times. This makes it necessary to develop technical measures to increase the cathode resource.

In this regard, the aim of this work is to analyze the operating conditions of the tungsten thermal cathode of plasma torch EDP-104A in the nitrogen plasma.

2. The device, characteristics and operation of EDP-104A plasma torch

The device, characteristics and operating characteristics of the plasma torch EDP-104A are described in [1–7]. The plasma torch in the assembly, its cathode and anode assemblies, the cathode and anode before and after operation are shown in figure 1.

(a) (b)

Figure 1. Plasma torch EDP-104A (a), its cathode and anode nodes, anode and cathode before and after operation (b).

The plasma torch has sufficient versatility. Its design combines single-chamber plasmatrons with a fixed and self-stabilizing arc length. Depending on the purpose of the plasma torch, power, power source, the output electrode (anode) is selected either with a constant diameter along the length, or stepwise. An arc is excited between the electrodes when the plasma torch is turned on. The gas stream blows the arc out of the interelectrode space and draws it along the axis of the output electrode. The output electrode is made of copper or non-magnetic steel. The internal electrode (cathode) is made of tungsten, zirconium, graphite and other materials, depending on the operating conditions and the type of gas. In environments of hydrogen and nitrogen, tungsten is usually used.

The melting point of tungsten, approximately 3500 °C, is quite sufficient to ensure significant electron emission at a temperature below the melting point. Tungsten is alloyed with lanthanum, yttrium, thorium to reduce the work function. When choosing a plasma-forming medium, the following factors must be considered:

- the ability to achieve high enthalpy values;
- the possibility of use as a reagent;
- inertness in relation to the target products;

- cost.

Enthalpy is one of the main parameters of the plasma flow, which largely determines the technological feasibility of the reactor. The comparison of the temperature dependence of the enthalpy of plasma-forming gases at atmospheric pressure shows that the enthalpy of molecular gases (nitrogen, hydrogen) at relatively low temperatures (4.103-1.104 K) as a result of the dissociation of molecules reaches high values and almost an order of magnitude exceeds the enthalpy of monatomic gases (argon, helium). The use of molecular gases allows high values of the thermal efficiency of plasmatrons to be achieved, which is 60-80% for hydrogen and nitrogen, while for plasmatrons operating on argon and helium, this figure is 40-50%. Along with this, the cost of argon and helium significantly exceeds the cost of hydrogen and nitrogen. For these reasons, the use of plasma-forming molecular gases seems to be more preferable.

Hydrogen and nitrogen in a number of processes can be used simultaneously as a heat carrier and a chemical reagent. The explosion safety of nitrogen, the simplicity of the design and the reliability in operation of the plasmatrons operating on it, for example, EDP-104A should also be noted.

Satisfactory operating life of the thermal cathode is achieved by choosing the material of the cathode insert (zirconium for oxidizing media, tungsten for inert media), applying a protective inert gas to the cathode, using a multi-position electrode assembly.

3. The conditions and performance analysis of the electrodes of plasma torch EDP-104A in nitrogen plasma

Studies of the conditions and performance of the electrodes of plasma torch EDP-104A in nitrogen plasma were carried out in [6, 8-10].

The life of the plasma torch, determined by the erosion of electrode materials, is its most important characteristic. Cathode and anode spots of electric arcs on cold electrodes are characterized by an extremely high level of heat flux density, reaching 104-105 MW/m^2. Such heat loads in the stationary mode are not able to withstand any of the known materials. To ensure an acceptable level of erosion of the electrode, the electrode sections of the arc are moved along the surface of the electrode by the action of aerodynamic or electrodynamic forces on them. When moving the electrode root, the surface of the electrodes undergoes cyclic thermal shock, as a result of which the crystal lattice "sways" and structural defects (cracks) of the electrode material arise, which leads to its mechanical destruction and decrease in thermal and electrical conductivity.

Therefore, the erosion rate of the electrodes is associated with physical processes in the electrode zoness of the arc discharge, on the electrode surface and inside the crystal lattice of the metal from which it is made. It is determined by such non-stationary processes as large-scale and small-scale arc bridging, the action of the external magnetic field on the arc column, and the aerodynamics of the gas flow in the plasmatron. Erosion for a cathode also depends on its diameter, design, plasma-forming gas composition, current strength, cyclic operation (number of plasmatron inclusions), quality of thermal contact of the surfaces of a tungsten rod and a copper cathode holder.

The erosion of the copper anode is determined by such factors as cooling conditions of the arc, current strength, magnitude of magnetic induction, and the protection of the surface of the anode with an inert or natural gas. The integral characteristic of the processes of electrode erosion, of course, which does not separately disclose the role of the occurring microprocesses, the presence of oxide films on the surface of the electrodes, thermal stresses in the metal, and the peculiarities of the motion of the electrode roots, is specific erosion, measured in kg/C. The appearance of the erosion of tungsten cathode and copper anode of the plasma torch EDP-104A is shown in figure 2 [6].

The cathode assemblies in the following design options have been investigated, tested in operation (figure 3): the rod cathode (a) in which a tungsten rod with a diameter up to 0.006 m is sealed in the water-cooled copper cage (cathode holder) with a given extension length (most often 0.003–0.005 m); the conical cathode (b), in which the tungsten rod with a diameter up to 0.006 m is pressed into a copper cage (cathode holder) cooled by water flush with its surface.

IOP Conf. Series: Materials Science and Engineering **866** (2020) 012019 doi:10.1088/1757-899X/866/1/012019

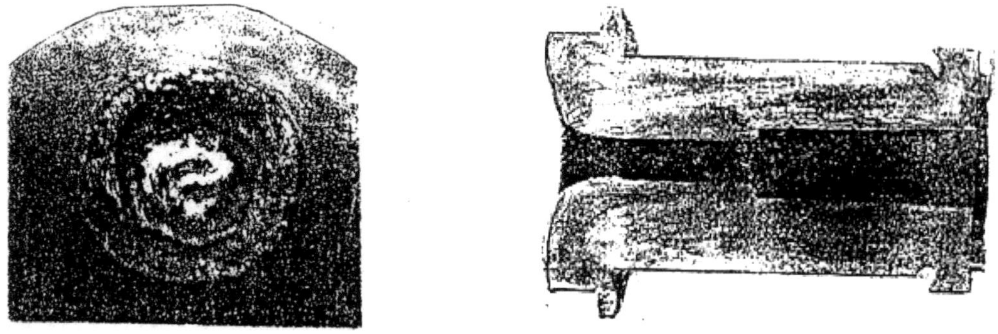

(a) (b)

Figure 2. Appearance of the tungsten cathode (a) and the copper anode (b), which worked in nitrogen at a current strength of 200 A for 100 and 400 hours, respectively.

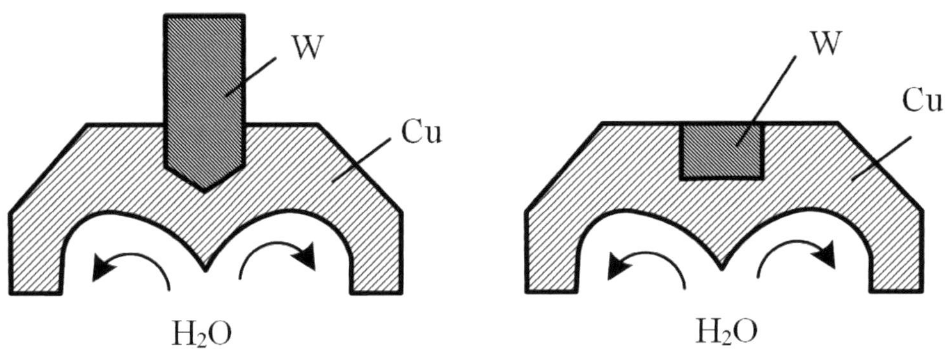

(a) (b)

Figure 3. Design options for the execution of cathode nodes.

According to [6], the surface temperature of a working rod tungsten cathode with a diameter of 0.003 m is about 4000 K (the melting point of tungsten is 3650 K), and the surface cooling rate when the current is turned off reaches 10^6 K/s. In the process, the molten zone is formed at the top of the cathode. With distance from the top, the surface temperature decreases. When working in nitrogen, a cathode spot is formed with a high level of current density and temperature, which leads to the formation, growth, and further decay of a molten tungsten bubble. Its occurrence is due to the expansion of the gas dissolved in tungsten. The specific erosion of the tungsten cathode when operating in nitrogen at a current strength up to 1 kA is about 10-12 kg/C. When the oxygen content in nitrogen is 0.1%, the nature of the destruction of the cathode is preserved.

When the oxygen content in nitrogen increases to 0.5%, particles of molten metal begin to fly out from the cathode surface, which sharply increases the erosion of the cathode. In this case, specific erosion reaches 10-9 kg/C. The conical cathode is characterized by more intensive cooling of the tungsten insert and a smaller working surface that undergoes erosion. According to [6], during the operation of the conical cathode, the end surface of the tungsten rod in the zone where the cathode spot of the arc is located is in the molten state.

Outside the melt, the gas evolution zone develops in the form of separate craters. Craters are formed due to the release of gases dissolved in tungsten through the molten surface. The specific

IOP Conf. Series: Materials Science and Engineering **866** (2020) 012019 doi:10.1088/1757-899X/866/1/012019

erosion of the tungsten cathode in nitrogen at atmospheric pressure, oxygen concentration up to 0.5% and current strength of 250–300 A lies in the range $(2 \div 5) \cdot 10^{-12}$ kg/C, and with an increase in oxygen concentration to 1.5% reaches $(2 \div 4) \cdot 10^{-12}$ kg/C.

The specific erosion of the copper cylindrical anode is also largely determined by the presence of oxygen in the working gas. So, at a current strength of 180 A in high-purity nitrogen containing about 0.001% oxygen, specific erosion is estimated at 10^{-12}-10^{-11} kg/C, and in nitrogen of technical purity (oxygen up to 0.5%) it increases by more than an order of magnitude.

4. Conclusion

The operating conditions analysis of the electrodes of plasma torch EDP-104A in nitrogen plasma was carried out. It is shown that the operating life of the plasma torch is determined by the erosion of the tungsten thermocathode, which at a current strength up to 1 kA and an oxygen content up to 0.1% vol. is approximately 10^{-12} kg/C and reaching 10^{-9} kg/C with an increase in oxygen content to 0.5%. Erosion of the thermal cathode is the result of the action of the electrode root on its working surface (repeated melting, evaporation and crystallization with the occurrence of thermal stresses and the formation of the cracks network in the depth of the cathode material) and oxidation of the cathode material by oxygen with the formation of tungsten oxides.

References

[1] Zhukov M F, Smolyakov V I and Uryukov B A 1973 *Electric Arc Gas Heaters* (Plasmatrons) (Moscow: Nauka) p 232

[2] Anshakov A S et al 1977 *Electric Arc Plasmatrons* (Novosibirsk: ITP SB AS USSR) p 87

[3] Zhukov M F 1980 *Electric Arc Plasmatrons* 1980 (Novosibirsk: ITF SB AN SSSR) p 84

[4] Zhukov M F et al 1992 *New Materials and Technologies and Extreme Technological Processes* (Novosibirsk: Nauka Sib. Brunch) p 183

[5] Zhukov M F et al 1994 *Thermal Plasma in the Technology of New Materials* (London: Cambridge Scientific Publishing House) p 580

[6] Zhukov M F et al 1995 *Plasmatrons. Research. Problems* (Novosibirsk: Publishing House of the SB RAS) p 202

[7] Galevsky G V et al 2008 *Basics of the Design of Plasma-metallurgical Reactors and Processes* vol 2 (Moscow: Flint: Science) p 228

[8] Galevsky G V and Rudneva V V 2018 *Nanomaterial Technology: Lecture Notes* (Novokuznetsk: SibSIU) p 106

[9] Galevsky G V, Rudneva V V and Polyakh O A 2018 *Fundamentals of Designing Electric Arc Plasmatrons: Learner's Guide* (Novokuznetsk: SibSIU) p 26

[10] Galevsky G V and Rudneva V V 2018 *Nanomaterial Technology: Practicum* (Novokuznetsk: SibSIU) p 29

IOP Conf. Series: Materials Science and Engineering **866** (2020) 012020 doi:10.1088/1757-899X/866/1/012020

Application of ladle electric steelmaking slag as a flux – additive for welding

A R Mikhno, N A Kozyrev, R E Kryukov, A A Usoltsev and A A Umansky

Siberian State Industrial University, 42 Kirova street, Novokuznetsk, 654007, Russia

E-mail: kozyrev_na@mtsp.sibsiu.ru

Abstract. The possibility of using ladle electric steelmaking slag as a flux – additive in the welding flux based on slag from silicomanganese production is considered. The effect of this flux – additive on the properties of the deposited metal is shown. The optimal composition of welding flux with the flux – additive providing a favorable appearance and properties of the weld metal was revealed.

1. Introduction

Recently, a number of studies have been carried out using technogenic wastes in the manufacture of welding and surfacing fluxes. A number of studies are devoted to improvement of the welding fluxes compositions for surfacing rolling rolls using new fluxes and flux – additives using industrial wastes from metallurgical enterprises (slags and sludges). In particular, new welding fluxes and flux – additives made using ladle electric-furnace slag, slag from the silicomanganese production have been proposed [1-9]. When applying such metallurgical wastes, the cost of welding and surfacing is reduced.

2. Results and discussion

In this paper, the possibility of using ladle electric steelmaking slag as a flux – additive in the manufacture of welding fluxes is considered.

Ladle electric steelmaking slag for the production of rail steel at EVRAZ ZSMK JSC with a chemical composition, wt. %: 1.26% – S; 0.02% – P; 1.31% – FeO; 36.26% – SiO_2; 36.19% – CaO; 0.22% – MnO; 11.30% – MgO; 6.17% – Al_2O_3; 0.28% – Na_2O; 3.34% – F; <0.12% – C. Production of flux – additives was carried out according to the scheme shown in figure 1.

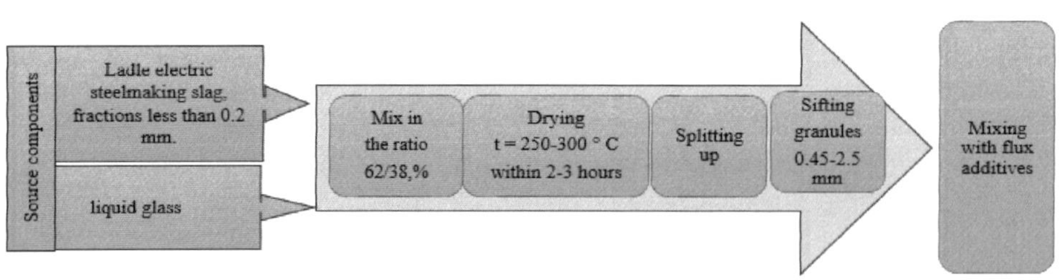

Figure 1. Scheme of flux – additive manufacturing.

IOP Conf. Series: Materials Science and Engineering **866** (2020) 012020 doi:10.1088/1757-899X/866/1/012020

Ready flux – additive with chemical composition, wt.%: FeO – 0.84; S – 0.98; P – 0.013; MnO – 0.06; MgO – 5.01; CaO – 35.47; Na_2O – 1.92; Al_2O_3 – 4.71; Cr_2O_3 – 0.04; K_2O – 0.11; TiO – 0.33 mixed with silicomanganese slag produced at the West Siberian Metallurgical Plant with the chemical composition, wt.%: S 0.15-0.17%, P 0.01%, F 0.28-0.76%, FeO 0.27-0.81%, CaO 22.85-31.70%, SiO_2 46.46-48.16 %, MnO 8.01-8.43%, MgO 6.48-7.92%, Al_2O_3 6.91-9.62%, Na_2O 0.26-0.36% K_2O to 0.6 2% in various ratios (table 1)

Table 1. The ratio of silicomanganese slag and flux – additive, %.

Flux marking	Flux – additive, wt%	Silicomanganese slag (SiMn), wt%
0	50	50
1	30	70
2	20	80
3	10	90
4	5	95
5	0	100

After mixing the flux with the flux – additive, the samples were surfaced on the plates made of 09G2S steel by flux core welding in the laboratory conditions using an ASAW-1250 welding tractor and Sv-08GA welding wire. After samples surfacing, judjing by their appearance it was revealed that a satisfactory quality of the deposited samples is achieved using not more than 30% of flux – additive. Then, X-ray spectral and metallographic analysis of deposited samples was carried out. The chemical compositions of deposited samples, slag crusts and fluxes are given in tables 2-4, respectively.

Table 2. The chemical composition of the deposited samples.

No.	Mass fraction of elements, %										
	Mn	Si	Ni	Cr	Mo	Cu	V	Nb	S	P	C
0	1.22	0.35	0.09	0.04	0.01	0.12	отс	0.011	0.036	0.008	0.04
1	1.25	0.38	0.08	0.04	0.01	0.13	0.002	0.011	0.033	0.009	0.04
2	1.22	0.41	0.09	0.04	0.01	0.13	0.001	0.11	0.031	0.010	0.04
3	1.30	0.42	0.09	0.04	0.01	0.13	0.003	0.010	0.024	0.011	0.04
4	1.30	0.44	0.08	0.04	0.01	0.13	0.002	0.010	0.019	0.012	0.04
5	1.16	0.43	0.11	0.05	0.021	0.14	0.007	0.003	0.019	0.012	0.07

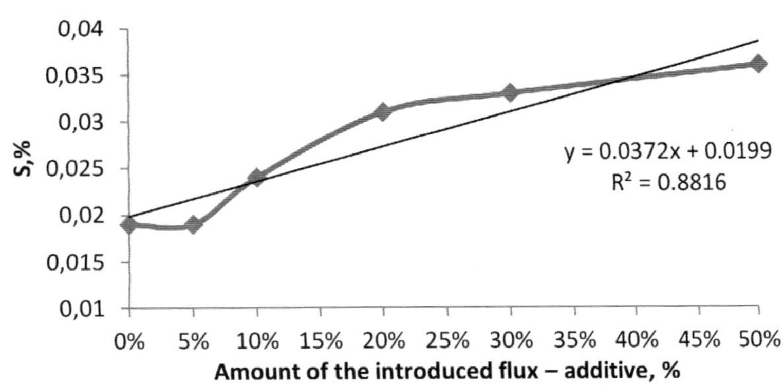

Figure 2. The effect of the amount of introduces flux – additive on the sulfur concentration.

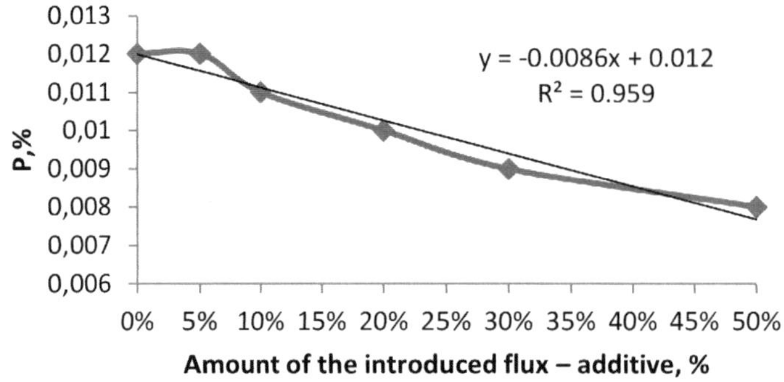

Figure 3. The effect of the amount of introduces flux – additive on the phosphorus concentration.

Table 3. The chemical composition of slag crusts.

No.	Mass fraction of elements, %													
	FeO	CaO	MnO	Al_2O_3	SiO_2	Na_2O	MgO	K_2O	S	P	F	Cr_2O_3	ZnO	TiO_2
0	1.66	34.48	3.90	5.59	44.61	2.78	4.87	0.14	0.46	0.012	1.66	0.036	0.005	0.017
1	1.69	33.28	5.34	6.05	44.55	2.05	4.64	0.11	0.33	0.011	1.24	0.032	0.005	0.015
2	1.78	33.10	6.36	7.23	43.13	1.19	4.38	0.088	0.23	0.012	0.83	0.034	0.004	0.012
3	1.62	32.47	7.04	6.63	43.39	0.89	4.24	0.081	0.20	0.011	0.68	0.033	0.004	0.011
4	1.54	32.42	7.61	7.87	44.28	0.59	4.28	0.11	0.16	0.011	0.53	0.033	0.003	0.014
5	1.69	32.35	7.78	6.59	42.50	0.30	5.55	0.12	0.21	0.011	0.37	0.04	0.012	0.018

Table 4. The chemical composition of the flux.

No.	Mass fraction of elements, %												
	FeO	CaO	MnO	MgO	SiO_2	Al_2O_3	K_2O	Na_2O	F	S	P	Cr_2O_3	TiO_2
0	0.74	35.64	2.58	4.92	43.33	5.19	0.17	3.99	1.99	0.73	0.013	0.041	0.21
1	0.57	34.03	4.62	4.58	43.80	5.09	0.14	3.28	1.63	0.54	0.012	0.036	0.17
2	0.42	32.06	6.90	4.03	46.20	6.85	0.082	1.40	0.89	0.34	0.011	0.024	0.11
3	0.42	33.12	7.28	4.52	45.10	6.33	0.082	1.40	0.87	0.32	0.011	0.029	0.11
4	0.38	32.71	7.59	4.34	45.04	6.60	0.081	1.07	0.77	0.28	0.012	0.031	0.10
5	0.50	31.34	7.97	5.74	46.09	6.61	0.01	0.40	0.45	0.33	0.011	0.05	0.07

Metallographic studies were carried out in the Material Science Center of SibSIU using OLYMPUS GX-51 optical microscope at × 100 magnification on microsections without etching. Assessment of non-metallic inclusions is carried out according to GOST 1778-70. The results of the study of non-metallic inclusion and their type are presented in figure 4 and table 5.

Table 5. Non-metallic inclusions in the weld seam zone.

Sample No.	Non-metallic inclusions, point	
	non-deformable silicates	point oxides
0	2b,3b, rare 4b	1a,2a
1	1b, rare 2b	1a
2	1b, rare 2b	1a
3	1b,2b	1a,2a
4	1b,2b	1a,2a
5	1b,2b	1a

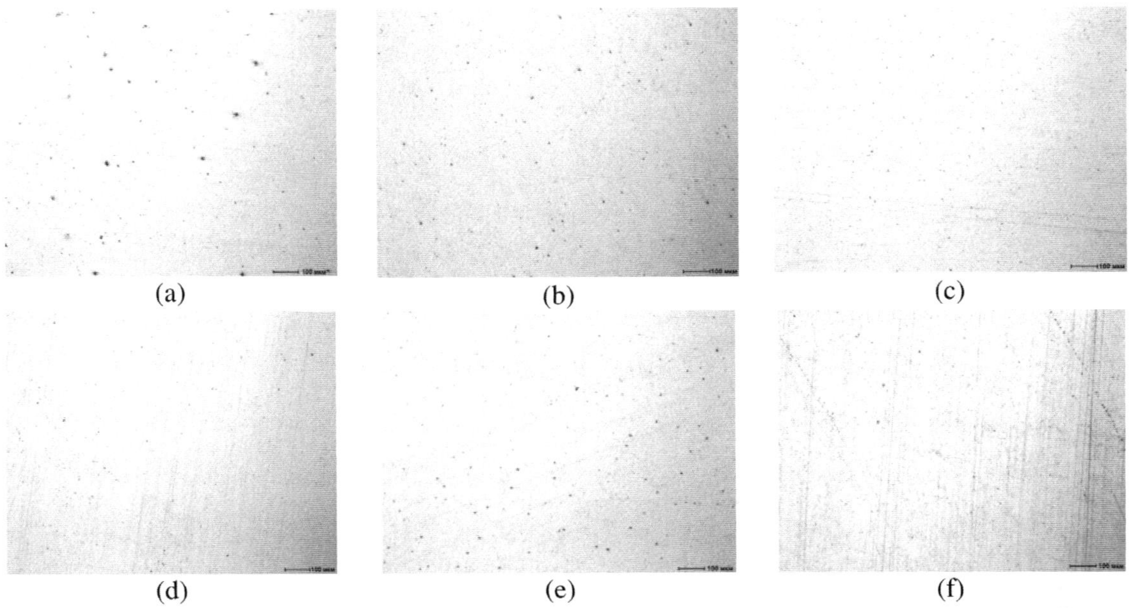

Figure 4. Non-metallic inclusions in the weld zone. Samples: a) 0; b) 1; c) 2; d) 3; e) 4; f) 5.

Analysis of the level of contamination by non-metallic inclusions in the deposited samples showed that when using more than 20% of flux – additives, the amount of non-metallic inclusions in the deposited layer increases.

3. Conclusion

When conducting the described studies, the possibility of using ladle electric steelmaking slag as a flux – additive was revealed. Studies showed that with the introduction of a flux – additive, the sulfur concentration in the deposited layer increases and the phosphorus concentration decreases. The study of non-metallic inclusions showed that it is better to use not more than 20% of flux – additive, since with a higher content of flux – additive, the number of non-metallic inclusions in the deposited layer increases, and the appearance of the deposited samples deteriorates.

References

[1] Volobuev Yu S, Surkov A V et al 2010 *Welding International Volume* **24(4)** 298–300
[2] Kozyrev N A, Kryukov R E et al 2017 *Ferrous Metallurgy. Bulletin of Scientific, Technical and Economic Information* **5(1409)** 85–89
[3] Kryukov R E, Kozyrev N A, Kozyreva O A and Usoltsev A A 2017 *IOP Conf. Series: Materials Science and Engineering* **253** 012007.
[4] Kozyrev N A, Kryukov R E et al 2018 *Ferrous Metallurgy. Bulletin of Scientific, Technical and Economic Information* **6(1422)** 55–66
[5] Kozyrev N A, Kryukov R E et al 2018 *Welding and Control* 168–182
[6] Kozyrev N A, Kryukov R E et al 2018 *Izv. Vuzov. Ferrous Metallurgy* **4** 274–279
[7] Kozyrev N A, Kryukov R E et al 2017 *Metallurgy: Technologies, Innovations, Quality* (Novokuznetsk: SibSIU) part 1 pp 288–292
[8] Kryukov R E, Kozyrev N A et al 2018 *XV International Congress of Steelmakers and Metal Manufacturers* (Moscow – Tula) pp 484–488
[9] Kryukov R E, Mikhno A R et al 2018 *Prospects for the Development of Metallurgy and Mechanical Engineering Using Completed Basic Research and R&D; FERROALLOYS* (Ekaterinburg) pp 326–330

IOP Conf. Series: Materials Science and Engineering **866** (2020) 012021 doi:10.1088/1757-899X/866/1/012021

Hardening of VK10KS hard alloy surface for solving tribological problems in machine-building

T N Oskolkova[1], A S Simachev[1], A M Glezer[2] and A R Fastykovskii[1]

[1]Siberian State Industrial University, 42 Kirova Street, Novokuznetsk, 654007, Russia,
[2]National Research Technological University "MISiS", 4 Leninsky Prospekt, Moscow, 119049, Russia

E-mail: oskolkova@kuz.ru

Abstract. Hard alloy with a gradient structure consisting of a viscous core made of VK10KS alloy and a more wear-resistant surface made of VK6-OM alloy is obtained by electrodischarge hardening alloying. In this case, the base of the alloy is able to absorb shock loads, and the surface can play the role of wear-resistant cutting elements.

1. Introduction

For stamped and boring tools, which are required to have an increased fracture energy, as well as a high wear resistance, there are other ways to increase their wear resistance in addition to coating hard alloy tools. These include the method of manufacturing a "multilayer" hard-alloy tool [1].

The scientists of All-Russian Research and Development Institute of Hard Alloys and Refractory Metals have developed a method for producing hard-alloy inserts for equipping a metal-cutting tool with a variable cobalt binder content, when the composition changes when pressing a blank for a plate: VK3–VK6–VK10–VK15–VK10–VK6–VK3 [2]. The obtained plate had a viscous core and a hard surface layer, which ensures its resistance to the processing of various materials.

However, this method for obtaining a multilayer hard alloy with a gradient structure is quite complex, energy-intensive and time-consuming, requires sophisticated equipment. Obtaining a multilayer hard alloy by this method is possible only on a flat surface. It is of interest to obtain hard alloys with a gradient structure – from a viscous core to a wear-resistant surface by the method of electrodischarge hardening by alloying (EDHA).

The advantages of the EDHA process are the ability to produce complex micrometallurgical processes in limited areas of the treated surface; extremely high adhesion strength of the applied coating layer to the treated surface; simplicity of the process; lack of distortion of the workpiece; small dimensions and low power consumption of equipment.

The essence of the EDHA process is that during a spark discharge, of the anode (electrode) erosion and transfer of erosion products to the cathode (part) occur. In this case, a layer is formed on the part having a changed composition and structure, which depend on the electrode material and the electrical mode of the installation. High discharge temperature (5000 ÷ 10000 °C), short duration of action ($10^{-6}...10^{-3}$ seconds), instantaneous cooling of heated sections with a mass of cold metal makes it possible to conclude that ultra-high-speed quenching occurs after EDHA. Along with this process, there is a contact or drop transfer of alloying elements from the hardening electrode to the treated surface when it comes in contact with it and subsequent diffusion resorption of these elements in the layer.

2. Results and discussion

In the framework of this work, a low-cobalt VK6-OM coating on the surface of VK10KS hard-alloy was obtained by the EDHA method [3 - 6]. The alloy VK6-OM in accordance with GOST 3884 - 74 was used as the alloying electrode. The alloys of the OM group (VK6-OM, VK10-OM, VK15-OM) were developed for cutting especially hard and abrasive materials. VK6-OM alloy of composition 91.9 WC + 6 % Co + 2 % TaC + 0.1% VC does not reduce strength, VC and TaC additives contribute to obtaining a fine-grained structure – they prevent the growth of WC phase grains, while VK6-OM alloy with its wear-resistant properties is not inferior to the most wear-resistant alloy of the WC – Co group (B 253), but significantly exceeds it in strength [1].

Thus, a gradient hard-alloy plate may consist of a viscous core that serves as a damper, well absorbs shock loads, is able to withstand significant bending stresses, and the surface layer has increased wear resistance compared to a viscous core. The creation of hard-alloys with a gradient structure should be considered as a new stage in the improvement of hard-alloys.

The surface treatment of VK10KS hard-alloy by the EDHA method was carried out using the portable installation UR – 121 manufactured by the Instrumental Plant "Koncern Podolsk" OJSC. The surface treatment of VK10KS hard-alloy was carried out using a double technological mode at the rate of 0.55 min/1 cm^2 in Norma mode, 3 + 1.0 min/1 cm^2 in Turbo mode.

The investigation of the structural features after processing and determining the depth of the surface hardening zone was carried out using OLIMPUS – GX 50 optical microscope and Philips SEM 515 scanning electron microscope.

Figure 1 shows the microstructure of VK10KS alloy with a hardened layer of VK6-OM after EDHA. A poorly etched layer was detected, the thickness of which reaches 20 μm.

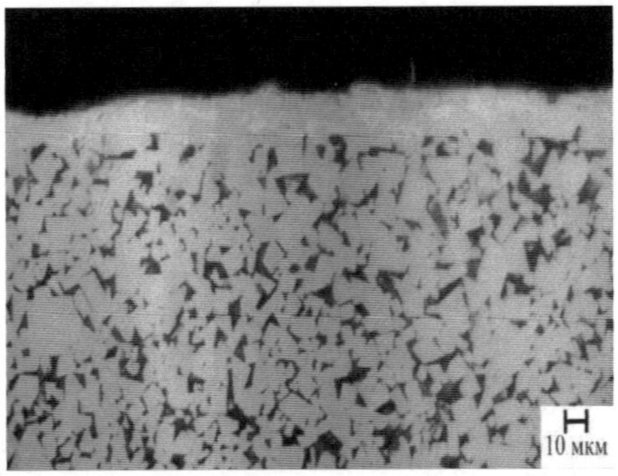

Figure 1. The microstructure of VK10KS alloy after EDHA coated with VK6-OM.

The results of scanning electron microscopy confirm the presence of a gradient structure on the VK10KS alloy: an increased content of tungsten in the surface layer and a lower content of cobalt (figure 2).

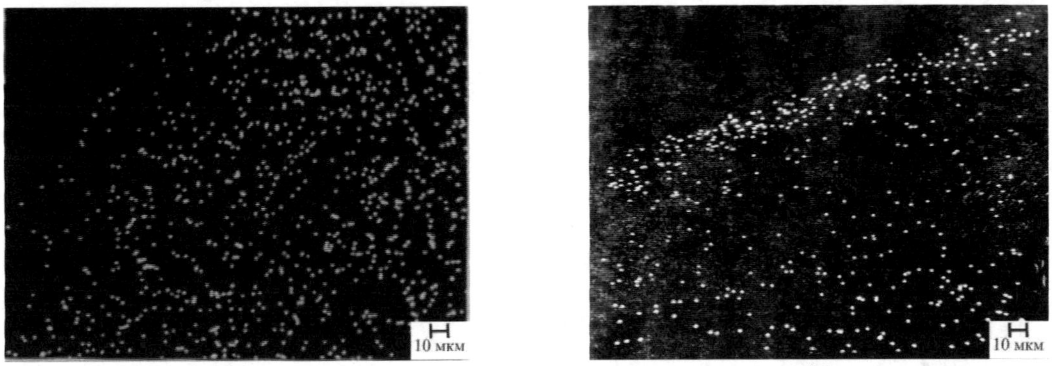

(a) (b)

Figure 2. Microstructure of VK10KS alloy after EDHA in characteristic x-rays:
a – cobalt, b – tungsten.

X-ray phase analysis showed that the surface treatment of the hard alloy by EDHA leads to a change in its phase composition (figure 3).

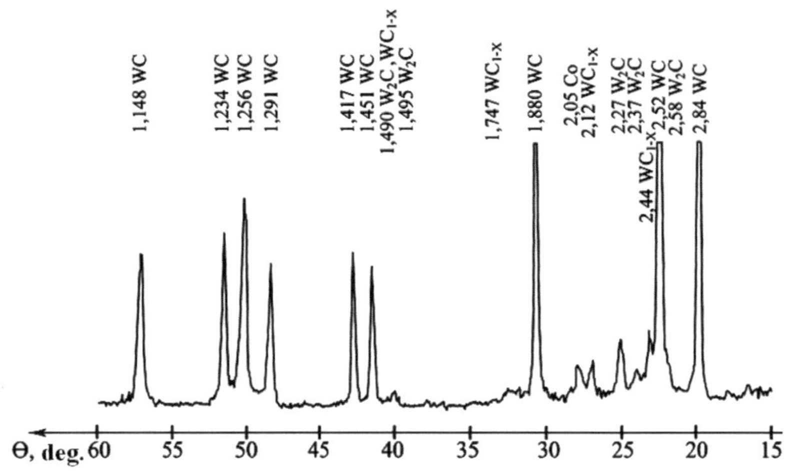

Figure 3. A fragment of the diffraction pattern of VK10KS hard-alloy after EDHA.

In addition to the existing tungsten monocarbide alloy WC and cobalt binder, which were present in the initial state in the hard alloy, a finely dispersed carbide of ditungsten W_2C and carbide WC_{1-x} are formed in the surface layer. The authors of [7] describe the positive role of divofram carbide in the surface layer of a hard-alloy WC + 20% Co after a high-energy impact on the surface, noting a higher W_2C hardness compared to WC hardness.

A change in the phase composition of the surface layer in the hard alloy was reflected in the indices of nanosolidness. The result of nanoindentation of the hard-alloy after EDHA is an increase in surface hardness to 22000 MPa (table 1).

Table 1. Properties of VK10KS alloy after hardening treatment.

Coating material	Nanohardness of the surface layer, MPa	Depth of the worn track, µm	Wear track area, µm^2	Coefficient of friction, µ	Roughness, R_a, µm
		Electrodischarge hardening by alloying			
VK6-OM	21649 ÷ 22000	10.8	941	0.23	1.83 ÷ 2.15
		VK10KS alloy in the initial state			
-	-	58	12921	0.41	0.41

Durability tests were carried out on a PC-Operated High Temperature Tribometer at room temperature. The wear of the samples was determined by measuring the depth and area of the track after testing, formed as a result of the action of a stationary diamond indenter on a rotating sample at a load of 3 N, the number of revolutions 4000, linear velocity of 2.5 cm/s. For comparison, similar tests were performed on the initial samples with the same parameters (figure 4). The depth and area of the wear track after testing, as well as the coefficient of friction after running-in, were measured using the Micro Measure 3D station high-precision measuring system with software. The analysis of the obtained profiles was carried out using the Mountains Map Universal sofrware, version 2.0.13. The results of tribological tests (figure 5, table 1) are presented as arithmetic mean values obtained from 10 tests in one processing mode on three samples.

(a) (b)

Figure 4. Appearance of tracks during tribological tests: a – VK10KS alloy coated with VK6-OM; b – initial alloy.

The wear of samples coated with VK6-OM after EDHA showed that the track depth is 10.8 μm, and that of the original sample, the track depth is 58 μm (figure 5). The cross-sectional area of the track of worn samples with and without coating is 941 μm^2 and 12921 μm^2, respectively (table 1). The friction coefficient (μ) for VK10KS alloy coated with VK6-OM is defined at 0.23, and for the initial alloy μ = 0.41 (table 1).

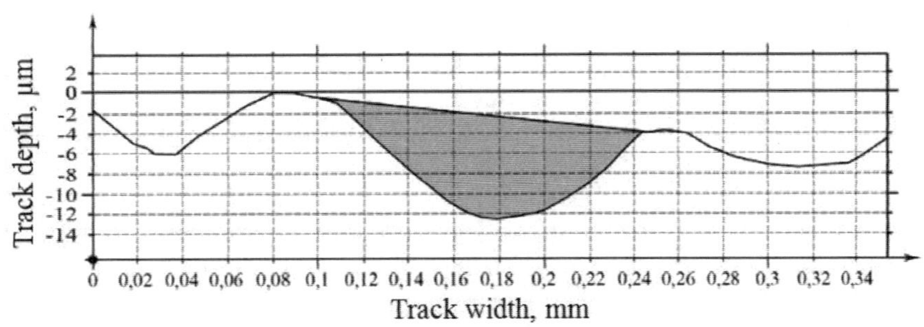

(a)

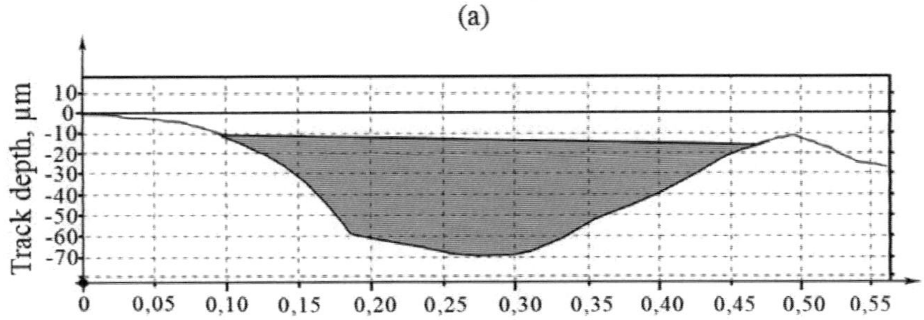

(b)

Figure 5. Profile tracks of samples wear: a – after EDHA; b – initial sample.

Since EDHA always leads to an increase in surface roughness [8, 9], an important factor is the study of microgeometry after hardening of VK10KS using this method, the results of which are presented in table 2. Hardening of the surface of VK10KS hard-alloy by EDHA slightly worsens the surface microgeometry, but keeps it within the technical requirements with recommended surface finish R_a = 2.5 μm [8].

IOP Conf. Series: Materials Science and Engineering **866** (2020) 012021 doi:10.1088/1757-899X/866/1/012021

Table 2. Treatment modes and microgeometry of the coating with a thickness of 20 μm.

Treatment		Treatment time, min	Roughness R_a, μm
Turbo	Norma 3		
80 %	20 %	1.10(0.2 Norma 3 + 0.9 Turbo)	2.15
60 %	40 %	1.20(0.2 Norma 3 + 1.0 Turbo)	1.90
50 %	50 %	1.40(0.4 Norma 3 + 1.0 Turbo)	1.87
40 %	60 %	1.55(0.55 Norma 3 + 1.0 Turbo)	1.83

3. Conclusion

Using the proposed EDHA method for coating the surface of a hard-alloy in comparison with existing methods makes it possible to obtain a hard-alloy plate with a gradient structure, for example, VK3–VK6–VK8– VK10–VK15–VK20. In addition, using the EDHA method, it is possible to restore the surfaces of various hard alloys after using the corresponding electrodes, this will extend the life of products equipped with these alloys, as well as save scarce materials (tungsten and cobalt).

References

[1] Panov V S, Chuvilin A M and Falkovsky V A 2004 *Technology and Properties of Sintered Hard Alloys and Products from Them* (Moscow: MISiS) p 464

[2] Anikin V N, Zolotareva N N, Kazantsev N I et al. 2007 *Patent of the Russian Federation* No. 2302925 Appl. 01.12.2005, Publ. 20.07.2007

[3] Oskolkova T N 2010 *Patent of the Russian Federation* No. 2401720 02.24.2009; Publ. 10.20.2010

[4] Oskolkova T N 2015 *IOP Conf. Series: MSE* **91** 012019

[5] Oskolkova T N and Glezer A M 2018 *Materials Science* **6** 21–31

[6] Oskolkova T N 2009 *Bulletin of Samara Scientific Center of RAS* vol 11 **5(2)** 320–23

[7] Tyurin Yu N, Kulkov S N and Kolisnichenko O V 2009 *Physical Engineering of the Surface* **3** 262–67

[8] Oskolkova T N 2011 *Hardening Technologies and Coatings* **10** 15–19

[9] Blinova T A, Pogonin A A and Boyko A F 2010 *Bulletin of Samara Scientific Center of RAS* vol 12 **1(2)** 301–303

IOP Conf. Series: Materials Science and Engineering **866** (2020) 012022 doi:10.1088/1757-899X/866/1/012022

Modeling and experimental study of the electrodes erosion of plasma torch EDP-104A in nitrogen plasma

G V Galevskiy[1], V V Rudneva[1] E G Orshanskaya [1], S G Galevskiy[2] and I Mishne[3]

[1]Siberian State Industrial University, 42 Kirova Street, Novokuznetsk, 654007, Russia
[2]St. Petersburg Mining University, 2, 21st Line, St Petersburg, 199106, Russia
[3]EST - Environmental System & Treatment

E-mail: kafcmet@sibsiu.ru

Abstract. Modeling and experimental studies of the electrodes erosion of plasma torch EDP-104A in nitrogen plasma were carried out. The resource of cathode and anode operation is determined by calculation. The conditions and performance of 178 cathodes were studied. The possibility of using natural gas to protect a tungsten cathode was investigated. The increase in the cathode life to the rated value was confirmed.

1. Introduction

Plasmatrons attract attention due to their effective implementation of chemical, metallurgical and other processes, create practically waste-free technologies, organize complex processing of raw materials, obtain materials with fundamentally new physical, mechanical and chemical properties, miniaturize industrial plants and automate them.

The created technological installations have high productivity. Advantages of electric arc plasmatrons allow their efficient use in industry:

1. The cost-effectiveness of converting electrical energy into heat by existing plasmatron designs, characterized by high values of electrical and heat efficiency.
2. Reliability and stability of the electric arc installation.
3. A large resource of electrodes, calculated in tens and hundreds of hours of operation, depending on the type of plasma torch, its power and the type of working gas.
4. A wide range of power – from hundreds of watts to several megawatts.
5. The ability to heat almost any gas, including reducing, oxidizing, inert, widely used in industrial technologies.
6. The simplicity of automation of the control mode of the electric arc.
7. Small size and rather small metal consumption.

The increasing scale of technical applications of electric arc devices require solution of a number of problems, associated, first of all, with the increase in the service life of the most heat-stressed elements – electrodes. Among them, the problems of increasing the enthalpy of the working gas and the inclusion of gases with complex composition in the technological processes, the need to increase the power consumption of the gas discharge should also be noted. The latter requirement is especially evident in the chemical and metallurgical industries, where the unit power of the plasma torch during long-term stationary operation should reach many megawatts. The implementation of these working conditions inevitably requires an increase in the current strength of the arc and, therefore, the use of materials with improved physical and mechanical characteristics, since an increase in the current strength leads to an increased erosion of the electrodes material (reducing the resource of their work).

The erosion rate of the electrodes is associated with physical processes in the electrode zones of the arc discharge, on the surface of the electrode and inside the crystal lattice of the metal from which it is made. It is determined by such non-stationary processes as large-scale and small-scale arc bridging, the action of an external magnetic field on the arc pillar, the aerodynamics of the gas flow in the plasmatron, and other factors. For example, bridging leads to the accumulation of damage in the surface layer of the electrodes.

In the last decade, as a result of theoretical and experimental studies, it was possible to improve significantly the quantitative indicators of specific erosion of tungsten thermal cathodes, reducing it to a record low value of 10^{-13} kg/C today with a current strength not exceeding 1 kA. Successes have also been noted in reducing the erosion rate of copper cooled cylindrical anodes with a movable supporting arc spot. Typically, the average values of specific erosion of such anodes are at the level of 10^{-9} kg/C at a current strength in the range 0.1^{-4} kA, pressure 10^5 Pa, and for a wide range of gases (air, nitrogen, oxygen, hydrogen).

The physical nature of the near-electrode processes is extremely complex, and their characteristics significantly depend on the material, shape and method of cooling the electrodes, pressure, temperature, nature of the flow and type of gas, as well as the method of organizing the arc discharge in a gas-discharge device. In addition, at high gas pressures (of the order of atmospheric and higher), the thickness of the electrode layers turns out to be very small, which at high current densities significantly complicates the experimental study of the electrode processes. Therefore, in recent years, theoretical methods for studying near-electrode processes have been increasingly used, along with experimental ones, which significantly expands the possibilities of elucidating their nature and determining the features of the process itself.

The aim of this work is the mathematical modeling of the electrodes erosion processes of plasma torch EDP-104A when they are in nitrogen plasma and the technological testing of natural gas to protect the tungsten cathode and increase its service life.

2. Modeling of the electrodes erosion processes of plasma torch EDP-104A when they operate in nitrogen plasma

Usually, the service life of the plasma torch is taken to be equal to the lower value of the service life of one of the electrodes. The experimental determination of the life of the electrodes is very time-consuming and expensive. In this regard, the life of the anode and cathode was estimated by calculation. For this, the complex profile of the anode worn part at the point of attachment of the electric arc behind the ledge was assumed to be similar to a triangle (figure 1 a), and the diameter of the crater formed in the cathode under the influence of the crater arc was equal to the diameter of the arc attachment (figure 1 b).

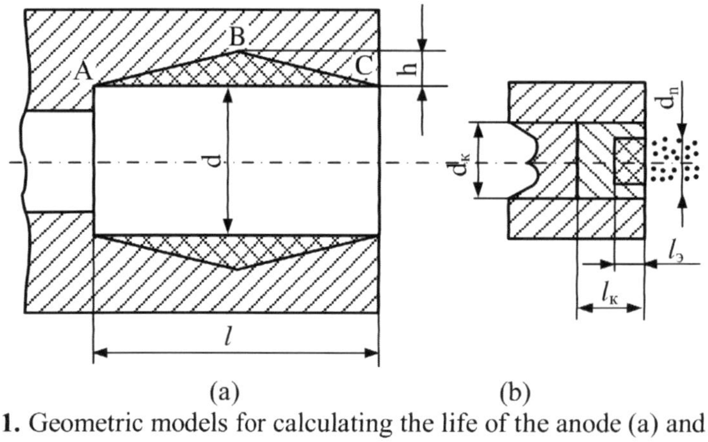

(a) (b)

Figure 1. Geometric models for calculating the life of the anode (a) and cathode (b).

The volume of the anode material removed as a result of erosion will be, m^3:

$$V_e = 0.5 \cdot h \cdot \pi \cdot l \cdot (d_2 + h). \quad (1)$$

Taking that l=3.6·10^{-2} m; d_2=12·10^{-3} m; h=0.2·10^{-2} m, then:

$V_e = 0.5 \cdot 0.2 \cdot 10^{-2} \cdot 3.14 \cdot 3.6 \cdot 10^{-2} \cdot (12 \cdot 10^{-3} + 0.2 \cdot 10^{-2}) = 1.58 \cdot 10^{-6}$ m^3.

The duration of continuous operation of the copper anode is:

$$t_a = \frac{\rho_e \cdot V_e}{G_e \cdot I}, \quad (2)$$

where ρ_e is the density of copper, ρ_e = 8.9 · 10^3 kg/m^3; G_e – specific erosion of the electrode, G_e = 4·10^{-12} kg/C; I – arc current, A.

$$t_a = \frac{8.9 \cdot 10^3 \cdot 1.58 \cdot 10^{-6}}{4 \cdot 10^{-12} \cdot 200} = 4883\text{h}$$

Define the cathode resource. We select tungsten as the cathode material. The diameter of the cathode tungsten insert [6], m:

$$d_c = \frac{1.3 \cdot 10^{-4} \cdot I^{0.5}}{1 - 2.15 \cdot 10^{-2} \cdot I^{0.5}}, \quad (3)$$

$$d_c = \frac{1.3 \cdot 10^{-4} \cdot 200^{0.5}}{1 - 2.15 \cdot 10^{-2} \cdot 200^{0.5}} = 3 \cdot 10^{-3}.$$

The length of the insert should be less than or equal to the diameter, l_c = 0.003 m is assumed. The permissible cathode erosion depth, m:

$$l_e = 0.3 \cdot l_c = 0.3 \cdot 0.003 = 0.1 \cdot 10^{-2} \text{ m}. \quad (4)$$

The diameter of the crater formed in the cathode under the influence of the arc is equal to the diameter of the arc attachment, m:

$$d_n = B \cdot I^{0.5}, \quad (5)$$

where B is a coefficient equal to 1.6·10^{-4} for nitrogen [1].

$d_n = 1.6 \cdot 10^{-4} \cdot 200^{0.5} = 2.3 \cdot 10^{-3}$ m.

The volume of material removed as a result of erosion will be, m^3:

$$V_e = \frac{\pi \cdot d_n^2 \cdot l_e}{4}, \quad (6)$$

$$V_e = \frac{3.14 \cdot (2.3 \cdot 10^{-3})^2 \cdot 0.1 \cdot 10^{-2}}{4} = 0.42 \cdot 10^{-8}.$$

By the formula (2), we calculate the time of cathode continuous operation, given that for tungsten ρ_e =19.34·10^3 kg/m^3, G_e=2·10^{-12} kg/C:

$$t_c = \frac{19.34 \cdot 10^3 \cdot 0.42 \cdot 10^{-8}}{2 \cdot 10^{-12} \cdot 200} = 56\text{h}.$$

Using the geometric model described above and methods for calculating the cathode life for specific erosion values (0.4 - 3.0)·10^{-12} kg/Cl, corresponding to an approximate oxygen content in technical nitrogen up to 1% vol., The obtained calculated dependence of the resource of the cathode operation on the specific erosion value is shown in figure 2. The nature of the dependence is confirmed by the erosion mechanism of a tungsten thermal cathode described in [1] in an oxygen-containing nitrogen plasma. Erosion of the cathode as a result of the action of the arc spot on its working surface (repeated melting, evaporation and crystallization with the occurrence of thermal stresses with the formation of a network of cracks in the depth of the cathode material) and its erosion as a result of oxidation with the formation of tungsten oxides. Apparently, the reduction in the erosion of the thermal cathode seems to be real due to the limitation of the oxygen content in technical nitrogen.

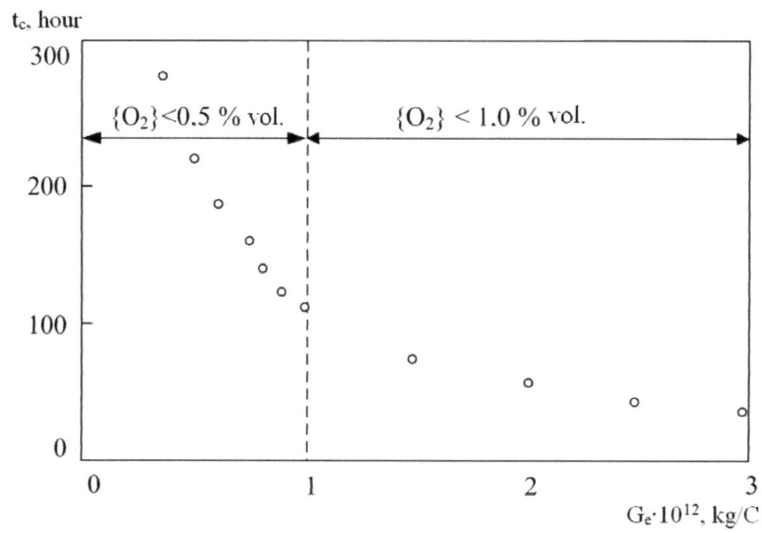

Figure 2. The calculated dependence of the cathode life on specific erosion.

3. Technological testing of natural gas to protect the tungsten thermal cathode of plasma torch EDP-104A

In the conditions of LLC Polimet, which implements plasma synthesis and plasma processing technologies in a three-jet plasma metallurgical reactor with a capacity of 150 kW using technical purity nitrogen as a heat carrier, the analysis of technological documentation that records the operation of the cathodes (installation and replacement dates) was carried out and the average resource for 2015, 2016, 2017, 2018 was determined. The statistical sample of cathodes was, pcs.: 2015 – 42; 2016 – 48; 2017 – 39; 2018 – 45. The oxygen content in the main nitrogen during this period varied within 0.7 – 1.6% vol. In this case, the same cathode could operate in a plasma-forming gas with different oxygen contents, which did not allow the dependence of the resource on the oxygen content to be revealed. The distribution of the average cathode resource by years is shown in figure 3. The average cathode resource was, hours: 2015 – 63, 2016 – 51, 2017 – 58, 2018 – 64.

As it can be seen that the average cathode resource is significantly lower than the declared passport value (100 h) and corresponds to approximate values of specific erosion (1.8 - 2.2)·10^{-12} kg/l.

The analysis of specialized scientific and technical literature has revealed two possible technological options for increasing the cathode life:

1) preparation of main nitrogen for use as a plasma-forming gas, including the removal of moisture and oxygen in electric column heaters with a chrome-nickel nozzle at a temperature of 500-600 °C, providing a residual oxygen content of not more than 0.01% vol. at a load (6 - 8)·10-4 m^3/m^2·s [2-4];

2) the use of a mixture of technical nitrogen and gaseous hydrocarbon as a plasma-forming gas, for example, natural gas introduced in an amount necessary for the binding of oxygen to carbon monoxide [5]. However, the first option was not widely used and mastered only at the laboratory level, and the second was not technologically confirmed in real production conditions.

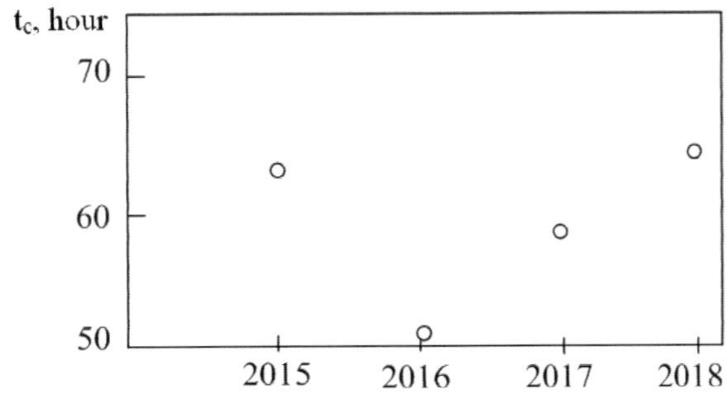

Figure 3. Distribution of the average cathode resource by years.

In this regard, in the conditions of Polimet LLC, a technological study of the cathode operation of plasma torches EDP-104A of a three-jet reactor with a capacity 150 kW in a heat carrier gas containing nitrogen of technical purity with natural gas additives containing 94.0-96.0% vol. methane at an arc current of 200 A. During the research, the oxygen content in technical nitrogen was (0.6 - 1.2) % vol. The cathode life was determined for three plasma-forming gas compositions: N_2 + 1.0% vol. CH_4, N_2 + 2.0% vol. CH_4, N_2 + 3.0% vol. CH_4. The average value corresponds to the stoichiometric for the reaction of the formation of carbon monoxide with an oxygen content of nitrogen of ~ 1.0 % vol. The results of the study are presented in figure 4 and generally confirm the technological feasibility of this approach: the average value of the resource increased to 105 hours and is achieved with a methane content of 2 to 3 % vol.

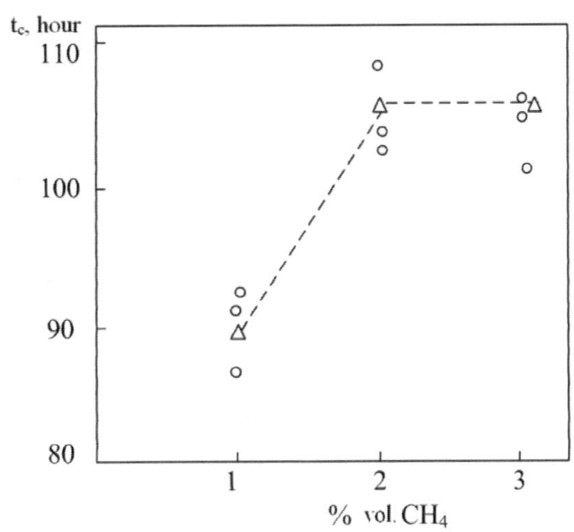

Figure 4. Dependence of the cathode life on methane content in nitrogen (○, Δ – practical and average values).

4. Conclusion

1) The modeling of electrodes erosion process of plasma torch EDP-104A during their operation in nitrogen plasma was carried out. Geometric models of a complex profile of cathode and anode worn parts are constructed. For a current strength of 200 A, an oxygen content in nitrogen approximately 1 % vol., specific anode erosion values of $4 \cdot 10^{-12}$ kg/C, cathode $2 \cdot 10^{-12}$ kg/C, the anode operating life is

4883 hours, i.e. virtually unlimited, the cathode resource is 56 hours, which is lower than the passport data (100 hours). The analysis of the calculated dependence of the cathode life on specific erosion shows that a satisfactory resource is 100 to 200 hours can be achieved with a specific erosion of (0.4 - 1) kg/C, which corresponds to an oxygen content of significantly less than 0.5% vol.

2) In the conditions of Polymet LLC, the survey of the conditions and indicators of 174 cathodes of plasmatrons EDP-104A for the period 2015 – 2018, achieved during operation in main nitrogen with an oxygen content of (0.7 - 1.6) % vol., was conducted. The average cathode life was 59 hours, which is significantly lower than the declared passport value (100 hours) and corresponds to approximate values of specific erosion (1.8 - 2.2)$\cdot 10^{-12}$ kg/C.

3) In the conditions of Polimet LLC, the study of the operation of the plasma torch cathodes EDP-104A of a reactor with a capacity of 150 kW was conducted in a heat carrier gas containing nitrogen, oxygen (0.6 - 1.2) % vol. and natural gas (methane up to 94.0-96.0 % vol.) (1-3)% vol. The increase in the average value of the cathode life up to 105 hours was confirmed when the methane content in the coolant gas is 2–3 % vol., which indicates the technological feasibility of introducing gaseous hydrocarbons into the composition of the plasma-forming nitrogen-oxygen mixture to bind oxygen to carbon monoxide.

References

[1] Zhukov M F et al 1995 *Plasmatrons. Research. Problems* (Novosibirsk: SB RAS) p 202
[2] Nozdrin I V, Rudneva V V and Galevsky G V 2014 *Chromide Boride – Nanotechnology, Properties, Application* (Saarbrücken (Germany): LAP LAMBERT Academic Publishing) p 233
[3] Nozdrin I V et al 2012 *Bulletin of Mechanical Engineering* **12** 78–83
[4] Nozdrin I V et al 2011 *Proc. of Int. Conf in Metallurgy: Technology, Management, Innovation, Quality* (Novokuznetsk: SibSIU) pp 119–123
[5] Nozdrin I V et al 2017 *Pat. of the Russian Federation* at PM No 107740, IPC H05B 7/18, H05H 1/24. No of appl. 2011112115 03/30/2017 publ. 08/10/2017

IOP Conf. Series: Materials Science and Engineering **866** (2020) 012023 doi:10.1088/1757-899X/866/1/012023

Development and application of a complex model of titanium diboride plasmosynthesis

G V Galevskiy[1], V V Rudneva[1], E G Orshanskaya[1] and K A Efimova[2]

[1]Siberian State Industrial University, 42 Kirova Street, Novokuznetsk, 654007, Russia
[2]Group of companies "Siberian Project", 11 Ordzhonikidze Street, Novokuznetsk, 654000, Russia

E-mail: kafcmet@sibsiu.ru

Abstract. An integrated multifactor mathematical model of plasmosynthesis of titanium diboride was developed. The model includes 3 submodels: "Plasma generation", "Evaporation of raw materials", "Boride formation". The model provides the implementation of multivariate research and engineering calculations of plasma processing technological indicators of various types of titanium-containing raw materials, industrial products and waste. The convergence of the predicted and practical results for various types of raw materials is 10.4-12.9%. In all cases, lower values of practical data are noted in comparison with the calculated ones, which is due to the influence of non-isothermal heat carrier gas.

1. Introduction

To solve design and technological problems and perform multivariate research and engineering calculations of effective plasma metallurgical processing of various types of titanium-containing raw materials, industrial products and waste, a complex model of boride formation has been developed.

The classical approach to construction of complex multifactorial deterministic model of a complex metallurgical process involves dividing it into component subprocesses and formation of the corresponding submodels based on the joint solution of the systems of mathematical equations describing them [1]. Given that the solution of the equations is carried out under the inevitability of a number of assumptions, the simulation results are predictive. Nevertheless, for the study of metallurgical processes, mathematical modeling is widespread enough, since with the appropriate computer software it allows for multivariate calculations and ensures the accumulation of information important for technological design.

2. Development of an integrated model

For the correct construction of a complex model, it is necessary to correctly pose the problems that are solved within the framework of each submodel in terms of their physical and chemical nature. Of the 7 components of plasma processing of the powder mixture of sequentially implemented technological processes (stages), three stages are the main ones that determine the quality of diboride as a commercial product. This is 1) generation of a high-enthalpy gas flow; 2) mixing powdered raw materials with plasma, its heating and evaporation; 3) the formation of a reaction boride-forming mixture, the formation of TiB_2 and the formation of its nanoparticles as a result of coalescence and coagulation processes. This determines the possibility of constructing a model based on three submodels: "Plasma generation" (1); "Evaporation of raw materials" (2); "Boride formation" (3).

Submodel 1 "Plasma generation" is based on the use of the following calculated dependences of the thermal and gas-dynamic characteristics of the plasma jets generated in plasmatrons and the plasma flow formed in the mixing chamber of the reactor:

$$U = \frac{11044}{10^{0.03M}} \left(\frac{G}{I \cdot d_1}\right)^{\frac{1}{3}} \tag{1}$$

$$\eta_n = [1 + 5.85 \cdot 10^{-5} \left(\frac{I^2}{G_g \cdot d_1}\right)^{0.27} \cdot \left(\frac{G_g}{d_1}\right)^{-0.27} \cdot \left(P_{N_2} \cdot d_1\right)^{0.3} \cdot \left(\frac{l_1}{d_1} + \frac{l_2}{d_2}\right)^{0.5}] \tag{2}$$

$$H_c = \frac{N_n \cdot \eta_n}{G_g} \tag{3}$$

$$T_c = f(H_c) \tag{4}$$

$$Re_c = \frac{\upsilon_c \cdot d_2 \cdot \rho_c}{\mu_c} \tag{5}$$

$$N = 3 \cdot P \cdot \eta_n \cdot \eta_{m.c} \tag{6}$$

$$H_n = \frac{N}{3G_g} \tag{7}$$

$$T_n = f(H_n) \tag{8}$$

$$Re_n = \frac{\upsilon_n \cdot d_3 \cdot \rho_n}{\mu_n} \tag{9}$$

The following notation is used in equations (1-9): N_n is the given power of the plasma torch, kW; U is the voltage across the arc, V; I – arc current, A; M is the molecular weight of nitrogen; G_g is the mass flow rate of the plasma-forming gas in the plasmatron, kg/s; d_1 and l_1 – diameter and length of the arc channel of the plasma torch, m; d_2 and l_2 – diameter and length of the ledge of the plasma torch, m; P_N2 – nitrogen pressure at the exit of the plasma torch, Pa; H^s and H_n are the specific enthalpy of the plasma jet and plasma flow, respectively, at the exit from the anode of the plasma torch and at the entrance to the reactor channel, kJ/kg; T_c and T_n are the mass-average temperature of the plasma jet and plasma stream, respectively, at the exit from the anode of the plasma torch and at the entrance to the channel of the reactor, K; η_n, $\eta_{m.c}$ – thermal efficiency of the plasma torch and mixing chamber, fractions of a unit; N is the net power supplied to the reactor, kW; Re_c, Re_n – Reynolds numbers determined for the conditions of the expiration of the plasma jet from the anode and the entry of the plasma flow into the reactor channel; υ_c, ρ_c, μ_c – velocity, density and viscosity of the plasma jet at the temperature T_c respectively m/s, kg/m^3, Pa·s; υ_n, ρ_n, μ_n – are the velocity, density and viscosity of the plasma flow at a temperature T_n, respectively, m/s, kg/m^3, Pa·s.

Equations (1, 2) were obtained by the scientific school of academician M.F. Zhukov [2] for the plasma-powered plasma torch EDP-104A based on the processing of experimental data using the theory of similarity. The submodel predicts the thermal and gas-dynamic characteristics of the plasma jets formed in the plasma torches and the plasma formed in the mixing chamber upon their collision, depending on the flow rate of the plasma-forming gas and the power supplied to the plasma torches.

Submodel 2 "Evaporation of raw materials" was proposed by A.L. Mosse and I.S. Burov [3] and is based on the joint solution of the equations of motion of raw materials, inter-component heat transfer and heat transfer of the plasma stream with the walls of the reactor, provides prediction of energy regimes of effective processing of various titanium-boron-containing raw materials. In this case, the dependence of the degree of evaporation K_p on the characteristics of the raw material (density ρ_p, specific heat C_p, thermal conductivity coefficient λ_p, temperatures of initial T_0, melting T_{pm}, evaporation T_{pr}, heat of fusion ΔH_m and evaporation ΔH_r, particle diameter d_p and particle shape coefficient $_{fp}$) and parameters of the reactor operation and the process (power supplied to reactor N, mass flow rate of plasma-forming G_g and transporting gas G_t, raw materials μ_r, diameters of the channels of the reactor D and lance D_0) are taken into account

$$K_p = f\,(\rho_p, C_p, \lambda_p, T_{pm}, T_{pr}, \Delta H_m, \Delta H_r, d_p, f_p, G_g, G_t, \mu_p, D \text{ and } D_0). \tag{10}$$

Submodel 3 “Boride formation” should be based on equations that take into account the influence of kinetic factors on the yield (content in products) of titanium diboride and equations describing the evolution of the dispersion of its particles in a plasma stream.

Kinetic equations should be formulated for the most likely boride formation reactions. However, the use of high-temperature kinetic calculations for processes involving the condensed phase in atmospheric pressure plasma using classical chemical kinetics methods is difficult due to the lack of information on the most likely boride formation reactions and the complete absence of data on their rate constants for the temperature range under study. In this regard, the macrokinetic approach seems to be more real, suggesting the choice of equations of the form obtained in [4] for the submodel

$$[TiB_2],[Ti_{free}],[B_{free}],[C][TiO_2] = f(K_p, T_0, T_3, \{H_2\}, [B], \{CH_4\}) \quad (11)$$

The equations describe the dependence of the titanium diboride content in plasma products on the degree of evaporation of the feed K_p, initial temperature T_0, quenching temperature T_3, the amount of boron in the charge [B], the hydrogen concentration in the plasma gas $\{H_2\}$, and the amount of the reducing agent $\{CH_4\}$ for TiO_2. The results obtained in studying the features of the boride formation process under the conditions of a plasma flow indicate a high probability of TiB_2 formation during crystallization of a titanium-boron-containing melt – a product of boriding by boron hydride of metal aerosol formed during the volume condensation of titanium vapor. This predetermines at least a two-stage review of this complex process.

However, the analysis of specialized scientific and technical literature indicates that a universal comprehensive model of the condensation process has not yet been developed, although theoretical options for predicting the dispersion of solid particles of condensation origin have been proposed.

The aforementioned and yet insurmountable shortcomings of the known mathematical models of the formation of nanoparticles under the conditions of a plasma flow prompted a number of researchers to study experimentally the temperature dependence of the average particle size during their enlargement [10-17].

The dependence obtained by the authors [17] for the synthesis in the plasma stream of TiB_2 from the reaction mixture (TiO_2+B+C_3H_8+C_4H_{10}) and the temperature range 2500–3200 K has the following form (d in m, T in K)

$$d = (1.46\pm0.025)\cdot10^{-5}\cdot T^{-0.741\pm0.165} \quad (12)$$

This temperature dependence is accepted as a calculated one for describing the evolution of the dispersion of TiB_2.

The structure of the mathematical model of boride formation processes is shown in figure 1. To simulate the boride formation process, a computer program “Modeling of the plasma synthesis of titanium boride” was created, which represents a complete product with the possibility of implementation on IBMPC computers running MS Windows and installed Microsoft Accesses. The program is registered in the joint fund of electronic resources “Science and Education” of the Russian Academy of Education (certificate on registration of electronic resource No. 21506 dated 12/07/2015).

3. Application of the model

Using the developed mathematical model, the parameters of plasma jets and flow, the conditions for the efficient evaporation of powdery raw materials and boride formation, and the dispersion evolution of boride nanoparticles, the results of which are presented in figures 2–6, were calculated. It can be seen that the efficiency of boride formation processes is directly determined by the conditions of evaporation of the raw material, boron gasification and methane pyrolysis.

4. Assessment of model adequacy

The accuracy of the developed model was evaluated by two methods:

- for option 1 ($Ti+B+H_2$) using a test sample obtained by processing in a plasma reactor a mixture containing PTM grade titanium powder with a grain size of 40 μm;
- for option 2 ($TiO_2+CH_4+B+H_2$) by comparing the calculated degree of evaporation of TiO_2 and the experimentally determined degree of its reduction in the conditions of a plasma flow; the degree of recovery was determined by the ratio of the actual and equilibrium concentration of carbon monoxide in the process gases.

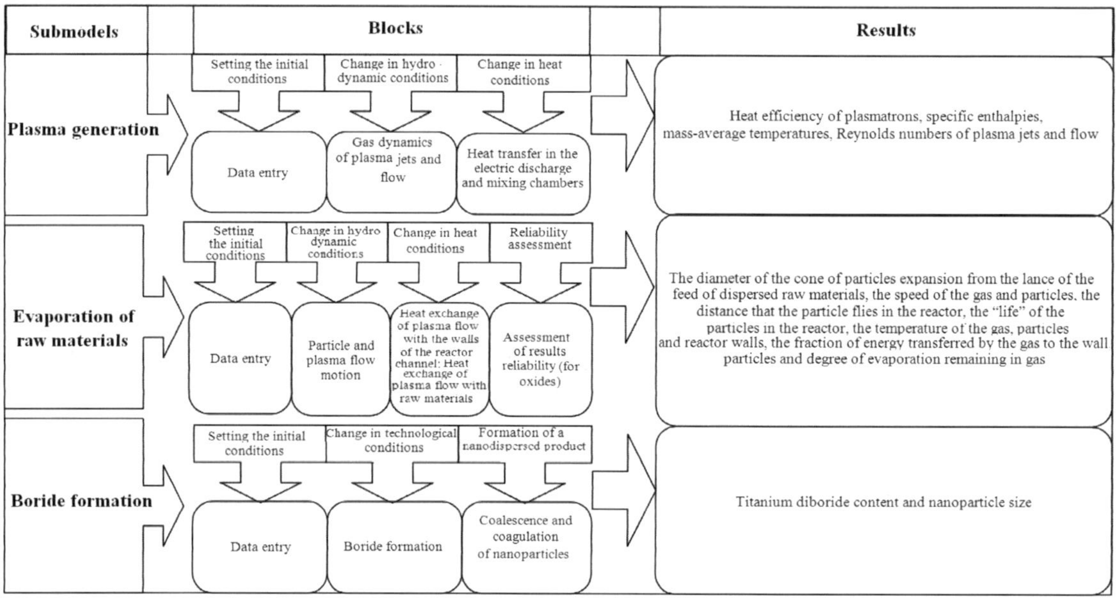

Figure 1. The structure of the mathematical model of boride formation processes.

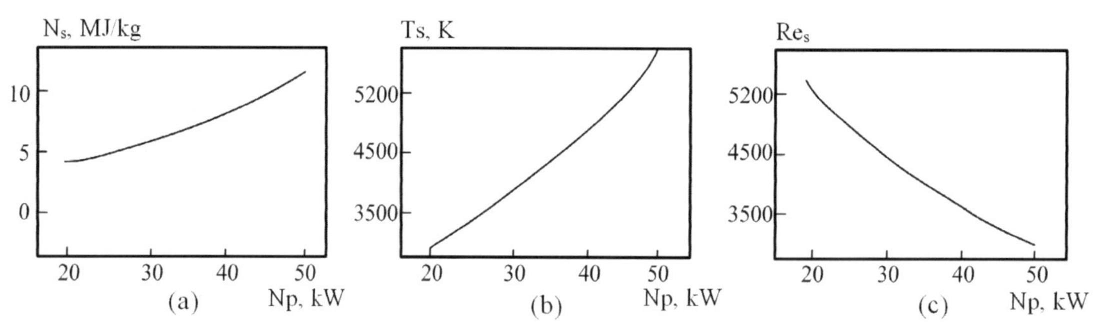

Figure 2. Dependence of specific enthalpy (a), mass average temperature (b) and Reynolds number (c) of a plasma flow at the entrance to the mixing chamber on the power of the plasma torch ($G_g = 3 \cdot 10^{-3}$ kg/s).

The accuracy assessment results for option 1 are shown in table 1, for option 2 in table 2. They confirm sufficient convergence of the predicted and practical results. In all cases, lower values of practical data are noted in comparison with the calculated ones, which is apparently caused by the influence of nonisothermal gas flow – the heat carrier.

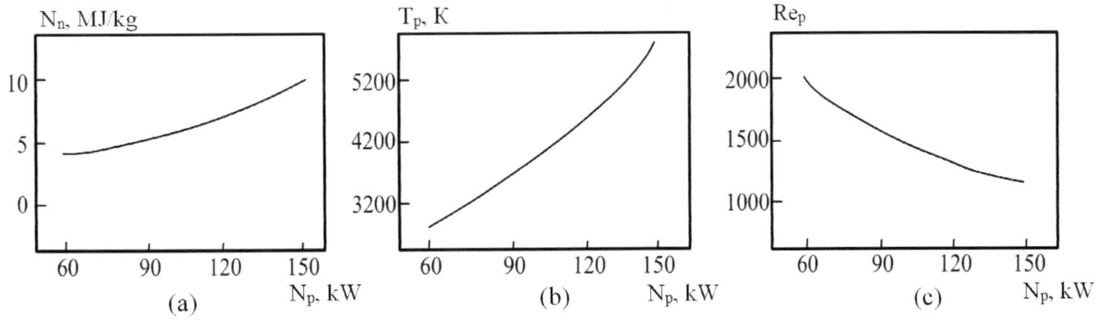

Figure 3. The dependence of specific enthalpy (a), mass average temperature (b) and Reynolds number (c) of the plasma flow at the inlet of the reactor channel on its power ($G_g = 3 \cdot 10^{-3}$ kg/s, D = 0.054 m).

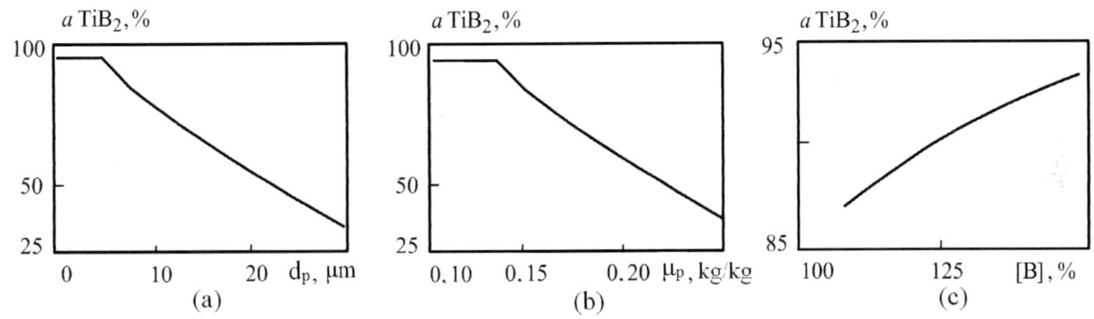

(a) μ_p = 0,14 kg/kgr, T_p = 5400 K, T_3= 2800 K, [B] = 112.5 %, {H_2} = 10 % vol.;
(b) d_p = 5 μm, T_p = 5400 K, T_3= 2800 K, [B] = 112.5 %, {H_2} = 10 % vol.;
(c) d_p = 5 μm, μ_p = 0,14 kg/kg, T_p = 5400 K, T_3= 2800 K, {H_2} = 10 % vol.

Figure 4. The dependence of the content (a) of titanium diboride in plasma products on the fineness of titanium powder (a), its mass consumption concentration (b) and the boron content in the charge (c) (Option 1 – (Ti+B+H_2)).

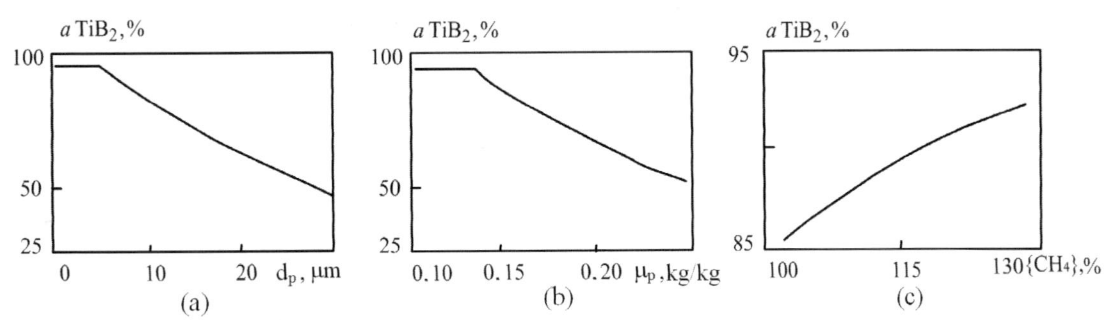

Figure 5. The dependence of the content of (a) titanium diboride in plasma products on the fineness of titanium powder (a), its mass consumption concentration (b) and methane consumption (c) (Option 2 – (TiO_2+CH_4+B+H_2)).

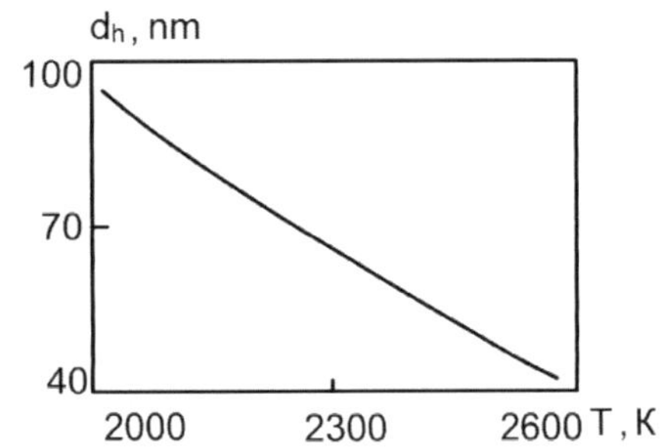

Figure 6. Evolution of the dispersion of titanium diboride nanoparticles in the plasma flow.

Table 1. Comparison of calculated and experimental results for option 1.

T_0, K	Titanium diboride content %							Deviation, %
	$y_{calc.}$	y_e^1	y_e^2	y_e^3	y_e^4	y_e^5	$y_{e\ av}$	
5400	36.6	34.9	32.1	31.2	30.5	30.9	31.9	12.9
5000	24.3	21.3	20.9	19.4	22.2	21.5	21.1	12.8

Table 2. Comparison of calculated and experimental results for option 2.

T_0, K	Evaporation rate, %	Degree of reduction α, %				Deviation, %
		α_e^1	α_e^2	α_e^3	$\alpha_{e\ av}$	
5400	100.0	88.1	90.9	89.8	89.6	10.4
5000	82.0	71.3	69.2	72.9	71.7	12.6

5. Conclusion

An integrated multifactor mathematical model of boride formation processes, that occur during the plasma metallurgical processing of titanium-boron-containing raw materials, was developed, including 3 submodels: 1) "Plasma generation", 2) "Evaporation of raw materials", 3) "Boride formation". The model has a block structure and allows the parameters of plasma jets and flow, conditions for efficient evaporation of powdery raw materials and boride formation, the evolution of the dispersion of boride nanoparticles to be calculated. A computer program was created for submodel 3 "Boride formation" (Certificate No. 21509 on registration in the joint fund of electronic resources "Science and Education" of the Russian Academy of Education). The effectiveness of the use of a complex mathematical model for performing research and engineering calculations of plasma processing technological indicators of various types of titanium-containing raw materials was confirmed.

References

[1] Tsymbal V P 2006 *Mathematical Mmodeling of Complex Systems in Metallurgy: a Textbook for Universities* (Kemerovo, Moscow: Kuzbassvuzizdat) p 431

[2] Zhukov M F et al 1995 *Plasmatrons. Research. Problems* (Novosibirsk: SB RAS) p 203

[3] Krasovskaya L I and Mosse A L 2000 *Plasma-chemical Processes in Three-jet Electric Arc Reactors* (Minsk: Institute of Heat and Mass Transfer n.a. A V Lykov of NAS of Belarus) p 196

[4] Galevsky G V, Efimova K A and Rudneva V V 2015 *Scientific and Technical Journal of SPbSPU* **2(219)** 141–150

[5] Solonenko O P et al 1995 *Thermal Plasma and New Materials Technology: In 2 Volumes* vol 2 (Cambridge: Cambridge Intersci) p 533

[6] Solonenko O P, Polak L S et al 1995 *Thermal Plasma Investigations and Design of Thermal Plasma Technologies* vol 2 (Cambridge: Cambridge Intersci) p 533

IOP Conf. Series: Materials Science and Engineering **866** (2020) 012024 doi:10.1088/1757-899X/866/1/012024

Technological basis of the plasma-metallurgical production of zirconium carbide

T I Alekseeva[1], G V Galevskiy[1], V V Rudneva[1] and S G Galevskiy[2]

[1]Siberian State Industrial University, 42 Kirova Street, Novokuznetsk, 654007, Russia
[2]St. Petersburg Mining University, 2, 21st Line, St Petersburg, 199106, Russia

E-mail: kafcmet@sibsiu.ru

Abstract. Based on the interpretation of the results of theoretical and experimental studies, a continuous technological process for zirconium carbide production in the plasma-metallurgical reactor, including plasma generation, plasma processing of zirconium-carbon-containing raw materials, the formation of ZrC, forced cooling and its separation from the process gases flow leaving the reactor, was developed.

1. Introduction

Zirconium carbide ZrC, investigated and put into circulation by the scientific school of the famous Russian material scientist G.V. Samsonov more than 50 years ago, is still in demand in the technology of various materials: cermet tool and construction materials, refractory and abrasive, for the modification of coatings [1 - 6]. The analysis of modern scientific and technical information reflects the trend of the transition from the use of coarse-grained zirconium carbide to micro- and nanocrystalline, which is due to the desire of scientists and technologists to achieve a qualitatively new level of operational properties of materials and coatings based on it.

The basis of modern production of zirconium carbide is the carbothermic method. The carbothermic method is implemented in several technological options and is used to obtain zirconium carbide when used as part of materials for abrasive processing, spraying and surfacing, protective coatings. However, the strategically important nanocrystalline zirconium carbide market segment is completely closed by foreign suppliers, including companies such as Nanostructured □ Amorphous Materials, Inc. (USA), Hefei Kaier Nanotechnology & Development Ltd. Co (China), NEOMAT Co (Latvia), PlasmaChem GmbH (Germany). This necessitates the development of Russian nanotechnology of zirconium carbide.

In this regard, the study and technological implementation of carbide formation processes during the plasma-chemical processing of zirconium-carbon-containing raw materials is an important scientific and practical task of great importance for the development of domestic technology of multifunctional zirconium compounds, as well as the effective solution of innovative problems of applied materials science.

2. Theoretical study

Theoretical research includes thermodynamic modeling of plasmosynthesis of zirconium carbide. Thermodynamic modeling of the synthesis processes was carried out with the aim of predicting the optimal conditions for the production of zirconium carbide (ratio of components and temperature), determining the equilibrium process indicators (the degree of conversion of raw materials to carbide, compositions of gaseous and condensed products), assessing the contribution to the carbide formation of gas-phase reactions that provide plasma conditions technologies for efficient processing of dispersed raw materials [7.8].

The objects of thermodynamic modeling were the C-H-N, Zr-C-H-N and Zr-O-C-H-N systems, which is due to the possibility of using zirconium and its dioxide, hydrocarbon feeds – methane, plasma forming gas – nitrogen as zirconium-containing raw materials. In the calculations, the temperature range 1000–6000 K was considered at a total system pressure of 0.1 MPa. The compositions of gaseous and condensed products necessary for analysis were calculated by the "constant method". The calculations were performed using the PLASMA computer simulation program for high-temperature complex chemical equilibria, which has an integrated database of interaction products for oxide, boride, carbide, and nitride-forming systems. The ratios of the starting components were set in accordance with the stoichiometry of the reactions of the formation of the target products and the parameters of the processing equipment (plasma gas flow rate).

The simulation results of the raw materials and heat carrier gas interaction for the conditions of a plasma-chemical reactor allow the parameters and implementation indicators of technological options for zirconium carbide production to be predicted. Technological options, parameters and indicators for the production of zirconium carbide are presented in table 1. The calculation took into account the possibility of raw materials loss in the reactor in the amount of 4% and plasma products in the capture system in the amount of 5%.

Table 1. Predicted parameters and indicators for zirconium carbide production.

Technological parameters and indicators	Technological options		
	Option 1 $Zr+CH_4+N_2$	Option 2 $ZrO_2+CH_4+N_2$	Option 3 $ZrCl_4+CH_4+N_2$
Consumption of plasma-forming gas (nitrogen), kg/s	$9 \cdot 10^{-3}$	$9 \cdot 10^{-3}$	$9 \cdot 10^{-3}$
The initial temperature of the plasma stream, K	5400	5400	5400
Quenching temperature, K	2800	2800	2800
Mass consumption concentration, kg of raw materials /kg of nitrogen	0.14	0.14	0.20
The degree of raw materials conversion to carbide	0.96	0.96	0.96
The carbide yield, %	92.0	92.0	92.0
Productivity, kg of zirconium carbide /h	1.1	4.29	2.45
The intensity of the process, kg /h·m^3	647	2523	1441
Consumption of zirconium-containing raw materials, kg/kg	2.73	0.82	2.29
Methane consumption, kg/kg	0.15	0.56	0.16
Consumption of heat carrier gas (nitrogen), kg/kg	21.43	5.73	16

3. Pilot study

Subsequently, experimental studies of carbide formation processes occurring in the nitrogen-hydrogen high-temperature flow during plasma processing containing zirconium dioxide – natural gas (methane) were carried out:

1. Based on the results obtained, a zirconium dioxide powder of TsRO grade according to GOST 21907-76, natural gas (methane), and technical nitrogen GOST 9293-84 were selected as raw materials for the production of zirconium carbide. It was established that the product of carbide formation is zirconium carbide ZrC.

2. The equations describing the dependences of the zirconium carbide content and free carbon (in %) in the products on the main technological factors were obtained for the technological option under study:

$[ZrC] = -109.72 + 0.0371\ T_0 + 0.0034\ Tq - 0.0687\{CH_4\}$,

$[ZrO_2] = 174.44 - 0.02995\ T_0 - 0.00236\ Tq$,

$[C_{free}] = -101.33 + 0.015\ T_0 + 0.0079\ Tq + 1.2487\ \{CH_4\} - 0.00019\ T_0\{CH_4\} - 0.00008\ Tq\{CH_4\}$,

$[N] = -7.092 + 0.0013\ T_0 + 0.00064\ Tq$,

IOP Conf. Series: Materials Science and Engineering **866** (2020) 012024 doi:10.1088/1757-899X/866/1/012024

(in the equations T_0 – the initial temperature of the plasma flow (5000 – 5400 K); Tq – quenching temperature (2000 – 2800 K); {CH4} – the amount of reducing agent (100 – 130% of the stoichiometrically necessary to obtain ZrC).

3. Peculiarities were determined and a probable mechanism of zirconium carbide formation according to the "vapor-crystal" scheme was proposed, presumably during the interaction of zirconium vapor and cyan; A generalized hypothetical carbide formation scheme has been compiled containing 2 zones: a high-temperature zone (6000 – 3500 K) of formation of the reaction mixture in which the processes of zirconium powder evaporation occur, and a lower-temperature (3500 – 2000 K) in which zirconium vapor is condensed, a significant decrease in hydrocarbons concentration and zirconium carbide formation.

4. A comprehensive physico-chemical certification of zirconium carbide was carried out. It was established that:

- the zirconium carbide content in the obtained products, not in contact with air is 94.2 – 93.61%, related impurities, %: zirconium dioxide 4.56 – 5.27 %, free carbon 1.32 – 1.12 %, nitrogen 1.87 – 2.12%;
- zirconium carbide obtained in the nanocrystalline state, is represented by faceted particles of a cubic shape in the size range from 10 to 40 nm.

Based on the interpretation of the results of theoretical and experimental studies, a continuous technological process for the production of zirconium carbide in the plasma metallurgical reactor was developed.

For the implementation of zirconium carbide plasmosynthesis, a hardware-technological scheme is proposed, shown in figure 1 [9-11], which includes the following operations and stages: 1) input control of raw materials and process gases; 2) preparation of ZrO_2 powder: storage, dosing; loading into a powder dispenser; 3) plasma processing; 4) cooling and partial deposition in the quenching and precipitation chamber of the gas stream leaving the reactor to the working temperature of the bag metal fabric filters (600-800 K), separation of zirconium carbide from it in the filters; 5) certification of nanocrystalline ZrC in chemical composition and dispersion, packing in sealed double plastic bags and sending to the warehouse of finished products.

The main quality indicators and technical and economic indicators for zirconium carbide production for the production conditions of custom-made batches are presented in table 2. The production cost and selling price were calculated in accordance with the recommendations [12]. The required investment for organizing the zirconium carbide production in 3 plasma-metallurgical reactors with a total capacity of 450 kW is 123.7 million rubles. At the same time, it is forecasted to achieve an annual output of 54 tonnes/year with a selling price of 35,142 rubles/kg ($ 576 / kg).

The prime cost of zirconium carbide is 20539 rub/kg and has the following structure, %: raw materials, materials, electricity 86; wages and social security contributions 3; maintenance and operation of equipment 4; general shop, general business, commercial expenses 7. The payback period for capital investments is four months.

A comparison of prices with price offers for nanocrystalline zirconium carbide of foreign companies, presented in figure 2, testifies to the competitiveness of the developed technological process.

Based on the interpretation of the results of theoretical and experimental studies, a continuous technological process for producing zirconium carbide in a plasma-metallurgical reactor was developed, including the preparation of raw materials, plasma generation, plasma processing of zirconium-carbon-containing raw materials, the formation of ZrC, its forced cooling and the release of technological gases from the reactor from the stream. The technological process provides, under operating conditions with a coefficient of equipment utilization of 0.7, a productivity of 18.16 tonnes/year per reactor at a selling price of 35142 rubles/kg.

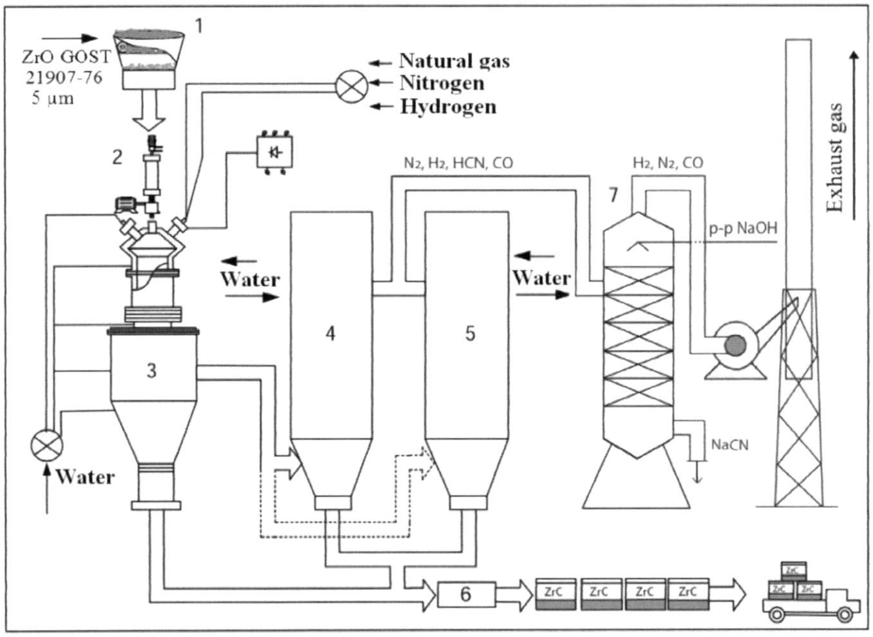

Figure 1. Process flow diagram for zirconium carbide production: 1 – rubbing of the powdered raw materials; 2 – loading of ZrO_2 powder into the dispenser; 3 – plasma treatment; 4-5 – cooling of the exhaust dust and gas flow and separation of the target product; 6 – collection, quality control and packaging; 7 – absorption option of exhaust gases neutralization.

Technical forecasts and suggestions for zirconium carbide use as part of functional protective coatings formulated on the basis of the formed ideas about its physicochemical properties features in the nanocrystalline state. In the conditions of Polimet LLC, the expediency of using zirconium carbide in the composition of corrosion-resistant protective coatings based on nickel instead of the used nanodiamonds was established.

Table 2. Key quality indicators and technical and economic indicators of zirconium carbide production.

Indicators	Value
Production area, m^2	300
Installed power, kW	450
Number of reactors, pcs	3
Equipment utilization ratio, the share of units	0.7
Schedule, number of shifts	2 shifts x 12 hours
Investments in fixed and current assets, mln. rub.	123.7
The specific gas flow rate of the coolant (nitrogen), t/t	5.73
Specific consumption of natural gas (methane), t/t	0.15
Specific consumption of zirconium-containing raw materials, t/t	0.99
Specific energy consumption, thousand kW·h/t	42.81
The content of the main phase (ZrC), %	92
Annual need for raw materials, t/year	
- zirconium dioxide ZrO GOST 21907-76	26.3
- природный газ (метан) - natural gas (methane)	4.9
Annual productivity, t/year	54.5
Planned cost, rub/kg	20539
Price (as of 05/01/2018), rub/kg	35142
Payback period for capital investments, years	0.3

IOP Conf. Series: Materials Science and Engineering **866** (2020) 012024 doi:10.1088/1757-899X/866/1/012024

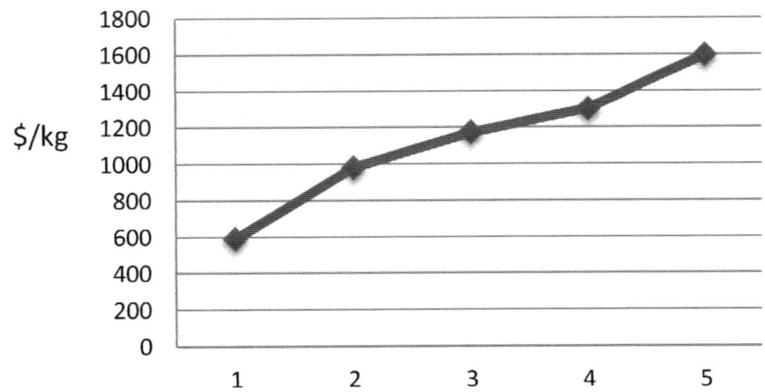

Figure 2. Global price level for nanocrystalline zirconium carbide: 1 – proposed technology; 2– "American Elements" (Los Angeles); 3 – "PlasmaChem GmbH" (Berlin); 4 – "NEOMAT Co" (Salaspils); 5 – "Nanostructured & Amorphous Materials, Inc." (Houston).

References

[1] Alekseeva T I, Galevsky G V et al 2017 *St. Petersburg Polytechnic University Journal of Engineering Science and Technology* **23(1)** 256–270
[2] Kosolapova T I 1968 *Carbides* (Moscow: Metallurgiya) p 300
[3] Xiang J Y, Liu S C et al 2011 *Journal of the European Ceramic Society* **31(8)** 1491–96
[4] Kuvshinov G G, Ermakova M A and Ermakov D Yu 2002 *Morphology and Texture of Silica Prepared by Sol–Gel Synthesis on the Surface of Fibrous Carbon Materials* (Novosibirsk: Boreskov Institute of Catalysis) p 432
[5] Jingjing Xie, Zhengyi Fu et al 2014 *Journal of the European Ceramic Society* **34(1)** 13.e1–13.e7
[6] Liliang Chen, Chihiro Iwamoto et al 2011 *RSC Advances* **1** 1083–88
[7] Hartmann U et al 2008 *The Charm of Nanotechnology Translation from It* (Moscow: BINOM. Laboratoriya Znaniy) p 173
[8] Rudskoy A I 2007 *Nanotechnology in Metallurgy* (St. Petersburg: Nauka) p 186
[9] Dubchak R V 1994 *Non-ferrous Metallurgy* **10** 28–33
[10] Galevsky G V, Kulagin N M and Mintsis M I 2004 *Aluminum Metallurgy. World and Domestic Production: Assessment, Trends, Forecasts* (Moscow: Flint: Nauka) p 280
[11] Galevsky G V and Rudneva V V 2018 *Nanomaterial Technology: Practicum* (Novokuznetsk: SibSIU) p 29
[12] Kerimov V E 2014 *Cost Accounting, Costing and Budgeting in Certain Sectors of the Manufacturing Sphere* (Moscow: Dashkov and K) p 384

IOP Conf. Series: Materials Science and Engineering **866** (2020) 012025 doi:10.1088/1757-899X/866/1/012025

Environmental monitoring of the reclaimed sites of JSC "EVRAZ ZSMK"sludge dump

A S Vodoleev[1], V A Androkhanov[2], M A Zaharova[1], O S Andreeva[3] and K I Domnin[1]

[1]Siberian State Industrial University, 42 Kirova Street, Novokuznetsk, 654007, Russia
[2]Institute of Soil Science and Agrochemistry SB RAS RF, 8/2 Academician Lavrentiev Ave., Novosibirsk, 630090, Russia
[3]Novokuznetsk Institute (branch) Kemerovo State University, 23 Tsiolkovskogo Street, Novokuznetsk, 654041, Russia

E-mail: botanik-egf@yandex.ru

Abstract. For conservation of technologically disturbed lands, various methods of reclamation are used to prevent their negative impact on the environment. At the same time, technological methods are used, including changing the technology of waste storage, changing the state of storaged products using briquetting and granulation, creating mechanical barriers that prevent the spread of dust or continuous coating with non-dusting material, and the physicochemical method – creation of a protective layer using foam or hydro-dusting in order to stabilize a dusty surface. The most environmentally effective is the biological method – creation of a protective vegetation cover. This method is of particular relevance due to the phytotoxicity of the technogenic substrate for sludge dumps and the difficulties of creating a stable vegetation cover. Taking into account the industrial and environmental characteristics of the waste in sludge dumps of JSC "EVRAZ ZSMK", an assessment of the physiological state was carried out as part of environmental monitoring and the chemical composition of plants growing in the reclaimed areas of this technogenic object was analyzed. A comprehensive assessment of the environmental safety of the use of sewage sludge from urban wastewater treatment plants on the substrate of the sludge storage facility as the main soil improver is given.

1. Introduction

Currently, in industrialized regions, as a result of the activities of industrial enterprises, technogenic ecosystems have formed that differ in the composition of the stored materials and the formation technologies. Basically, they are represented by various dumps and open-pit mines that are formed during minerals mining, waste from the concentration of sinter plants, tailings, sludges and ash dumps of thermal power plants. Technogenic ecosystems are formed by technical means and are fundamentally different from natural ecosystems that previously functioned in this territory. These differences are due to the special forms of relief, the properties of the stored substrates, and the almost total absence of biota [1]. Therefore, to restore ecosystems to their natural state, it is necessary to carry out a set of reclamation measures [2].

By the type of the work performed, measures for the restoration of technologically disturbed lands can be divided into:

- landscape reclamation (land reclamation) associated with the restoration of the disturbed landscape (terrain, soil and vegetation);

- environmentally friendly (environmentally protective), associated with the elimination and neutralization of the harmful effects of industrial waste on the living environment (soil, water, air).

The implementation of these measures can reduce the negative consequences of waste disposal and begin the formation of already natural-technogenic complexes on the site of disturbed lands, and it is very important to organize environmental monitoring in the reclaimed areas already at the initial stages of reclamation in order to assess the effectiveness of the work done and the condition of the reclaiming ecosystems.

Sludge dumps and tailings, as part of the technogenic landscape, occupy a special place in the framework of studies on the restoration of disturbed lands, in particular, in the development of ways and methods for their biological restoration. The main emphasis in choosing the direction of reclamation of sludge collectors is aimed at technological, mechanical and physico-chemical methods of combating dust, which are developed and recommended for technologically disturbed lands in metallurgical production.

When reclaiming sludge and tailing dumps, the following methods are used to combat surface dusting:

- Technological method. This method includes changing the storage method according to particle size or physico-chemical composition, as well as changing the composition and condition of the dumped products using briquetting and granulation.
- Mechanical method: creation of barriers to prevent the spread of dust or a continuous coating of non-dusting material.
- Physico-chemical method: creation of a protective layer using foam or water dedusting in order to stabilize the dusty surface.
- Biological method: creation of a protective layer from plants. This method is of particular relevance in connection with the phytotoxicity of the technogenic substrate of sludge and tailing dumps and the difficulties of creating a sustainable vegetation cover.

Biological reclamation of sludge dumps is limited by several factors:

1. The absence or critical lack of mineral nutrition.
2. Unsatisfactory water-physical properties due to the predominance of sand fractions in the beach area having high filtration properties.
3. Easy deflationary ability of particles, which complicates the fixing of seeds and seedlings of plants.

Modern physicochemical methods do not give a complete picture of the ecological situation in a particular area, so there is a need to use biomonitoring data and conduct bioindication studies. A change in the chemical composition of the habitat is accompanied by a mandatory change in the composition of the biota. Biological indication of environmental pollution may include methods for assessing the vital state of plants (physiological), the structures of the emerging phytocenoses (phytocenotic), migration paths and transformations of chemical elements entering the plants (biochemical).

2. Characteristics of the studied object

Sludge dumps is a technogenic facility for central collection, accumulate and settle the solid phase of pulps, sludge and wastewater from industrial wastes of JSC “EVRAZ ZSMK”, West Siberian Central Heating and Power Station, CPP “Kuznetskaya” and return of clarified water to consumers. The sludge dump has been in operation since 1964. Since the beginning of operation, more than 100 million tonnes of waste have been disposed.

The total area is more than 300 hectares with a dam height of about 25 m. On all sides, the sludge storage facility is surrounded by a dam composed of a mixture of coarse screening of coal processing waste and converter slag. The dam is of trapezoidal sectional structure, the outer slope of which is terraced. In total, two terraces are formed on the slopes of the dam, the flat surface of which is used for laying technological roads.

IOP Conf. Series: Materials Science and Engineering **866** (2020) 012025 doi:10.1088/1757-899X/866/1/012025

The width of the terraces ranges from 10 to 40 meters. The sides of the terraces are currently not fixed – their steepness is determined by the angle of repose – and for this reason are subject to destruction by water-erosion and deflation processes. A view of the sludge storage facility from space is presented in figure 1.

Figure 1. Sludge storage facility of JSC "EVRAZ ZSMK", satellite view.

At the same time, these terraces play a protective role, protecting the dam from destruction by water erosion and deflation processes, although at the same time, their edges and steep slopes continue to collapse quite intensively. To prevent the destruction of the edges of the entire surface of the terraces, various methods of fixing the surface, in particular, methods of biological reclamation, are required. These methods should be aimed at improving the physical and chemical properties of the substrate, creating conditions for the development of phytocenoses, which in the end result should lead to the formation of a solid turf and surface fixation.

3. Assessment of the physiological state of plants at the sludge dump

The physiological parameters of woody plants (balsamic poplar, ash-leaved maple, goat willow, small-leaved elm) were studied during the growing season (July) located in the lower part of the second terrace of the reclaimed plot No. 5 in the morning from 9 a.m. to 12 p.m. The temperature was in the range of 20 – 22 °C, illumination from 25000 to 42000 lux. The state of stomata, the intensity of photosynthesis, and the chlorophyll content of experimental plants were studied [3].

The studies have shown that the state of stomata in the plants indicates sufficient water supply for their tissues, therefore water, as a participant in photochemical light reactions in this habitat, is not a limiting factor. The content of photosynthetic pigments is able to ensure the absorption of carbon at an average level, although the level of photosynthesis intensity does not reach its maximum. The reasons for this discrepancy may be the lack of nutrients available to plants, nitrogen and phosphorus, which provide plastic and energy metabolism with their carbon nutrition. The previously performed agrochemical characterization of the dam rocks of the sludge dump at JSC "EVRAZ ZSMK" site confirms this assumption, while the content of mobile forms of potassium is quite sufficient, which ensures normal water exchange of plants and is confirmed by stomatal control.

The results obtained on the physiological parameters of woody plants indicate their adaptive capabilities, despite the high daily temperature, moisture deficiency, and the lack of optimal mineral nutrition. The correlation between the content of chlorophyll in the leaves of the studied plants and the intensity of photosynthesis is more pronounced during the July drought.

The introduction of non-traditional soil improvers – wastewater sludge (WWS) of urban treatment facilities into the technogenic substrate of the sludge storage facility of JSC "EVRAZ ZSMK" is accompanied by an increase in the content of organic substances and fractions of physical clay, which increases the sorption capacity of the formed organomineral substrate (technozem). In this case, a slight increase in the content of heavy metals is recorded. This is due, on the one hand, to an increase in their WWS content in the substrate, and, on the other hand, to an increase in the mobility of metals in the dam substrate, as a result of increased humidity and organic matter content. Subsequent decomposition of the organic matter of technozems leads to a decrease in sorption ability and an increase in the bioavailability of metals. The introduction of WWS is accompanied by a change in the pH value, which leads to a decrease in the mobility of metals as a result of complex formation, but conditions may arise for the formation of soluble organometallic complexes.

4. Chemical analysis of plant material in the sludge dump

When WWS is added to the sludge dump, the content of the main nutrient elements in the surface layer of the reclaimed areas increases: carbon – up to 12.3%, nitrogen – up to 0.92%. At the same time, the level of pollution with heavy metals decreases several times: vanadium – from 3 (medium) to 2 (low), tin and mercury – from 2 (low) to 1 (permissible). The indicators for arsenic, copper, nickel and zinc (low level), as well as for lead, chromium, cobalt, cadmium and fluorine (acceptable level) remain unchanged.

According to the content of mobile forms of heavy metals (copper, zinc, lead and nickel), the obtained mineral-organic substrate of the sludge dump is assigned to the second, low level of pollution. The content of cobalt and chromium is lower or corresponds to the permissible level (MACp = 6.4 mg/kg).

The analysis of chemical composition of the biomass of herbaceous plants is an indicator of the possible removal of toxic elements from the root layer. According to the results of studies, the excess of the MAC of arsenic, lead, zinc, nickel, copper and antimony in slag+WWS was found. Comparison of the content of these elements in plants showed that arsenic was not detected in plants (cadmium, selenium, beryllium were also not found). The content in the plants of the experimental plots, mg/kg: lead – 0.3–1.4, zinc - 64–400, nickel – 0.7–1.2, copper – 3.7–9.2, vanadium – 0,7–2.7, manganese – 148–274, with a gross content of these elements in technozems ("rock + WWS" and "slag + WWS"), respectively, mg/kg: lead – 29 and 40, zinc – 239 and 191, nickel – 32 and 22, copper – 67 and 68, vanadium – 157 and 157, manganese – 563 and 7565. These data show that zinc is a potential hazard in terms of the removal of toxic elements. The taking of heavy metals by plants is maximum in the first years of the placement of WWS, over the years their concentration decreases and becomes comparable with the background indicators of the control plant material.

By microbiological characteristics, pathogenic microflora was not found in the studied material and corresponds to SanPiN 4630-88 [4, 5]. The introduction of WWS increases the level of microbial population to ecological and trophic diversity, therefore, WWS, as ameliorants, are classified as "polluted" in terms of "coli titer".

5. Conclusion

Comprehensive monitoring of the environmental safety of WWS use on the rocks of the sludge storage facility of JSC "EVRAZ ZSMK" showed that this unconventional soil improver cannot be

IOP Conf. Series: Materials Science and Engineering **866** (2020) 012025 doi:10.1088/1757-899X/866/1/012025

considered as a source of additional pollution during reclamation work. The content of nutrients in the formed technozems during the placement of WWS is increased, which ensures survival and sustainable growth of vegetation. The selection of the species composition of plants is determined by their resistance to extreme conditions and soil-improving properties. This allows a stable phytocenosis in the face of adverse effects of metallurgical production wastes to be formed and helps to improve the environmental status of the suburban area of Novokuznetsk.

References

[1] Dvurechensky V G and Androkhanov V A 2017 *Living and Biocos Systems* 20
[2] *GOST 17.5.3.04-83 General Requirements for the Reclamaition of Disturbed Lands*
[3] Vodoleev A S, Androkhanov V A and Klekovkin S Yu 2007 *Soil Improvers: Reclamation Aspect* (Novosibirsk: Nauka) p 148
[4] *SanPiN 2.1.7.1038-01 Hygienic Requirements for the Arrangement and Maintenance of Landfills for Municipal Solid Waste*
[5] *SanPiN 4630-88 Sanitary Rules and Standards for the Protection of Surface Waters from Pollution*

IOP Conf. Series: Materials Science and Engineering **866** (2020) 012026 doi:10.1088/1757-899X/866/1/012026

Study of the characteristics of iron-containing waste from steelmaking and rolling production

A E Anikin[1], G V Galevskiy[2], V V Rudneva[2] and E V Nozdrin[2]

[1]Group of companies "Sibirsky proyekt", 11 Ordzhonikidze Street, Novokuznetsk, 654000, Russia
[2]Siberian State Industrial University, 42 Kirova Street, Novokuznetsk, 654007, Russia

E-mail: kafcmet@sibsiu.ru

Abstract. A study of the physicochemical characteristics of mill scale and dehydrated sludge from gas purification of oxygen-converter production was made. Their chemical, phase, particle size distribution and density are determined. It was established that the total iron content in the scale is 73.3%, in the sludge – 41.2%. The technological expediency of their use in metallization processes is determined. In this case, preliminary briquetting in a mixture with a carbon reducing agent is necessary.

1. Conclusion

Currently, the following technological options for metallization of various raw materials are known [1-4]:

1) Production of partially metallized materials for blast furnaces.

2) Production of metallized product (sponge iron) in solid form for melting in steelmaking units (temperature 500-1000 °C).

3) Production of metallized product in a plastic state (production of critical iron) for various purposes, including as an option of pyrometallurgical enrichment of refractory, poor and complex ores (temperature 1100-1400 °C).

4) Obtaining of liquid metal (cast iron or intermediate) for remelting in steelmaking furnaces (temperature above 1200-1400 °C).

The main solid-phase metallization processes are Midrex, HyL, Danarex, Finmet, etc. [5-9], and the main liquid-phase iron reduction processes are Corex, Finex, Romelt, Ausmelt, HIsmelt, TECHNORED, etc. [2 - 4] .

Metallization of fine-grained and powdery oxy-iron-containing wastes is an important area in metallurgy. According to the World Steel Association [10], the production of direct reduction iron over the past 10 years has grown in the world by 18%, in Russia – by 66%, which indicates an extremely dynamic development of this area. Metallization is especially important for mini-factories due to the lack of sinter production in their structure, which allows metallurgical enterprises to process rolling mill scale, sludge from blast furnace and steelmaking and other iron oxide-containing wastes. The urgency of the problem is due to a significant number of mini-factories in the world – about one thousand [11], and, accordingly, a significant amount of generation of oxy-iron-containing waste. According to forecasts, in 2020 more than 50% of global steel production will be provided by mini-factories. In the territory of the former USSR, 7 mini-factories have already been built: in Russia

(Komsomolsk-on-Amur, Kaluga, Saratov, Rostov and Tyumen regions), in Belarus (Zhlobin) and in Moldova (Chisinau) [11, 12].

To study the metallization processes of iron-oxide-containing technogenic raw materials, their physicochemical certification is necessary. As an object of study, mill scale and gas treatment sludge from the oxygen-converter shop No. 1 of EVRAZ ZSMK JSC (BOF sludge) were selected. These materials were also studied previously by a number of authors [13–16], but the data presented are either contradictory or insufficiently complete.

2. The purpose of the research

The aim of the present work is to study the physicochemical characteristics of mill scale and dehydrated sludge from a gas purifier of basic oxygen steelmaking (chemical, particle size and phase composition, density) – a promising technogenic oxide-iron-containing raw material for producing metallized briquettes demanded in steel production.

3. General characteristics of materials and methods for their research

Rolling scale is formed as a result of secondary oxidation of the surface layer of the metal when heated before rolling. When heating 1 tonne of steel, approximately 25-30 kg of scale (2.5-3.0%) is formed [8]. According to the World Steel Association [5], up to 50 million tonnes of mill scale are formed annually in the world, and up to 1.5 million tonnes in Russia.

At the construction site of EVRAZ ZSMK JSC, approximately 200 thousand tonnes of mill scale are formed per year. According to [6], on average, 1.0-1.5 million tonnes of rolled steel are produced at one metallurgical mini-plant; therefore, approximately 25-45 thousand tonnes of mill scale are formed. Scale in sinter production is widely used as an iron-containing component of sinter charge.

Sludges from basic oxygen steelmaking are formed during wet gas treatment of converter gases. Sludge refers to rich (Fe_{tot}=55÷67%) or relatively rich iron (Fe_{tot}=40÷55%). With wet gas cleaning, 10-30 kg of sludge is formed per 1 tonne of smelted steel (1-3%) [9]. Consequently, approximately 14.0–38.0 million tonnes of BOF sludge is generated in the world, and 0.5–1.5 million tonnes in Russia. Dehydrated sludge is most widely used in sintering as an iron-containing component of sinter charge. The specific consumption of sludge can reach 200 kg/t of sinter. However, a significant drawback of this technology is the fact that the content of zinc and lead increases in the resulting agglomerate, which is unacceptable due to the significant influence of these impurities on the masonry of blast furnaces.

In this regard, a number of alternative technologies are proposed, such as a scheme for heat treatment of sludge in a rotary furnace, a scheme for treating sludge with spent nitric acid pickling solution at a neutralization station, and a combined two-stage scheme with solid-phase metallization and zinc removal from the sludge and further liquid phase recovery of partially metallized intermediate. But some metallurgical plants, in particular EVRAZ ZSMK JSC, dump some of the sludge into the sludge storage facilities. So, about 2.7 million tonnes of iron in the composition of iron-containing sludge is accumulated in the sludge storage facility of EVRAZ ZSMK JSC.

It should be noted that the gas treatment sludge from different BOFs varies significantly in properties due to the fact that different smelting technologies are used, and there are three technologies for removing and purifying exhaust converter gases:

1) systems working with air suction through the gap between the converter and the boiler-cooler and the complete combustion of carbon monoxide in the latter, i.e. with air flow coefficient $\alpha > 1$;

2) systems operating without access of air to the gas path and without afterburning of carbon monoxide, i.e. when $\alpha < 0.15$;

3) systems operating with partial afterburning of carbon monoxide in a converter gas cooler, i.e. for $1 > \alpha > 0$.

When choosing between sludges of BOF-1 and BOF-2, EVRAZ ZSMK JSC settled on sludge BOF-1 due to the higher content of total iron in it. When analyzing the physicochemical properties of the above materials, the following characteristics were determined: chemical, particle size and phase compositions, as well as density.

The chemical composition was determined according to GOST R 52939-2008. The mass fraction of total iron was determined according to GOST 23581.18-81 (1987 Edition, amendment No. 1 of May 1987), the mass fraction of iron (II) oxide was determined according to GOST 23581.3-79 (1986 edition, amendment No. 1 dated July 1982 and No. 2 dated January 1986), metallic iron – according to GOST 23581.11-79 (1986 edition with amendment No. 1 dated January 1986), arsenic – according to GOST 23581.8-79 (1986 edition as amended by No. 1 of January 1984 and No. 2 of January 1986), phosphorus by GOST 23581.19-91, sulfur by GOST 23581.20-81 (1987 edition as amended by No. 1 of May 1987), copper – in accordance with GOST 23581.6-79 (1986 edition with amendment No. 1 of January 1984 and No. 2 of January 1986).

The particle size distribution was determined according to GOST-27562-87, the phase composition was determined by x-ray analysis. Density determination was carried out according to GOST 26732-88.

4. Results and discussion

The results of determining the chemical composition of mill scale and sludge are shown in table 1. From table 1 it can be seen that mill scale has a higher total iron content (73.3%) than BOF-1 sludge (41.18%). It should be noted that the total iron content in the BOF -2 sludge is about 35%. The following picture is observed in the distribution of iron over oxides: in mill scale, FeO accounts for 75.5%, and Fe_2O_3 – 20.86%; in the sludge, on the contrary, FeO accounts for 4.69%, and Fe_2O_3 – 53.67%.

Table 1. The chemical composition of mill scale and sludge.

Chemical composition, %	Scale	Sludge
$Fe_{tot.}$	73.30	41.18
$Fe_{red.}$	Not def.[a]	54.08
Fe_2O_3	20.86	53.67
FeO	75.50	4.69
SiO_2	2.24	1.47
Al_2O_3	0.20	0.12
CaO	0.24	20.59
MgO	0.24	0.36
K_2O	Not def.	0.13
Na_2O	The same	0.10
TiO_2	»	0.04
Mn	»	0.74
MnO	0.66	Not def.
P	0.019	0.15
Cr_2O_3	Not def.	0.016
V_2O_5	The same	0.04
$S_{tot.}$	0.036	0.21
SO_3	Not def.	0.07
$S_{sulfide.}$	The same	0.18
BaO	»	< 0.01
Ni	»	0.012
Cu	»	0.03
Zn	»	0.28
Pb	»	0.09
As	»	0.0025
$C_{tot.}$	»	4.32
$C_{sol.}$	»	0.82
LOI	»	16.57

[a] Was not defined

Thus, mill scale is less oxidized than sludge. The low content in mill scale (less than 3%) of iron-free compounds should also be noted. In the sludge, on the contrary, the CaO content reaches 20.59%. In addition, in mill scale, the sulfur and phosphorus contents are 0.036% and 0.019%, respectively, against 0.21% and 0.15% in sludge. However, the slurry contains elements and compounds that are alloying in the production of steel. These are compounds of manganese, nickel, vanadium, chromium. The sludge also contains up to 4.32% of total carbon, including up to 0.82% of solid carbon. Thus, both mill scale and sludge are of interest as oxide-iron raw materials for the metallization process.

The study results of the particle size distribution of scale and sludge are shown in table 2. From table 2 it is seen that the sludge is a finer dispersed material than mill scale.

The predominant mineral of sludge is magnetite. Hematite, wustite, calcite, feldspar are also present. The data obtained are in good agreement with the data of chemical analysis. The true density of mill scale was from 4.6 to 4.9 g/cm^3, converter sludge from 3.5 to 5.0 g/cm^3. Thus, the density of mill scale and sludge CCC is comparable.

The analysis of the properties of the specified oxide-iron raw materials allows the feasibility of its use in the metallization process to be concluded. Moreover, the use of mill scale is more preferable due to the higher content of total iron in it.

Table 2. Granulometric composition of mill scale and sludge.

Size class content, %	Scale	Sludge
> 2.5 mm	24.7	7.4
1.6-2.5 mm	18.3	3.6
1.0-1.6 mm	20.3	3.8
0.63-1.0 mm	6.1	5.6
0.315-0.63 mm	12.3	8.5
0.16-0.315 mm	11.5	6.9
0.10-0.16 mm	3.7	4.0
0.063-0.10 mm	1.5	4.8
0.05-0.063 mm	1.4	1.9
0.032-0.05 mm	0.2	9.0
0.016-0.032 mm	–	22.3
0.008-0.016 mm	–	10.3
< 0.008 mm	–	11.9
Total	100.0	100.0

5. Conclusion

The chemical, phase, granulometric compositions and density of mill scale and dehydrated sludge for gas treatment of BOF No. 1 of EVRAZ ZSMK JSC were investigated. The total iron content in the scale is 73.3%, in the sludge – 41.2%. The content of FeO and Fe_2O_3 in the scale is 75.5% and 20.9%, in the sludge 4.7% and 53.7%. It was established that sulfur and phosphorus content is lower in the scale – 0.036% and 0.019%, respectively, against 0.21% and 0.15% in the sludge. Sludge also contains 20.6% CaO, 4.3% of total carbon. The granulometric compositions of scale and sludge are significantly different: the content of the particle size of classes +1.0 mm and -0.016 mm in the scale is 63.3% and 0%, in the sludge – 14.8% and 44.5%. The true density of the scale is 4.6 - 4.9 g/cm^3, sludge – 3.5 -5.0 g/cm^3. The obtained results confirm the technological feasibility of using fine-grained scale and powdery sludge in metallization processes, including their preliminary briquetting in a mixture with a carbon reducing agent, for example, brown coal semi-coke, supplied in the form of a fine-grained product of class 0 – 3 mm.

References

[1] Yusfin Yu S and Pashkov N F 2007 *Metallurgy of Iron* (Moscow: Metallurgiya) p 320
[2] Lungen KH B, Peters M and Shmele P 2010 *Ferrous Metals* **9** 52–66
[3] Roshchin V E and Roshchin A V 2013 *Electrometallurgy and Steel Metallurgy* (Chelyabinsk: SUSU) p 572

[4] Timofeeva A S, Nikitchenko T V and Timofeev E S 2014 *Extraction of Ferrous Metals from Natural and Technogenic Raw Materials* (Stary Oskol: TNT) p 303
[5] Kazakov R A et al 2013 *Metallurg* **2** 30–33
[6] Nokhrina O I, Rozhikhina I D and Khodosov I E 2017 *Proc. of the XX Int. Sci. and Pract. Conf. Metallurgy: Technologies, Innovations, Quality* (Novokuznetsk: SibSIU) part 1 pp 130–134
[7] Timofeeva A S et al 2014 Bulletin Ferrous metallurgy 2 35–39
[8] Nokhrina O I, Rozhikhina I D, Rybenko I A and Khodosov I E 2016 *Steel in Translation* **46(4)** 237–244
[9] Panishev N V et al 2014 *Theory and Technology of Metallurgical Production* **2(15)** 22–26
[10] *Steel Statistical Yearbook 2016. Brussels: Worldsteel Committee on Economic Studies* https://www.worldsteel
[11] Yaroshenko Yu G, Gordon I M and Khodorovskaya I Yu2012 *Energy-efficient and Resource-saving Technologies of Ferrous Metallurgy* (Ekaterinburg: UIPTs) p 670
[12] Tsirulnikov E V 2013 *Ferrous Metals* 9 33–35
[13] Makhortova Yu V and Bondar A S 2009 *Sources of Generation and Physicochemical Properties of Rolling Waste* http://masters.donntu.org
[14] Dobrovolsky I P, Rymarev P N and Safina T A 2008 *Bulletin of Chelyabinsk State University* **17** 31–35
[15] Edilbaev A I et al 2013 *Coke and Chemistry* **9** 20–27
[16] Nokhrina O I, Rozhihina I D and Hodosov I E *JOP Conference Series: MSE* **91** 1–8

IOP Conf. Series: Materials Science and Engineering **866** (2020) 012027 doi:10.1088/1757-899X/866/1/012027

Energy generation from wood waste

A K Solovyov[1] and G K Koshkina[2]

[1]Siberian State Industrial University, 42 Kirova Street, Novokuznetsk, 654007, Russia
[2]Kuzbass State Technical University, 28 Vesennaya Street, Kemerovo, 65000, Russia

E-mail: ak100752@yandex.ru

Abstract. The reconstruction of the heating system of the administrative and amenity building of "Tekhnologii Retsiklinga" LLC is proposed by replacing the heat energy of electric boiler with the heat energy produced by a wood waste boiler.

1. Introduction

In many regions of the Russian Federation, biofuel boilers successfully operate, although their percentage is very low (1-2%) and there are no plans concerning the complete transition to renewable energy sources in the country. However, even several biofuel boiler houses can serve as examples for enterprises, both in their own region and in the whole country for the efficient use of biomass [1]. If we compare our country, for example, with Finland, in Finland wood chips are preferable to fuel oil and coal in the vast majority of regions of the country, and in Russia, even in areas rich in forests, expensive coal and oil products are imported.

The idea of using wood waste as fuel has already been realized in numerous designs of pellet boilers. But pellets are quite expensive to manufacture, at least $ 7-8 per tonne, so today chip boilers are becoming more relevant. In this case, fuel costs are reduced by approximately 40-50%. In addition, the use of wood chips in boiler rooms allows the problem of wood waste utilization to be solved.

Chip fuel is environmentally friendly fuel. Its ash content is less than 3%. In the process of burning this fuel, the release of carbon dioxide into the atmosphere is equal to the amount absorbed by the plant during its growth. The low cost of wood chip makes it a worthy alternative to pellets and briquettes. But wood chip as a heat source is attractive only in areas located close to guaranteed sources of its formation. Transportation of wood chip over long distances is unprofitable due to its small specific gravity, compared with pellets and pressed fuel briquettes.

Thus, wood chip are affordable, inexpensive and energy-intensive raw materials suitable for heat and electricity. The calorific value of 100 kg of dry wood chip is equivalent to the combustion of about 40 kg of coal or 50 kg of dry peat.

2. Schematic diagram of a boiler of KTU series

Currently, a significant amount of wood waste delivered by "EVRAZ ZSMK" JSC to the territory of the slag processing workshop of "Tekhnologii Retsiklinga" LLC is received for further processing (1400 tonnes). In addition, in "EVRAZ ZSMK" JSC there are approximately 500 m^3 of branches and cut trees, as well as foundry models – about 1500 tonnes. Every year, the municipal enterprise of the city collects branches and cuts of trees in an amount of more than 5000 m^3.

An electric boiler is used to heat the administrative and amenity building. The main disadvantage of such heating system is the high tariffs for electricity. For example, during the heating season from

IOP Conf. Series: Materials Science and Engineering **866** (2020) 012027 doi:10.1088/1757-899X/866/1/012027

September 2017 to April 2018, 742000 kW were spent on heating the administrative and amenity building, which is 2 619 260 rubles.

In order to reduce heating costs, it was proposed to reconstruct the heating system of the administrative and amenity building by replacing the heat of an electric boiler with the heat energy of a wood waste boiler. For the normal operation of the boiler and the prospect for expansion of the company, as well as for the possible prospects of introduction of heat generation into the market, it was decided to purchase a boiler unit KTU-750 manufactured by "Teploresurs" LLC. Figure 1 shows a schematic diagram of KTU-750 boiler.

The design feature of the furnace is an arch vault allowing highly moist fuel up to 55% to be burnt and it is made with one furnace front, which provides maintenance of the grate and loading of lump fuel (cuttings, firewood, etc.). The design of the boiler KTU-750 provides a window for the mechanized supply of bulk fuel (sawdust, wood chips, bark, peat, etc.).

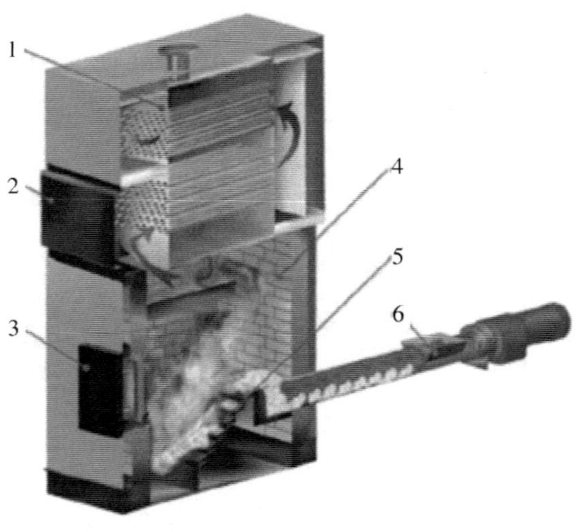

Figure 1. The basic scheme of boiler KTU-750: 1 – heat exchanger; 2 – cleanout of the heat exchanger; 3 – furnace door; 4 – burner; 5 – grate; 6 – fuel supply mechanism.

A special design of the furnace is the inclined and horizontal grate, in combination with an air distribution device, it is possible to use both bulk and lump fuel as fuel without additional boiler modernization.

For ease of operation, two cleanouts are provided – on both sides of the furnace, and a viewing window for visual inspection. The heat-insulated furnace cover increases the efficiency by reducing heat loss through the furnace walls.

The blower fan, which is supplied with the boiler, provides air supply to the undergrate space, afterburning of fuel, and the air flow, in turn, being heated when the walls of the furnace are cooled, passes through the grate and the fuel layer gets involved in the main combustion.

The heat exchanger with smoke tubes. The heat carrier (water) is pumped into the heat exchanger, flue gases wash the inner walls of the pipes where the heat exchange takes place. A safety group is installed on the heat exchanger of KTU-750 boiler, which includes two safety valves, a pressure gauge and a ball valve.

Safety valves are used to release excess steam (water) from the boiler when the pressure in it is higher than the calculated one. The manometer serves to control the pressure in the heat exchanger. The ball valve is designed to bleed the vapor-air mixture from the heat exchanger. At the heat exchanger branches, glasses are provided for direct and return water temperature sensors. A fitting with a ball valve mounted on it is embedded in the heat exchanger to remove water when the boiler stops.

The heat exchanger of the boiler is equipped with a cleanout, which provides access to the heat exchange surfaces of the smoke pipes at the time of the work on cleaning the heat exchanger from ash deposits. The heat exchanger is thermally insulated from the environment, which eliminates the loss of heat energy and increases the efficiency of the boiler. From the operational hopper-dispenser using the fuel supply mechanism – 6 (screw conveyor or hydraulic pusher), the fuel enters the burner of the boiler – 4. The fuel is burned on the inclined grate – 5, on which loose (sawdust, shavings, wood chips, peat, crop waste) and lump wood is burnt.

In the process of burning in the combustion chamber, the fuel goes through three stages (three combustion cycles):

Stage 1 – drying of the fuel. In boiler houses installed at wood processing enterprises, in most cases, wood chips and sawdust with high humidity obtained after sawing up fresh trunks are used as fuel for boilers. In plywood factories, wood is pre-steamed in process water pools before treatment, in this case the fuel has high humidity. At the first stage of the fuel movement in the boiler furnace, sawdust and wood chips from the fuel tray are squeezed directly into the furnace, where they are distributed along the width of the furnace front, on the inclined grates. Moisture evaporation occurs because of the heating effect of the red-hot furnace lining and the radiant energy of combustion reflected from the vault screen of the furnace.

Stage 2 – primary oxidation and release of CO. Layered combustion occurs at relatively low temperatures of 700-750 °C on the inclined and horizontal grates with forced separate blowing.

Stage 3 – CO afterburning in the flare of the boiler furnace. Afterburning takes place directly in the boiler furnace, in the air supply zone (oxidation activator) at a temperature of 900-950 °C, which is most optimal for the carbon oxidation reaction to occur. Air intake is made from the air gap between the lining and the outer lining of the boiler. This helps to reduce heat losses and increase efficiency. At each combustion stroke in proportion to the fuel supply, a dosed supply of oxidizer (atmospheric air) by several fans (blow and afterburning) is organized. After the residual carbon is completely burned, the ash falls through the grate to the ash pan, from where it is manually (or mechanically, depending on the design) removed through the cleaning hatch directly in the mode of operation of the boiler.

A heat exchanger is a special construction for transferring heat energy from a heated coolant to a colder one. In the design of boilers of KTU series, a heat exchanger of a gas-tube type is used. It is a "barrel" with smoke pipes passing through it, piercing it from the front of the boiler to the back with a 180 ° rotation of gases according to the two-way scheme. Flue gases passing through the pipes heat the coolant circulating inside the heat exchanger and washing the outer walls of the pipes. The rate of passage of flue gases is determined by the level of vacuum in the combustion chamber and is regulated by a smoke exhauster.

The optimum speed of movement of the heat agent in the heat exchanger is automatically regulated and is provided by the pumping station. A heat-insulating lining is used at the inlet of the heat exchanger to protect the jacket from high-temperature gases.

Flue gas anti-explosion valves and safety group valves for heat carrier – hot water are used. The coolant is supplied through the inlet, and the outlet through the output collector. The collectors are equipped with coolant temperature sensors. Information from the sensors is fed to the control panel, allowing the fuel supply to the furnace to be adjusted, thereby keeping the specified coolant parameters. The heat exchanger housing is lined with a sheet profile.

The fire pipe design of the heat exchanger minimizes the need for chemical water treatment (CWP) and allows the heat exchanger to be serviced directly in the operation mode of the boiler. Fuel can be supplied either automatically or manually. High-moisture wood waste, milled peat (relative humidity up to 55%) without preliminary drying, dry wood waste can serve as fuel for boilers of this series.

This boiler equipment has a total length of 5.6 m, a width of 1.4 m, a height of 3.750 m. The equipment will be installed exactly in the middle of the building, where the distance from the walls of the building will be 2.3 m for the comfortable operation of this equipment according to SNiPII-35-76. The building has two openings:

- the first opening – two meters high and one meter wide for the passage of personnel;
- the second opening is three meters wide and four meters high for unloading wood chips by road transport.

3. The main stages of the technological process for the production of wood chips

The main stages of the technological process for the production of wood chips at "Tekhnologii Retsiklinga" LLC are:

- waste sorting, preparation of wood raw materials for processing;
- pre-processing, preparation of wood at the site, delivery to the receiving hopper of the production line;

- production of wood chips, storage of finished products and loading into the boiler.

The raw materials for the production of wood chips are wood waste delivered by "EVRAZ ZSMK" JSC to the territory of the slag processing workshop of "Tekhnologii Retsiklinga" LLC. Considering the peculiarities of the incoming raw materials and the fact that the raw materials for production are waste with a humidity of at least 20%, high-temperature dryers are not required for their drying, therefore, the problem of safe removal of moisture can be solved in another way. For example, by "kinetic" removal of water from the surface of wood particles during their fine grinding. The heating medium necessary for this, in which atmospheric air acts, occurs directly in the mill body as a result of the temperature separation of the air flow similar to the Ranque-Hilsch Effect.

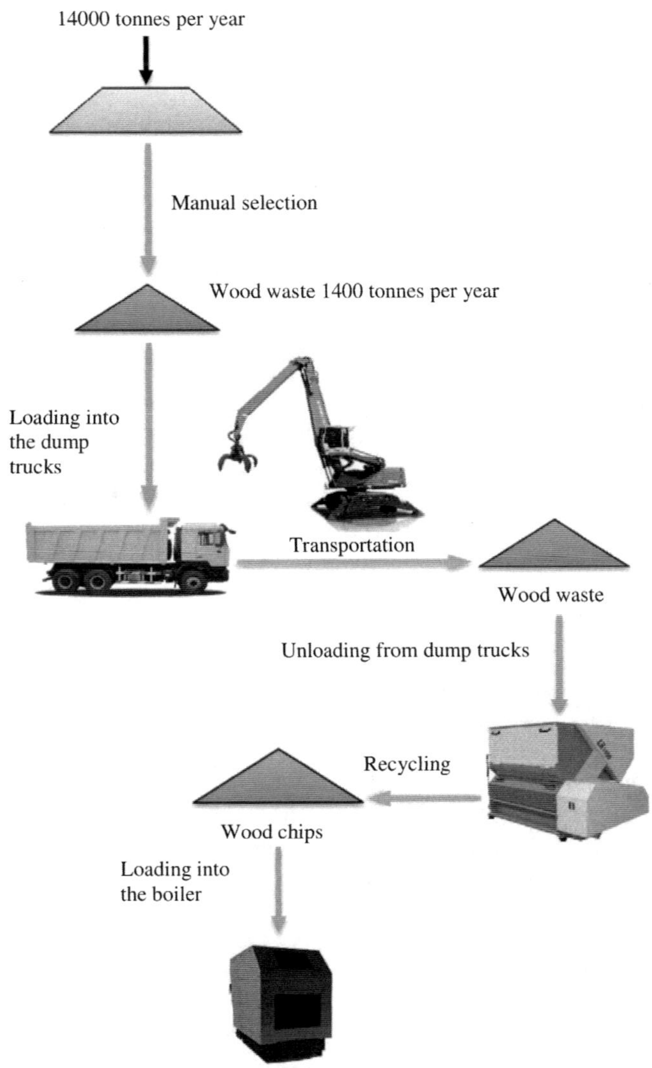

Figure 2. Scheme of wood chips production from wood waste.

This technology is implemented in SCARABEY small-sized automated lines, which are equipped with combined mills and heaters of the S.A.M.P. 2012 model. This equipment was specially designed for the production of wood chips from raw materials with humidity up to 30% in automatic mode, without the use of fire-hazardous "fire" drying methods. The SCARABEY line most fully meets the requirements of small industries.

For pre-processing and grinding of secondary waste wood of oversized dimensions, a twin-shaft shredder PPM 1 and a chip mill are used. The main purpose of the shredder is the grinding and crushing the bulky waste – wooden pallets, logs, lumber waste. The chip mill crushes large wood chips and other wood wastes (branches, knots, sawmill waste) to produce chips of smaller fractions that can be used as fuel for a boiler room. Figure 2 shows a diagram of the production of wood chips from wood waste.

4. Conclusion

The competitiveness of wood chips in comparison with other alternative solid fuels underlines the success of wood chips in the biofuel market and its stable growth rate. When implementing this measure, the expected annual profit of the enterprise will be 917008 rubles, the payback period of the project is 3.16 years.

References

[1] Golovkov S I, Koperin I F and Naydenov V I 1987 *Energy Use of Wood Waste* (Moscow: Lesnaya Promyshlennost) p 224

IOP Conf. Series: Materials Science and Engineering **866** (2020) 012028 doi:10.1088/1757-899X/866/1/012028

Prediction of carbon dioxide emissions from fuel combustion

V V Sterligov

Siberian State Industrial University, 42 Kirova Street, Novokuznetsk, 654007, Russia

E-mail: p.s.1976@bk.ru

Abstract. An objective method for estimating carbon dioxide emissions using specific emissions per 1 kJ of fuel energy is presented. To unify the calculations of the calorific value of gaseous and solid fuels, it is proposed to use a regression calculation model for both coal and gas. The applicability of this approach is proved and on its basis the calculation of possible CO_2 emissions at a given unit power.

1. Introduction

In existing assessments of the environmental situation in the world, pessimistic forecasts are increasingly dominating. And it refers not only to foreign media. In Russia, and even more importantly for us – in Kuzbass, materials appear that predict the disastrous ecology of coal Kuzbass. Written by non-professionals, such publications may provoke a negative reaction from untrained readers.

Meanwhile, the situation is not so tragic and during this period the objectively best information is needed, which, recognizing the difficult situation, would search for and suggest the ways to solve this problem. The basic document created by the international community on climate warming, the so-called "Framework Convention on Climate Change" (UNFCCC) [1], in paragraph 2, contains a precautionary statement that urges mankind to seek funds, even if they are still scientifically and not represented. This determines the relevance of any work aimed at solving the global problem of the greenhouse effect.

In another document created for the implementation of the UNFCCC, the so-called the "Kyoto Protocol" [2], in Art. Section 5 contains the definition of the main measure to combat warming caused by the greenhouse effect, which is energy saving leading to a reduction in greenhouse gas emissions and which can be achieved by direct reduction in the consumption of fuel and especially coal. In our opinion, this is a primitive and even vulgar solution due to the extensive factor. But since society needs energy to function in all areas of life, it is more rational to develop and introduce new ways to use fuel that would give reduced emissions of carbon dioxide – the main representative of greenhouse gases.

The activity of the international society on these energy-ecological problems has not yet reached Russia. It should be said that not all atmospheric experts know that in Russia, as long as the whole world, a fee for carbon dioxide emissions was introduced. But if you look at one of the most authoritative sources on energy-environmental problems [3], then it shows the payment for emissions of SO_2, NO_x, etc., but it does not exist for CO_2.

The Kyoto Protocol also noted the need to introduce accounting for CO_2 emissions, but so far practically nothing was done in Russia in this direction. Since the country has committed itself to continue to reduce CO_2 emissions by signing the Paris Agreement in December 2015 [4], it is necessary to prepare to fulfill these obligations.

2. The unified model for calculation of gas calorific value

This paper presents an attempt to simplify the determination of CO_2 emissions based on the elemental composition (i.e., on the content of individual chemical elements) as opposed to the calculation of the chemical composition, the determination of which is laborious and not always available at any enterprise.

Earlier [5] we proposed the idea of using specific emissions per unit of energy (c). The value of c is defined as follows:

$$c = \frac{m_{CO_2}}{Q_{H}^{P}}, \text{ kg/kJ}. \quad (1)$$

Therefore, it is important to determine quickly and efficiently the values of Q_{H}^{P}. Obviously, the model for solid and liquid fuels has its own advantages for the dtermination. In the future, solid fuel will be called coal, and gaseous – gas.

It is hypothesized that the heat from combustion of gaseous fuels and carbon dioxide emissions can be calculated using a simpler model used for coal.

For gaseous fuels from the theory of combustion [5] the equation for determining the calorific value is known:

$$Q_{H}^{P}=39.847CH_4+107.43H_2+122.63CO+63.790C_2H_4+590.59C_2H_4+912.81C_3H_8, \text{ kJ/m}^3, \quad (2)$$

where CH_4, H_2, CO, C_2H_4, C_3H_8 is the percentage content of fuel components.

It should be noted that this equation is deterministic, i.e. defined in strict accordance with the laws of physics and chemistry. Its additive structure is determined by the absence of interaction during the combustion of various components. Numerical coefficients take into account the heat effect of complete oxidation.

For solid and liquid fuels in the source [5] you can find different formulas, but the most used is the so-called “Mendeleev’s formula”:

$$Q_{H}^{P}=399C^{p}+1030\,H^{p} - 109(O^{p}-S^{p})- 25W^{p}, \text{ kJ/kg}, \quad (3)$$

where C^p, H^p, O^p, S^p, W^p – the content per working mass of individual chemical elements.

This is a regression equation that establishes a relationship based on statistical materials. The additive character is also explained by the independent burning of individual elements. Numerical coefficients are obtained by statistically processed observation results.

Unlike the first model, where it is clearly known which element in combination with another is a substance, the amount of which needs to be known exactly, as well as the thermal effect of its oxidation, the second model does not imply knowledge of elements combinations, it does not contain information about their thermal reactions, etc. For this model, there is no need for chemical gas analysis.

3. Hypothesis proof

For this purpose, calculations were carried out for three flammable gases according to the periodic formula for solid fuel. The calculation results are presented in table 1.

Table 1. Comparative calculation of the calorific value of burnt gases by the formula for solid fuel.

Fuel (gas)	Mass composition, %	Calorific value Q_{H}^{P}, MJ		
		according to the equation (2)		according to the equation (3)
		MJ/m^3	MJ/kg	MJ/kg
Methane CH_4, ρ=0.714 kg/m^3	C=75.00 % H=25.00 %	35.847	51.776	51.206
Propane CH_4,	C=81.81 %	912.281	46.969	46.929

ρ=1.96 kg/m³	H=18.19 %			
Natural gas (volumetric composition) CH_4 =85 %, C_2H_4=10 %, CO=2 %, ρ=0.7585 kg/m³	C=71.05 % H=24.12 % N=4.34 % CO=0.49 %	37.094	48.900	48.927

The base value in the calculation of Q_{H}^{P} was the value determined by the adopted for gas equation (2), MJ/m³. By dividing this value by density, the quantity Q_{H}^{P} MJ/kg was determined, which was compared with the calculation results by the proposed method using the elementary mass composition of the gas and equation (3) for solid fuel.

For all three cases, the approximation error was determined by formula (4):

$$\Delta Q=\frac{(Q_{\text{H}}^{\text{P}})_1-(Q_{\text{H}}^{\text{P}})_2}{(Q_{\text{H}}^{\text{P}})_1}\cdot 100\ \%. \tag{4}$$

For the base, we took the value Q_{H}^{P} calculated by formula (1).

For all three cases, the error was: ±1.0%, which is an acceptable error level for the proposed simplified methodology for calculating Q_{H}^{P} and is comparable to the error when using basic equations (1) and (2).

4. Methodology for determining emissions

Since the hypothesis of the applicability of the elementary composition for calculation of the calorific value can be accepted for gaseous fuels, it can be used to estimate the carbon dioxide emission potential and determine the specific emission value m_{CO_2} per unit of released energy, i.e.

$$m_{\text{CO}_2}=\frac{M_{\text{CO}_2}}{Q_{\text{H}}^{\text{P}}}. \tag{5}$$

This expression is the key one, and on its basis the following algorithm is proposed for the existing heat engineering unit.

1. The production capacity of the unit G, t/h is determined. As a rule, it is known even at the design stage, therefore, the proposed methodology for CO_2 emissions can be used both for forecasting emissions and for studying existing units;

2. The required amount of energy is calculated on the basis of regulatory or practical data according to the performance of unit G:

$$N = eG\ \text{kW}, \tag{6}$$

where e is the normative specific indicator of energy consumption per unit of production, J/t (unit/s),

G – unit capacity, t/h (unit/s).

The choice of fuel is carried out if there is the possibility of different options, and its consumption is determined B, kg/s (t/h):

$$B=\frac{N}{Q_{\text{H}}^{\text{P}}}\ \text{kg/s (m}^3\text{/s)}; \tag{7}$$

here Q_{H}^{P} – is the heat from fuel combustion, kJ/kg (kJ/m³);

4. The composition of the fuel (coal) is determined by the mass of carbon M_C

$$M_C = B\cdot \%\text{C kg/s}; \tag{8}$$

By formula (2) we determine the value of Q_{H}^{P}.

In the case of using gas fuel, the carbon mass is calculated by summing the contributions of all carbon-containing fuel components, i.e.

$$M_C = \sum_{i=1}^{n} M_{C_i}, \quad (9)$$

which is possible with knowledge of the fuel composition.

6. The calculation is based on the material balance of combustion of any carbon-containing fuel component, for example

$$C+O_2=CO_2;$$
$$CO_2+1/2O_2=CO_2;$$
$$CH_4+2O_2=CO_2+2H_2O, \text{ etc.}$$

From these reaction equations it follows that for any hydrocarbon with complete combustion (and in the existing heat engineering unit it is certainly necessary to achieve complete combustion of fuel) 1 atom C generates 1 CO_2 molecule; mass ratio m_{CO_2}/m_C = 44/12 = 3.67 = const. This makes it possible to determine the predicted carbon dioxide emission:

$$M_{CO_2} = 3.67 \cdot M_C, \text{ kg/s.} \quad (10)$$

To determine the carbon content (% by mass in the fuel) it is enough to burn the fuel and determine the carbon content in the fuel and its mass from the CO_2 content in the combustion products.

5. Conclusion

Using the proposed methodology, it is possible not only to predict the amount of greenhouse gas CO_2 emissions, but also the amount of fees for these emissions, as well as to abandon the costly analysis of gas composition and determine CO_2 emissions based on the fuel and energy balance of the enterprise.

References

[1] *United Nations Framework Convention on Climate Change* (New York) (05/09/1992)
[2] *Web site for consultation on the practical study of the Kyoto Protocol mechanisms* http://www.carbonmarketsolutions.com
[3] Lisienko V G 2003 *Chrestomathies on Energy Conservation* (Moscow: Teploenergetik) p 688
[4] *Website of the Paris Agreement on Climate Change* http://www.ipcc.ch (Paris) (12/15/2015)
[5] Sterligov V V, Kozlova N E and Abdalyk T E 2017 *Bulletin of the Siberian State Industrial University* **2(20)** 60–64
[6] Semikin I D and Averin S N 1965 *Fuel and Fuel Economy of Metallurgical Plants. Textbook for Students of Metallurgical Specialties at Higher Education Institution* (Moscow: Metallurgiya) p 392

IOP Conf. Series: Materials Science and Engineering **866** (2020) 012029 doi:10.1088/1757-899X/866/1/012029

Mathematical modeling of the process of carbon bubbles formation in the interaction of a gas jet and a melt bath in the oxygen converter

V V Solonenko, E V Protopopov, V D Sarychev, A Yu Granovskii and M V Temlyantsev

Siberian State Industrial University, 42 Kirova Street, Novokuznetsk, 654007, Russia

E-mail: sarychev_vd@mail.ru

Abstract. The article discusses the issues of modeling the processes of interaction of gas jets and metal melt during blowing of a converter bath. The authors developed and implemented using the finite element method a mathematical model based on a direct numerical solution of the Navier-Stokes equation without using a turbulence model. This model helped to study the dynamics of the velocity field and the boundary of the penetration of the jet into the metal. According to the results of mathematical modeling, a mechanism for the formation of carbon monoxide bubbles is suggested.

1. Introduction

In the top-blown converter, a jet of oxygen at a supersonic speed is directed by means of a lance to the bath of molten metal. In this case, two processes are realized: the formation of a cavity on the surface of the bath due to the dynamic pressure of the jet and the formation of a vortex flow in the volume of whole bath. The size, shape and boundary of the cavity are important parameters that determine the decarburization of cast iron in the bath, accompanied by the formation of carbon monoxide, which can react with oxygen in the upper space of the converter to form carbon dioxide.

This latter process is usually called the afterburning reaction, the exothermic effect of which is at twice the decarburization reaction [1]. Therefore, the creation of optimal conditions for the afterburning of carbon monoxide in the cavity of the converter is an actual scientific and practical task [2].

Experimental studies on the problem of a gas jet collision with a liquid surface, starting from [3] in the 60's to later experiments [4–6], were carried out using reduced cold models in which oxygen was replaced by air and molten steel was replaced by water. In these experiments, the integral dimensions and surface shape were investigated, while the local deformation of the cavity boundary was not considered.

The modeling of deformations of the free bath surface and the flow in a metal under the influence of jets was considered in numerous works, for example, in [7–12]. In [8], a physical and mathematical model of the instability in the behavior of the bath boundary was proposed. Model [9] describes the fluctuations of the melt level. In [10], a numerical study of the interaction between jets and a slag-metal bath is presented using a three-phase volumetric model of a liquid. Numerical results show that the cavity profile and the slag/metal/gas interface remain unstable as a result of the propagation of surface waves, which is probably the main factor determining the generation of metal droplets and their initial spatiotemporal distribution. Recent reviews [11, 12] present data on the interaction of an oxygen jet with a molten bath based on the volume of fluid method (VOF) using one or another turbulence model. Turbulence models use turbulent viscosity, which significantly exceeds molecular viscosity, which allows calculating mixing processes over a long period of time in the melt pool and in the upper space of the converter. However, this leads to a loss of flow detail. If to refuse the use of turbulent viscosity, important details of the process can be noted, for example, the formation of bubbles at times of one tenth of a second.

The works on the phenomena occurring in the upper space of the converter and on afterburning are mainly empirical in nature [13], where it was found that the afterburning coefficient in the exhaust gases, defined as the ratio of the concentration of carbon dioxide to the total concentrations of CO and CO_2, increases with the increase in height of the lance. An increase in the temperature and carbon content in the bath leads to lower values of this indicator.

2. Problem statement and solution method

Let us consider the interaction of a gas jet of the converter moving with speed u_0 and a bath of melt with the physical characteristics presented in table 1.

The motion of gas and melt is described by the Navier-Stokes equation for a viscous incompressible medium:

$$\begin{gathered} \rho\frac{\partial \mathbf{u}}{\partial t} + \rho(\mathbf{u}\cdot\nabla)\mathbf{u} = -\nabla p + \mu\Delta\mathbf{u} + \mathbf{F}_g \\ \nabla\cdot\mathbf{u} = 0 \end{gathered} \tag{1}$$

Where $\mathbf{u}$ is the velocity vector, p – pressure, $\mathbf{F}_g$ – volumetric gravity, ρ – density, μ – dynamic viscosity.

This system was solved numerically in a flat coordinate system using the finite element method. The finite element mesh is shown in figure 1. A conservative Level Set method was used to track the interface between the melt and gas. The method consists in calculating the scalar function for the entire computational domain:

$$\rho\left(\frac{\partial\phi}{\partial t} + \nabla(\phi\mathbf{u})\right) = \chi\left[\varepsilon\nabla\cdot\nabla\phi - \nabla\cdot\left(\phi(1-\phi)\frac{\nabla\phi}{|\nabla\phi|}\right)\right] \tag{2}$$

Where ε is the parameter determining the thickness of the transition layer and equal to half the size of the mesh cell, χ is the stabilizing parameter equal to the maximum speed achieved in the calculation area. In the transition zone, the density, viscosity, and electrical conductivity are approximated:

$$\begin{gathered} \rho = \rho_g + (\rho_m - \rho_g)\phi. \\ \mu = \mu_g + (\mu_m - \mu_g)\phi \end{gathered} \tag{3}$$

The surface tension force is calculated by the formula:

$$\begin{gathered} \mathbf{F}_{st} = \nabla\cdot\left[\left(\gamma\left(\mathbf{I} - \mathbf{n}\mathbf{n}^T\right)\right)\delta\right], \\ \mathbf{n} = \frac{\nabla\phi}{|\nabla\phi|}, \quad \gamma = |\phi(1-\phi)||\nabla\phi| \end{gathered} \tag{4}$$

Where $\mathbf{I}$ is the identity matrix, $\mathbf{n}$ – the normal vector to the surface, γ – the surface tension coefficient, δ – the Dirac delta function, which is not equal to zero only on the contact surface.

Table 1. Geometric and physical characteristics.

Designation	Value	Description
H	5.254 m	Estimated converter height
H_m	2.27 m	Melt layer height
D	7.42 m	Converter diameter
U_0	500 m/s	Gas velocity
ρ_g	0.23 kg/m^3	Gas density
μ_g	$0.73\cdot 10^{-4}$Pa·s	Dynamic gas viscosity
ρ_m	8136 kg/m^3	Melt density
μ_m	0.0058 Pa·s	Dynamic melt viscosity
D_{out}	50 mm	Lance nozzle diameter
T_0	1812 K	Initial temperature
σ	1 N/m	Surface tension coefficient

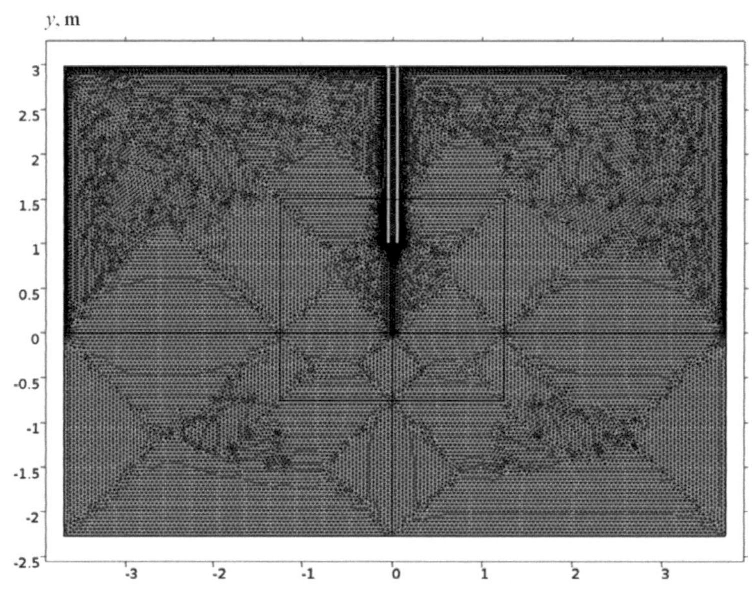

Figure 1. The calculated finite element mesh.

2.1. *Initial and boundary conditions*

At the initial moment of time, the melt is at rest, and the gas flows at a speed u_0. On the walls of the tuyere the adhesion condition is set:

$$\mathbf{u} = 0 \tag{5}$$

Figure 2 presents the results of calculations of the velocity field of jet penetration into the metal for various time instants, from which the instability of the interface follows. The development of instability leads to the formation of metal droplets (Figure 2 a-c) and the gas bubble inside the liquid (figure 2 c). This suggests a mechanism for the formation of carbon monoxide bubbles. The bubble formed due to instability, the oxygen of which interacts with carbon (a heterogeneous reaction), turns into carbon monoxide. Under the influence of Archimedean forces, this bubble rises to the surface and bursts. Then the upward hydrodynamic flow carries carbon monoxide into the upper space of the converter.

IOP Conf. Series: Materials Science and Engineering **866** (2020) 012029 doi:10.1088/1757-899X/866/1/012029

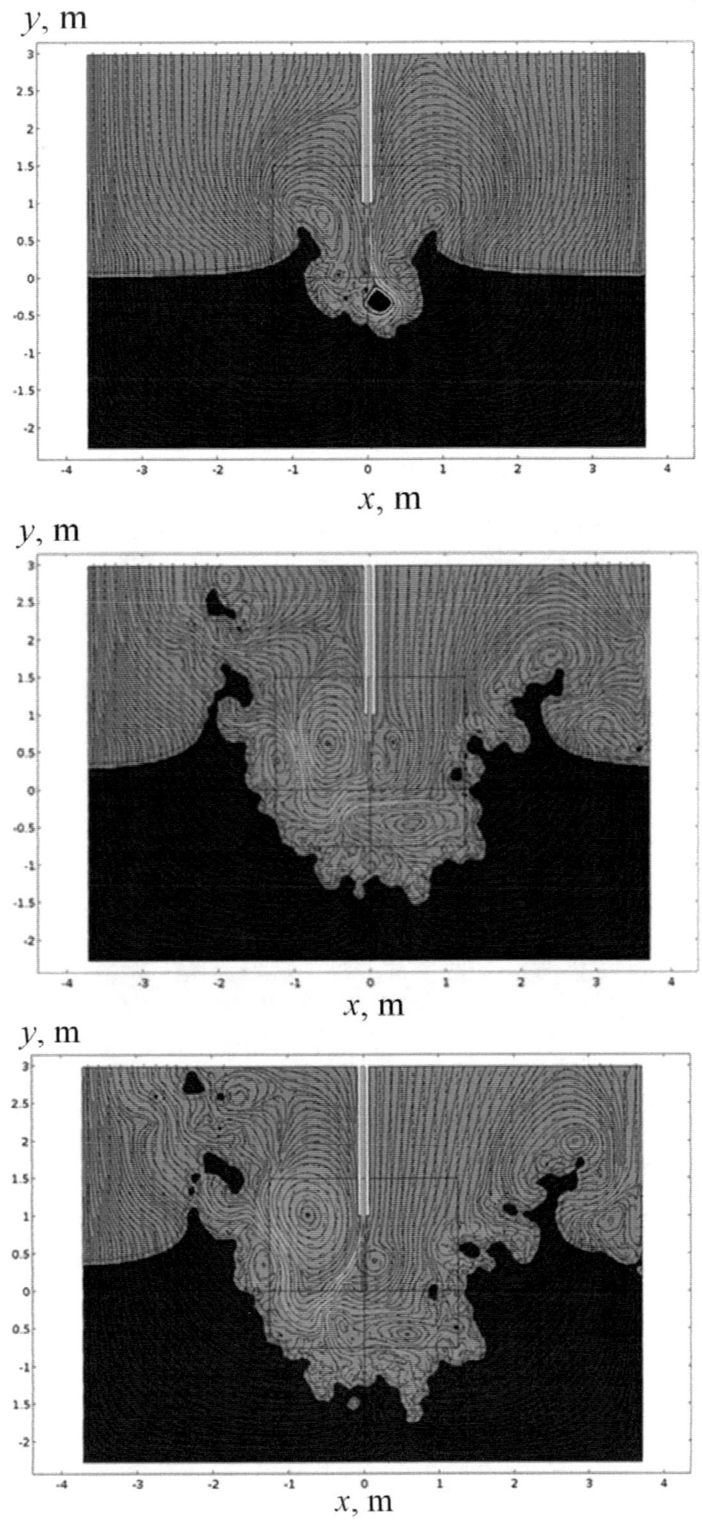

Figure 2. Speed fields during the jet penetration into the metal.

IOP Conf. Series: Materials Science and Engineering **866** (2020) 012029 doi:10.1088/1757-899X/866/1/012029

3. Conclusion

The direct method of numerical simulation of the Navier-Stokes equations solved the problem of the interaction of a supersonic gas jet with a liquid metal. The distributions of the velocity field and the geometry of the gas-liquid interface for various time instants are obtained. The obtained interface form proves that hydrodynamic instabilities are realized on it, leading to the formation of metal particles in the gas and gas bubbles in the metal. The analysis of the hydrodynamic situation made it possible to propose a mechanism for the formation of carbon oxide bubbles.

References

[1] Doh Y et al 2013 *Metall and Material Trans B* **44** 653
[2] Protopopov E V et al 2015 *Izv. Vuzov. Ferrous Metallurgy* **58(4)** 248–257
[3] Banks R B and Chandrasekhara D V 1963 *J. Fluid Mech.* **15** 13–35
[4] Nordquist A, Kumbhat N, Jonsson L et al 2006 *Steel Res Int.* **77** 82–90
[5] Hwang H Y and Irons G A 2012 *Process Metall. Mater. Process. Sci.* **43** 302–315
[6] Mordasov M M, Savenkov A P and Chechetov K E 2016 *J. of Technical Physics* **86(5)** 20–29
[7] Miloshvich Kh and Rychkov A D 1998 *Computational Technologies* **3(6)** 54–62
[8] Olivares O, Elias A, Sanchez R, Diaz-Cruz M and Morales R 2002 *Steel Res.* **73** 44–51
[9] Odenthal H, Kempken J, Schluter J and Emling W 2007 *Iron Steel Technol.* **4** 71–89
[10] Li Q, Li M, Kuang S et al 2015 *Metall and Materials Trans B.* **46** 1494
[11] Cao L, Wang Z, Liu Q and Feng X 2018 *ISIJ International* **58(4)** 573–584
[12] Ersson M and Tilliander A 2018 *Wiley in Steel Research International Steel Research International* **89** 1700108
[13] Huber J-C, Lehmann J and Cadet R 2008 *Revue de Métallurgie* **105(3)** 121–126

IOP Conf. Series: Materials Science and Engineering **866** (2020) 012030 doi:10.1088/1757-899X/866/1/012030

Diversification of the production and processing of pulverized wastes and ores based on the SER jet-emulsion unit

V P Tsymbal, A A Rybushkin, I A Rybenko, V I Kozhemyachenko, S N Kalashnikov and L A Ermakova

Siberian State Industrial University, 42 Kirova str., Novokuznetsk, 654007, Russia

E-mail: rybenkoi@mail.ru

Abstract. The principles of creating a self-organizing jet-emulsion reactor (SER) are considered, the role of dissipative structures in solving problems of controlling the metal chemical composition is shown. The technological scheme of the unit is presented, in which it is possible to process pulverized materials with a wide range of changes in chemical composition, including sludge, scale, poor pulverized ores and tailings.

1. Introduction

In the period from 1992 to 2001, thanks to the fruitful cooperation of scientists of SibSIU with specialists of the West Siberian Metallurgical Plant, a fundamentally new jet-emulsion metallurgical process and a unit was developed and implemented as a large-scale pilot plant in the second converter shop, which allows pulverized ore materials and waste in one stage without agglomeration to be processed [1, 2]. The SER process is characterized by low energy consumption and a small specific volume of the unit.

These results were achieved through the use of the basic principles of the theory of self-organization (synergetics) [3, 4] and some physical effects for their implementation, such as intense dispersion of the mixture with counter jets, gas-dynamic channel locking and others [1, 5].

The combination of these factors and measures made it possible to create dynamic dissipative structures in the aggregate, which led to a sharp decrease in the specific aggregate volume (10-15 times) and provided significant control flexibility, including the ratio of the rates of reduction and oxidation reactions, which made it possible to directly produce metal with a fairly wide range of carbon content [2].

Currently, several dozen new units for direct reduction are known in the world, but in none of them, including COREX, it is fundamentally impossible to obtain an iron-carbon alloy with a carbon content below 2%, due to the fact that the processes in them are close to a state of equilibrium, and it is impossible to control separately the composition of the metal and slag.

The article shows that the simultaneous solution of the problem of controlling the chemical metal composition and reducing the unit specific volume is achieved by constructive and mode organization of dynamic dissipative structures [2, 5], which play the role of unique "chambers" that were planned to be created in the initial designs of continuous steelmaking units.

2. Basic principles and solutions

The following principles and decisions were laid at the basis of the constructive implementation of the SER process and unit [6]:

IOP Conf. Series: Materials Science and Engineering **866** (2020) 012030 doi:10.1088/1757-899X/866/1/012030

- creation of a large reaction surface and a two-phase working mixture (gas suspension or emulsion), that is, the transfer of the process to the field of gas dynamics, which allowed high speeds of physical and chemical processes to be obtained;
- organization of the forced movement of the working (reaction) mixture in a closed system under pressure, which made it possible to create a significant deviation from thermodynamic equilibrium and at the same time solve the problem of internal transport of reaction products through all series-connected devices;
- the use of a nonlinear dependence of the flow velocity of a two-phase medium on gas content to create a pressure potential due to the effect of gas-dynamic locking of the connecting channel;
- creation of dissipative structures that are significantly deviated from thermodynamic equilibrium, which made it possible to obtain great opportunities for controlling the chemical composition of metal and slag.

3. Design and technological features of the SER unit

The unit with a very small specific volume and energy intensity was created using the approaches and principles described above [1, 5]. The technological scheme and a brief description of the SER unit (self-organizing jet-emulsion reactor) are presented in figure 1.

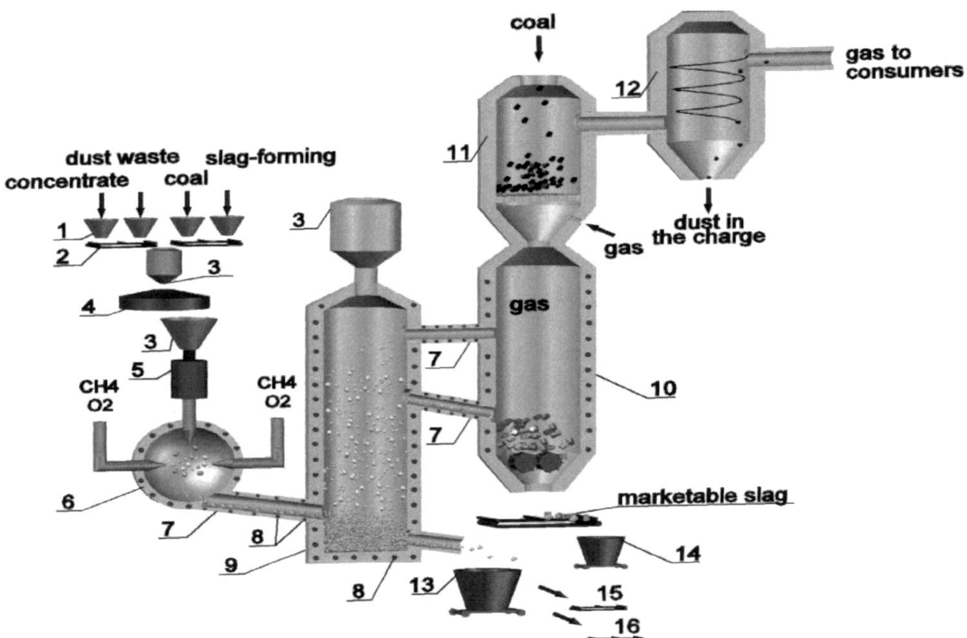

Figure 1. The technological scheme of the SER unit.

The basis of the technological scheme of the mini-module is [6]: the charge supply system 1–5, the reactor-oscillator 6, the connecting channel with gas-dynamic self-locking 7, the refining settler 8, simultaneously playing the role of the wet gas cleaning of the first stage, the receiver 9, as well as the skull cooling system 10, a channel 11 for issuing a gas-slag emulsion and a channel 12 for transferring part of the gas, a slag receiver 13 with a granulator 14, a system for heat recovery in a fluidized bed or for the transformation of flue gases into synthesis gas 17 and a gas purification system 18. The high pressure potential created in the oscillator reactor 6, as well as the complete isolation of the process from the atmosphere, allow the reaction products to be pushed through all energy recovery devices behind the main technological unit without the use of high-temperature flow factors [6].

IOP Conf. Series: Materials Science and Engineering **866** (2020) 012030 doi:10.1088/1757-899X/866/1/012030

Due to the high intensity of the gas-dynamic processes occurring in the unit, the flow of an aggressive two-phase medium, the protection of all elements of the unit is carried out on the basis of circulation skull cooling. To solve this problem, a specialized automated design and calculation system was developed that can also perform diagnostic functions when implementing this cooling system and maintaining the skull [1].

4. Development and implementation status

The process and the unit were implemented as a large-scale pilot plant in the second converter shop of ZSMK on the vacant site of cast iron overflow (figure 2). In total, more than 50 people took part in the creation of the installation and conducting experiments.

The experienced metallurgists-practitioners made a team of steelmakers under the leadership of V.P. Tyutyulnikov and played an important role in improving the design of the installation and conducting unique experiments, one of which is shown in figure 3.

In this experiment, a high-temperature two-phase jet was "shot" from a spherical reactor onto a lined area, which provided valuable data on the physical and chemical properties of particles emitted from the first reactor, chemical reaction rates, etc., as well as the identification of mathematical models.

Figure 2. General view of the pilot plant.

Figure 3. High temperature jet from the first reactor.

IOP Conf. Series: Materials Science and Engineering **866** (2020) 012030 doi:10.1088/1757-899X/866/1/012030

5. Development and implementation status

Several new low-energy-intensive technologies were developed and tested experimentally, including: direct reduction of pulverized ores and waste (sludge and mill scale) without agglomeration, production of manganese alloys from lean pulverized ores, separation of titanium-magnetite concentrates into iron and conditioned titanium slag. The process was patented in the main metal-producing countries: USA, Japan, South Korea, European patent, Russia [7].

The developed process and unit has a certain degree of versatility. It is possible, by changing the blowing mode and afterburning degree of the reducing fuel, to process pulverized materials with a wide range of changes in chemical composition, including poor dusty ores and dressing tailings.

In addition, the SER unit can be used as a gasifier for pulverized coal fractions, but the most economical option is to use a mixture of coal and pulverized iron-containing waste. This facilitates the process of maintaining a stable layer of slag-metal emulsion, inside which fuel is burned, and also reduces the consumption of gaseous oxygen through the use of oxygen from iron oxides.

Besides, it becomes possible to use excess temperature, in the case when the priority is gasification of coal to produce synthesis gas.

6. Diversification of production based on new technologies

In order to implement one of the possible ways to diversify production, technical proposals have been developed for the implementation of a complex of two SER plants in the free areas of the molds shop of West Siberian Metallurgical Plant.

The development of technical proposals was a logical continuation of the design studies related to the testing of a pilot plant at ZSMK and the further steps for the introduction of a continuous metallurgical process. The proposals were aimed at developing the layout of technological complexes, the design of the main equipment and their aggregate acquisition.

In the mold shop there are certain reserves for the use of equipment (primarily hoisting-and-transport) and free production facilities. In addition, in terms of production category and spatial parameters, the mold shop fully corresponds to the intended purpose.

The issues of infrastructure energy supply, auxiliary facilities were worked out, a preliminary expert assessment of the economic efficiency of construction was carried out.

The possibility of constructing a complex for a small-tonnage installation of a continuous metallurgical process in the mold shop was considered in the current production environment and new technological processes planned for real projects. The design materials substantiated the technological and constructive prerequisites for the placement of two technological modules of the SER type in the main building of the molds workshop:

- mini-module with an annual capacity of 30000 tonnes, designed to debug new technologies and obtain manganese alloys from poor dust-like ores and concentrates;
- a technological module with a capacity of 300000 tonnes, designed to receive the original charge stock, and with the further development of other steel products.

Figure 4 shows the layout of equipment for implementing the technology in development of the above concept, designed to:

- obtaining from dusty materials a primordial (alloyed) charge billet for electric arc furnaces (first stage);
- small shaped casting with special properties by using volume crystallization;
- a small profile, including pipes of small diameter, by continuous drawing up through a mold with a liquid metal coolant (with further development of the technology).

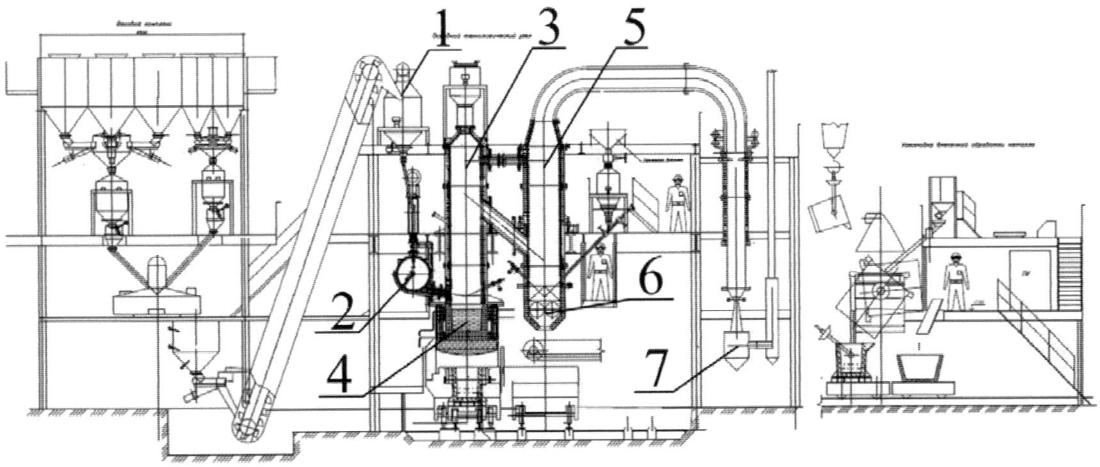

Figure 4. Equipment layout: 1 – charge loading system, 2 – spherical reactor-oscillator, 3 – vertical column reactor (refining sedimentation tank), 4 – induction-heated receiver, 5 – slag collector, 6 – slag granulator, 7 – gas purification system

6.1. Basics of technology

They are considered on the example of obtaining a billet stock or other steel products from ore and dusty metal-containing waste – sludge and scale.

A continuous metallurgical process is implemented in the form of a sequence of technological operations taking place in reactors connected to an in-line unit line (figure 5).

6.2. Equipment layout

The oscillator reactor, the refining settler, the unit for secondary metal processing, and the ingot casting machine represent the main production unit. The oscillator reactor and the refining settler are tied up with technological platforms and supports, which serve as independent load-bearing metal structures. The ball-shaped reactor-oscillator is mounted in suspension to the site at around 7100 mm with a mark in the center of the ball + 5600 mm. At the upper point of the reactor sphere (in the center), the pipe path of the bulk material supply system is connected. The bulk material supply system includes a weighing hopper (weighing funnel) for receiving and weighing the prepared charge, an activator for dosed delivery of the charge from the funnel, a vibrating feeder for loading the charge into the screw batcher.

The vertical screw metering unit is designed for gas-tight feed of the charge into the reactor. The metering unit is designed in such a way that the transported charge is compacted in the pipe channel, forming a kind of gas shutter that prevents the release of process gases from the emulsion reactor.

The refining sump is made with reference to the emulsion reactor at a distance of 1900 mm from the vertical axis. With a stiffness belt, it rests on a work platform at the level of +3900 mm and, in turn, bears the supports of the platforms +5100 mm, +7100 mm, +8300 mm.

IOP Conf. Series: Materials Science and Engineering **866** (2020) 012030 doi:10.1088/1757-899X/866/1/012030

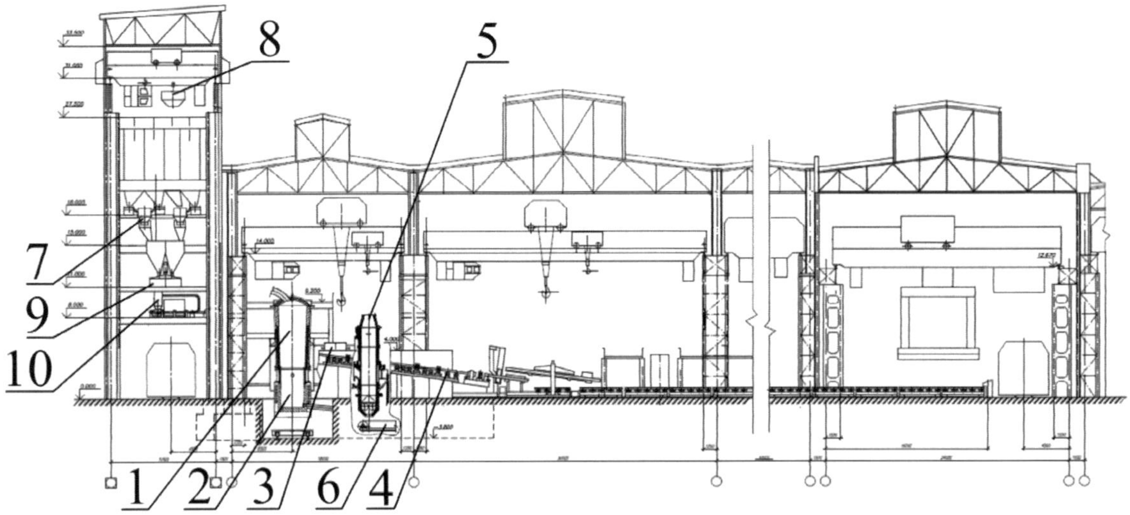

Figure 5. Cross section of the layout diagram: 1 – recovery zone of the column reactor, 2 – oxidation zone and receiver, 3 – connecting channel, 4 – metal casting machine, 5 – slag receiver, 6 – granulator, 7 – hopper block, 8 – bridge crane, 9 – mixer, 10 – vibration feeder.

A boiler-cooler leaves the refining sump from under the dome zone through the branch pipe. The tube-like panel of the boiler from the upper horizontal level (axis mark + 9100 mm) is refracted vertically down to the mark of +3000 mm and then goes to gas treatment.

The lower part of the sump is separated from the zero level of the floor by a distance of 3700 mm in order to freely approach the cart with the electric furnace or induction furnace installed to receive the liquid intermediate. An electric furnace (or induction furnace) acts as a unit for out-of-furnace metal processing. The casting machine, which is a chain conveyor with molds, is designed for casting a liquid metal product into ingots.

6.3. Construction conditions

According to an expert evaluation of the columns of the main building in the mold shop, the piles have a bearing capacity margin, which allows them to be used to absorb the corresponding additional design loads when expanding the production of the shop. The metal frame of the building makes it quite simple to mount new ones, as well as strengthen existing metal structures. A cross-section of the layout diagram is shown in figure 5.

6.4. The purpose of the production program

The purpose of the production program consists in the processing of dusty iron-containing wastes, primarily of current production. Table 1 shows the approximate amount and chemical composition of these materials.

Table 1. The composition of the pulverized waste from the current production.

Waste	Quantity, t/year	Fe_2O_3	FeO	$Fe_{tot.}$	SiO_2	Al_2O_3	CaO
Scale	150000 (including 60 000 scale with fuel-oil residue)	16.6	65	67	1.5	0.4	0.5
Sludge (gas purification)	100 000	15	62	65	2.0	0.04	7.2
Blast furnace dust	15 000	39	14	38	8	3	6.4

The possible savings from processing pulverized wastes of only current production using an example of such a plant as ZSMK is equivalent to refusing to purchase iron ore raw materials for 15 days in the context of one year. Cost recovery will be 1.5 years after the start of production. It is necessary to emphasize the feasibility of orientation, at the first stage of the implementation of the production program, to obtain the original charge for electric arc furnaces.

The fact is that, despite the increasing difficulties in selling finished steel, the market for scrap metal remains unsaturated, and the cost of scrap is at the level of pig iron. In Russia, in recent years several mini-factories have been built and are planned to be built, in connection with which, the scrap deficit is growing. The exception is perhaps only the Ural region, where there are many old closed or reconstructed factories.

The problem of contamination of recycled scrap with non-oxidizable impurities (copper, nickel, etc.) becomes more and more acute, which does not allow smelting of some steel grades or worsens the quality of the metal. In this regard, direct reduction iron (pellets or briquettes), which is much more expensive than scrap, has to be added to the charge of electric steel furnaces, and there is a danger of spontaneous combustion during long-distance transportation. In addition, when using briquettes, energy costs and the amount of slag increase significantly.

These drawbacks are deprived of the original charge stock, which is planned to be obtained from sludge and scale using the proposed technology [1], in addition, it is characterized by low energy costs and capital intensity in comparison with the known direct reduction processes [8].

So, for example, according to the results of European studies and comparative analysis of technologies aimed at reducing carbon dioxide emissions, it is argued that the replacement of the technological route "blast furnace – oxygen converter" with direct reduction processes based on the use of natural gas and steel smelting in electric arc furnaces for the reduction CO_2 emissions is probably economically inefficient "[9]. Table 2 compares the technical and economic indicators of the SER process with the closest and most advanced analog in the world – the COREX unit [10].

Table 2. Comparative analysis of technical and economic indicators of the SER and COREX processes.

Indicators	COREX	SER	Advantage
Energy intensity, GJ/t	29	15 – 17	in 1.7 times
Specific volume, t/m^3 day.	1.1	11	in 10 times
Capital costs, $/t year	350	120 –150	in 2.5 times

Given the above analysis of the effectiveness of direct recovery processes existing in the world, it can be argued that the implementation of the first stage of the program presented in the draft will open up opportunities for wider dissemination of the above technologies, including through the sale of licenses [7]. The unit with a capacity of 30 thousand tonnes may be of interest to foundries of machine-building plants in connection with the need to replace obsolete metallurgical production, as well as dilute scrap contaminated with non-oxidizable impurities, with original metal.

And units with a capacity of 300 thousand tonnes, taking into account the possibility of incorporating relatively small capital costs and quick payback into existing buildings, can become the basis for the creation of full-cycle mini-metallurgy, including inside reconstructed factories.

7. Conclusion

When creating an SER unit, an attempt was made for the first time in world metallurgy to use some ideas of the theory of self-organization (synergetics), such as the principles of submission and least forcing, a large deviation from thermodynamic equilibrium, etc. [6]. To implement these principles, it was possible to apply a number of physical effects, such as the dispersion of the mixture with oncoming gas jets, the creation of a self-organizing reactor-oscillator due to the use of the critical expiration of the two-phase flow and feedback on the change in the gas content of the reaction products, the lower flow of the working mixture from the reactor-oscillator to the column reactor, organization of forced movement and internal pneumatic transport of a two-phase working mixture [6]. The combination of these factors and measures made it possible to create dynamic dissipative structures in the unit (instead of physical zones), which led to a sharp decrease in the specific volume

IOP Conf. Series: Materials Science and Engineering **866** (2020) 012030 doi:10.1088/1757-899X/866/1/012030

of the aggregate (10-15 times) and provided significant control flexibility, including the ratio of the rates of reduction and oxidation reactions, which made it possible to directly obtain a metal with a fairly wide range of carbon content [6].

The project is presented for the placement and equipment layout for the implementation of the technology, first of all, obtaining the original charge from sludge and scale with fuel-oil residue for electric steel furnaces.

For more information on the status of the development of the SER process, please visit: http://www.sibsiu.ru (Scientific and technical developments – Self-organizing jet-emulsion reactor).

References

[1] Tsymbal V P, Mochalov S P, Rybenko I A et al 2014 *SER Process – Metallurgical Jet-emulsion Reactor* (Moscow: Metallurgizdat)

[2] Tsymbal V P, Mochalov S P and Shakirov K M 2015 *Springer Journals* **59** 119–25

[3] Nikolis G 1979 *Self-organization in Nonequilibrium Systems* (Moscow: Mir)

[4] Haken G 1985 *Synergetics. Hierarchy of Instabilities in Self-Organizing Systems and Devices* (Moscow: Mir)

[5] Tsymbal V, Olennikov A et al 2016 *Sustainable Industrial Processing Summit & Exhibition SIPS* **1** 214–27

[6] Tsymbal V P, Rybenko I A et al 2018 *Proc. of the II Int. Sci. and Pract. Conf. Dedicated to the 90th Anniversary of Y.G. Yaroshenko* (Ekaterinburg: UrFU) pp 177–84

[7] Tsymbal V P, Mochalov S P, Rybenko I A et al 2009 *Pat. of the Russian Federation* No. 2371482 appl. 12.03.2007, publ. 10.27.2009

[8] Degel R, Freling K et al 2016 *Ferrous Metals* **4** 40–49

[9] Meijer K, van der Stel J, Zeilstra C et al 2015 *Proc. METEC and 2nd ESTAD* (Düsseldorf Juni 2015) 15–19

[10] Lungen Kh, Knop K and Steffen R 2007 *Black Metals* **2** 13–25

IOP Conf. Series: Materials Science and Engineering **866** (2020) 012031 doi:10.1088/1757-899X/866/1/012031

Modelling of conditions of manganese reduction with carbon, silicon and aluminum

I A Rybenko[1], O I Nokhrina[1], I D Rozhikhina[1], I E Proshunin[2] and M A Golodova[1]

[1]Siberian State Industrial University, 42 Kirova Street, Novokuznetsk, 654007, Russia
[2]JSC EVRAZ ZSMK, Novokuznetsk, Russia

E-Mail: golodova_ma@mail.ru

Abstract. The paper contains the research results of the process of manganese oxides reduction by various types of reducers – carbon, silicon and aluminum, the purpose of which is definition of a possibility of the use of carbon-bearing material (coke, silicon carbide), ferrosilicon and aluminum by production of alloys of manganese and direct alloying of steel in a ladle by manganese and definition of optimal conditions of implementation of these processes.

1. Introduction

Metallurgy is one of the resource-intensive industries. The need to reduce the energy and material intensity of metallurgical products and increase their operational properties requires both the improvement of traditional metal production schemes and the development of new technologies.

The solution to the problems of improvement of the existing and development of new metallurgical technologies is connected, first of all, with studies of high-temperature processes in complex thermodynamic systems with physicochemical transformations in the equilibrium and nonequilibrium conditions. Since experimental studies, as a rule, are quite expensive, and often difficult to implement, in these conditions a computational experiment is of great importance, which allows the states and processes to be analyzed and conclusions about the behavior of the objects under study based on model concepts to be drawn. In this regard, it is urgent to create methods and instrumental modeling systems that allow for multivariate calculations, research and solve optimization problems, which will open up new opportunities for the development and design of advanced metallurgical technologies.

2. Methods of research

The solution to the problem of determining the conditions for the reduction of metals from oxides was carried out using thermodynamic modeling methods based on the calculation of equilibrium states in model thermodynamic systems [1–3]. The advantage of these methods is the possibility of calculating the equilibrium state of any process based on only the fundamental laws of thermodynamics. The calculation of thermodynamic equilibrium allows the fundamental possibility of obtaining certain substances to be found out, the isolation of which is the main one when solving the problem of assessing the ultimate limit state, and to determine the range of permissible parameter values.

When implementing thermodynamic modeling, we used the Terra software package developed at Moscow State Technical University, which allows, on the basis of the maximum entropy principle, the equilibrium composition of a multicomponent, heterogeneous thermodynamic system for high-temperature conditions [1,2] to be determined, as well as an instrumental system in the form of a set of programs and databases data allowing multivariate calculations created by the author [4]. The thermodynamic modeling technique was successfully applied to solve the problems of studying the reduction processes of iron, nickel, vanadium, barium, and strontium [5–9].

An elementary system is formed by setting the moles numbers of its constituent components. If necessary, the compositions of condensed solutions are formed. When two thermodynamic parameters are selected, multivariate calculations of equilibrium compositions are carried out depending on the thermodynamic parameters or the flow rates of the initial materials.

The study of the metal reduction process was carried out by calculating the equilibrium compositions in systems, the input stream of which consists of a combination of $Me - O - C - Fe - Si - Al$ elements and is represented by a set of substances $lMeO - nC - mSi - kAl - pFe$. Depending on the values of the parameters l, k, n, m and p, one or another initial composition of the system was formed.

The set of substances that can be formed at a given elemental composition of the mixture was determined as a result of numerical simulation for a selected temperature range and various thermodynamic states. From the complete list of possible substances, only those substances were selected the concentrations of which exceeded the value of 10^{-4} mol/kg of the mixture.

3. Results of the research

The study of the carbon-thermal reduction of manganese in the $MnO_2 - C$ system in the absence of iron showed that the reduction of manganese begins at temperatures above 1723 K with a carbon consumption of more than 1.5 moles (figure 1). At the same time, the process of evaporation of manganese begins. At a temperature of 1723 K with an excess of carbon, Mn_7C_3 manganese carbide is present in the system, which disappears with the increase in the temperature. Complete reduction of manganese occurs at a carbon consumption of 2 moles.

Calculations in the $MnO_2 - Si$ system showed that the reduction of manganese by silicon is feasible over the entire specified temperature range. The research results are presented in figure 2, from which it follows that the complete reduction of manganese occurs at a silicon molar rate of 1 mol. This value corresponds to a maximum content of manganese in the system of 47%, which decreases with an increase in the consumption of reducing agent due to dilution with excess silicon.

The results of calculations of the joint reduction of manganese with carbon and silicon in the $MnO_2 - C - Si$ system revealed the predominant effect of carbon on the reduction process, which occurs at temperatures above 1773 K (figure 3). The dependences of the process parameters on the carbon consumption at various silicon consumption at a temperature of 1823 K are shown in figure 4, from which it follows that the maximum content of reduced manganese corresponds to a carbon consumption of 1 mol with a minimum silicon content in the system.

With the increase in the carbon, the manganese content decreases due to an increase in the concentration of silicon carbide (figure 5). The content of gaseous manganese at this temperature is negligible and does not exceed 2.5%.

As a result of calculations of the process of manganese reduction by carbon in the $MnO_2 - SiO_2 - C$ system, it was revealed that the addition of 0.5 moles of silicon oxide does not change the qualitative nature of the dependences obtained for the $MnO_2 - C$ system, but only leads to a decrease in the reduced manganese content from 48 to 37.8% due to dilution with silicon oxide.

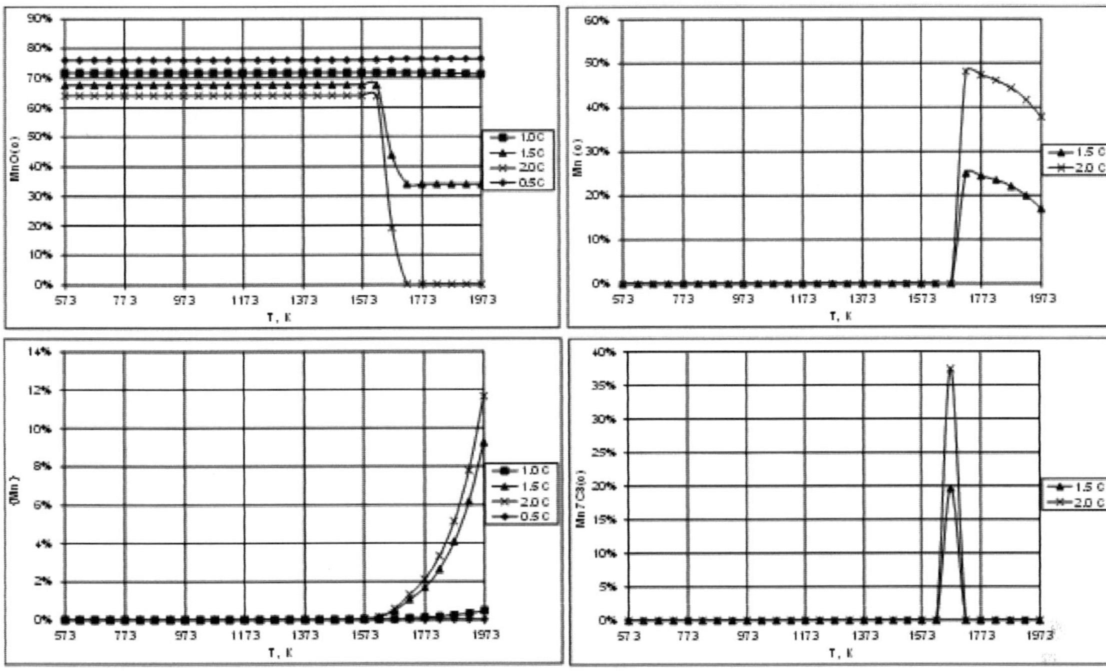

Figure 1. Relations of equilibrium compositions in the $1MnO_2 - nC$ system with temperature at n=0.5; 1.0; 1.5; 2.0 mole.

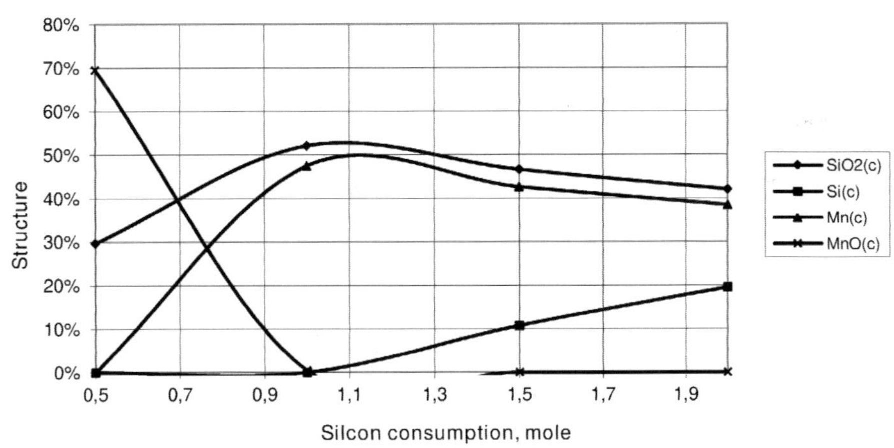

Figure 2. Relations of equilibrium compositions in the $MnO_2 - Si$ system with silicon consumption.

Similar results were obtained in the study of the $MnO_2 - SiO_2 - C - Si$ system. Manganese reduction also occurs at temperatures above 1773 K. The predominant reducing agent is carbon, the best results correspond to the consumption of carbon 1 mol and silicon 0.5 mol. The content of reduced manganese at these parameters is lower (30%) than in the $MnO_2 - C - Si$ system (47%) due to dilution with silicon oxide.

The addition of CaO to the $MnO_2 - SiO_2 - C - Si$ system also did not significantly affect the nature of the dependences. The maximum degree of reduction of manganese corresponds to the temperature of 1773 K and a consumption of 1 mol of carbon and 0.5 mol of silicon. An increase in the amount of

IOP Conf. Series: Materials Science and Engineering **866** (2020) 012031 doi:10.1088/1757-899X/866/1/012031

CaO in the system leads to a slight decrease in the amount of reduced manganese due to dilution (figure 6).

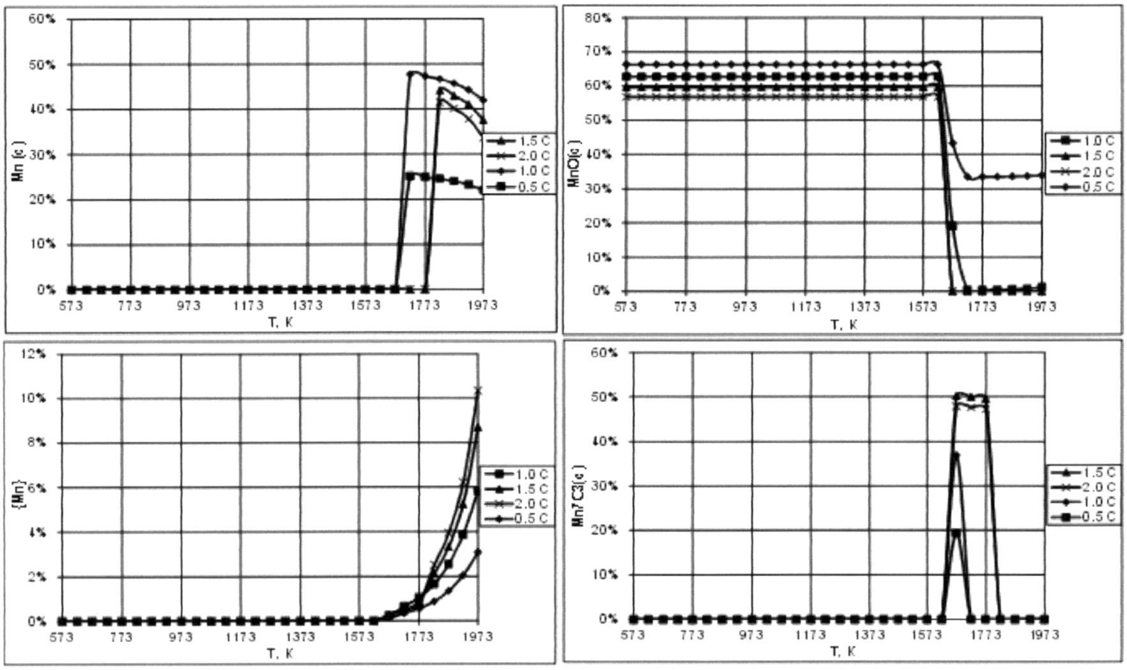

Figure 3. Relations of equilibrium compositions in the $1MnO_2 - 0.5Si - nC$ system with temperature at n = 0.5; 1.0; 1.5; 2.0 mole.

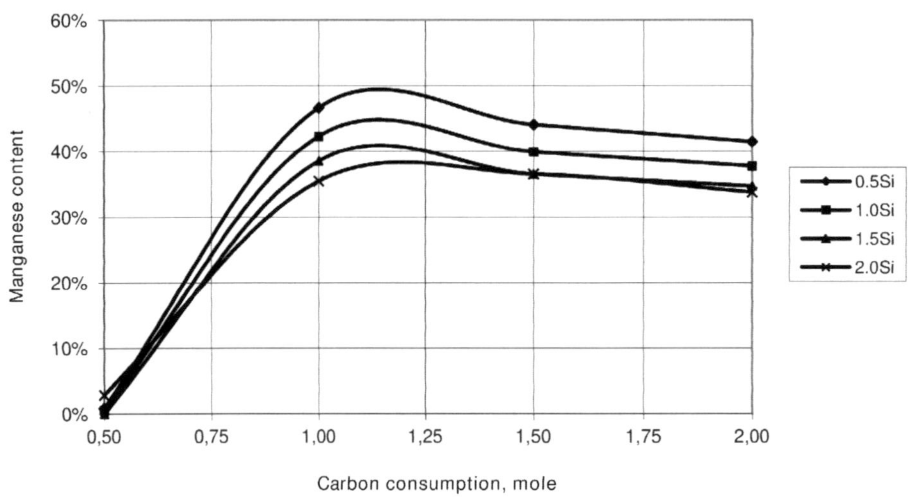

Figure 4. Dependence of reduced manganese content in the $MnO_2 - Si - C$ system on carbon consumption at different silicon consumption.

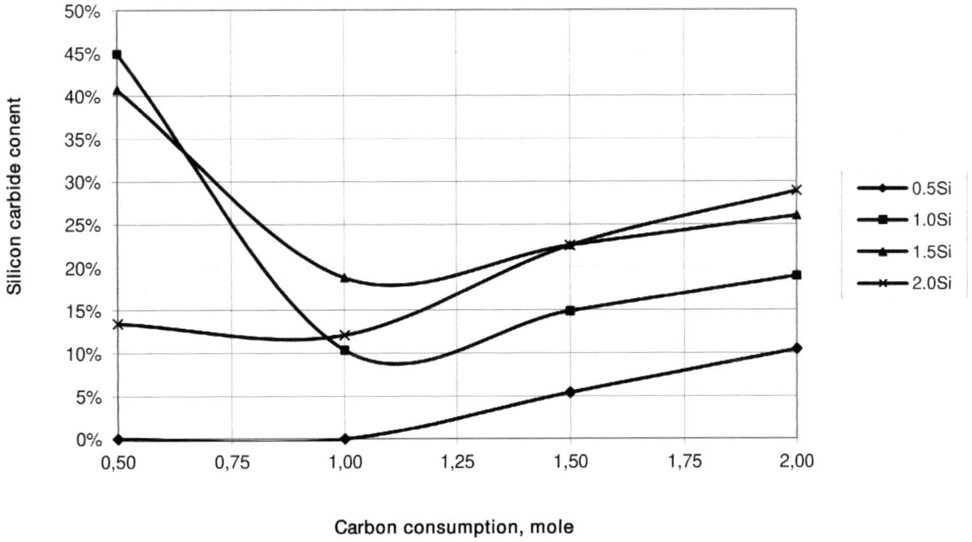

Figure 5. Dependence of silicon carbide content in the $MnO_2 - Si - C$ system on carbon consumption at different silicon consumption.

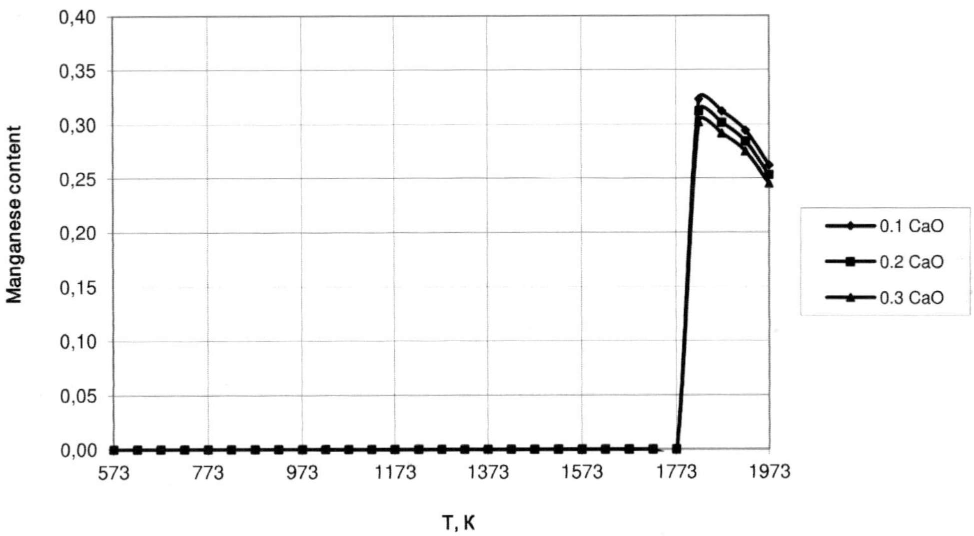

Figure 6. Dependence of reduced manganese content in the $1MnO_2 - 0,\ 5SiO_2 - 1C - 0.5Si - nCaO$ system on temperature at different CaO flow rates.

The results of calculations of the processes in the $MnO_2 - Al$ system showed that manganese is reduced over the entire specified temperature range. The dependences of the equilibrium compositions on the consumption of aluminum are shown in figure 7.

IOP Conf. Series: Materials Science and Engineering **866** (2020) 012031 doi:10.1088/1757-899X/866/1/012031

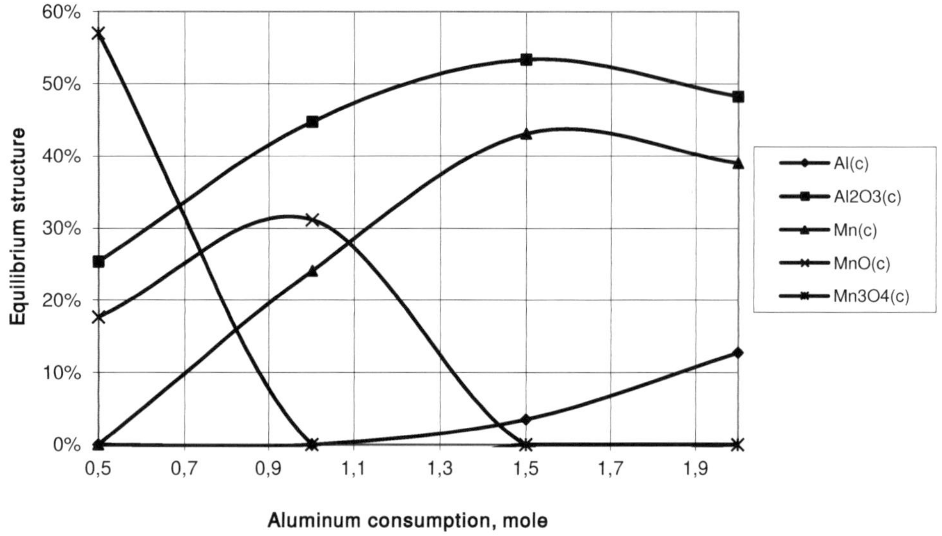

Figure 7. Dependence of equilibrium compositions in the $MnO_2 - Al$ system on aluminium consumption.

The maximum degree of manganese reduction corresponds to the consumption of aluminum 1.5 mol. The results of studies in the $MnO_2 - Si - Al$ system using the Terra program showed that manganese is reduced over the entire specified temperature range. The optimal total consumption of reducing agents, providing a maximum content of reduced manganese of 43%, is 1.5 mol.

An increase in the expenditures of silicon and aluminum leads to their accumulation in the system and a decrease in the concentration of reduced manganese (figure 8).

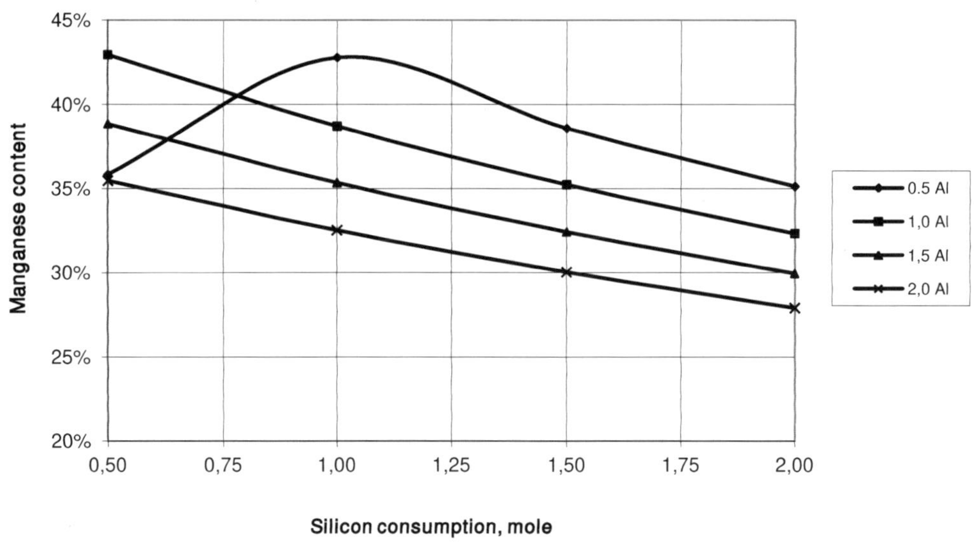

Figure 8. Dependence of reduced manganese content in $MnO_2 - Si - Al$ system on silicon consumption at different amount of aluminum.

The addition of CaO to the $MnO_2 - Si - Al$ system did not change the nature of the dependences; the maximum content of reduced manganese also corresponds to the total consumption of reducing agents of 1.5 moles in a given temperature range. An increase in CaO consumption leads to a slight decrease in the concentration of manganese in the system due to dilution.

The analysis of the results of studies in the $MnO_2 - SiC$ system showed that complete reduction of manganese is possible at any temperature in a given range with 1 mol of silicon carbide in the system. At higher flow rates, silicon carbide accumulates in the system, thereby reducing the concentration of reduced manganese. At temperatures below 1823 K, the reduced manganese is in the form of Mn_7C_3 carbide, and at higher temperatures, it is in its pure form in the condensed and gas phases.

4. Conclusion

Thus, it can be concluded that any of the described reducing agents or a combination thereof in certain ratios can be used as reducing agent when using oxide manganese containing alloy smelting and steel processing materials. Naturally, for the targeted use of silicon and aluminum in steel processing, a prior thorough deoxidation of the metal and slag is required to reduce the overall oxidation of the metal-slag system.

Acknowledgments
The study was carried out with the financial support of the Russian Foundation for Basic Research and Kemerovo Region – Kuzbass within the framework of scientific project No. 20-48-420001/20.

References

[1] Trusov B G 2005 *III International Symposium in Combustion and Plasma Chemistry* (Almaty: Kazak Iniversiteti) p 52
[2] Belov G V and Trusov B G 2013 *Thermodynamic Modeling of Chemically Reacting Systems* (Moscow: MSTU n.a. N.E. Bauman) p 96
[3] Vatolin N A et al 1994 *Thermodynamic Modeling in High-temperature Inorganic Systems* (Moscow: Metallurgiya) p 352
[4] Rybenko I A 2016 *Application of the Methodology and Instrumental System for Calculating Metallurgical Processes to Develop the Theoretical Foundations of Resource-saving Technologies: Monograph* (Novokuznetsk: SibSIU) p 187
[5] Golodova M A et al 2016 *Bull. Ferrous Metallurgy* **10** 27–34
[6] Kichigina O Yu 2010 *Izv. Vuzov. Ferrous Metallurgy* **10** 20– 23
[7] Nokhrina O I 2010 *Izv. Vuzov. Ferrous Metallurgy* **10** 17–20
[8] Dmitrienko V I et al 2012 Izv. Vuzov. Ferrous Metallurgy **4** 27–29
[9] Nohrina O I et al 2016 *Izv. Vuzov. Ferrous Metallurgy* vol 59 No 4 237–244

Analysis of the main trends in the development of the production of grinding balls in Russia and abroad

A A Umansky[1], A V Golovatenko[2] and A N Prudnikov[1]

[1]Siberian State Industrial University, Novokuznetsk, Kirova 42, 654007, Russia
[2]EVRAZ Consolidated West Siberian Metallurgical Plant, 16 Kosmicheskoye Shosse, 654043, Russia

E-mail: umanskii@bk.ru

Abstract. The article presents the results of the analytical review of the main trends in the grinding balls production in Russia and foreign countries. It is shown that grinding balls of domestic production have lower performance characteristics in relation to foreign counterparts, which is due to the technological features of their production, including the use of unalloyed steels with high hardenability and rejected billets of steels for other purposes (in most cases, rail steels), non-optimal technology of heat treatment. Based on the analysis, a list of promising areas for the development of domestic production of grinding balls that contribute to the competitiveness of this type of product in the domestic and foreign sales markets is determined. These areas include: increase in the share of rolled balls from alloy steels and cast balls from alloy cast irons, improvement of the technology of balls heat treatment and usage of new specialized steel grades for their production.

1. Introduction

Currently, over 2 billion tonnes of raw materials are crushed in various industries annually. At the same time, drum mills are mainly used for grinding, and balls, cylinders made of steel and cast iron are used as the grinding medium. The global production and consumption of grinding media is about 3 million tonnes per year. At the same time, the main consumers are the USA and Canada (about 20%), CIS countries (about 16%), China, and Australia.

The most common type of grinding media is grinding balls, due to their significant competitive advantages relative to other types of grinding media [1]. In particular, the specific consumption of grinding balls is 20–40% lower compared to cylpebs, which is explained by a much lower probability of a split during operation due to the absence of "stress edges" (angles) [2]. There is an improvement in the quality of grinding of crushed raw materials and an increase in the productivity of mills [3].

The volume of consumption of grinding balls in Russia is estimated at 250 thousand tonnes per year, while the main consumers of this type of product are the mining, metallurgical and cement industries (figure 1).

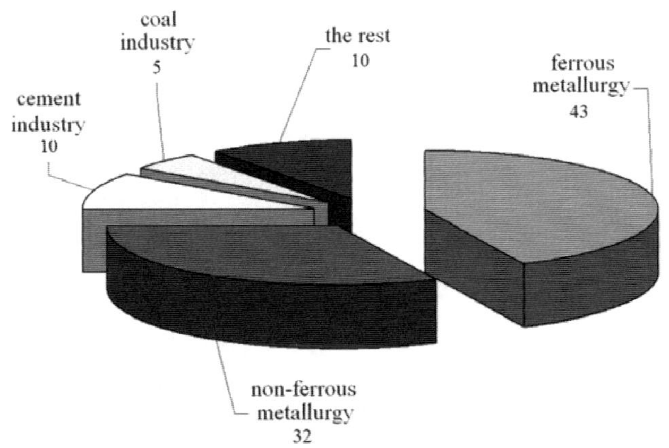

Figure 1. Consumption pattern of grinding balls in Russia.

The severe operating conditions of grinding balls, consisting in periodic shock loads, abrasive wear, determine the following requirements for their quality [4]: high hardness and strength, increased resistance to abrasion, resistance to splitting and chipping. Increase in the service life of the balls is an important task, since the cost of grinding media is about 30% of the cost of the grinding process. In addition, in the event of premature wear and cracking of the balls, not only their specific consumption increases, but also the quality of the crushed raw materials deteriorates due to the ingress of particles of chopped grinding balls into it.

The increase in the operational properties of balls is achieved by a combination of two methods: improvement of chemical composition of cast irons and steels used for the balls production and the use of special heat treatment modes. The article presents the analysis of existing trends in the development of ball production in these directions.

2. Results and discussion

The analysis of the structure of grinding balls production in Russia shows (figure 2) that the largest share is made of steel balls produced on specialized ball-rolling mills by screw rolling. High-performance ball-rolling mills are available at West Siberian (JSC "EVRAZ ZSMK") and Nizhny Tagil (JSC "EVRAZ NTMK") metallurgical plants, as well as at Guryevsky metallurgical plant – figure 2. Forging and stamping methods for the steel balls production are practically not used, due to their significantly lower (2-8 times) productivity and increased metal losses (10-15%).

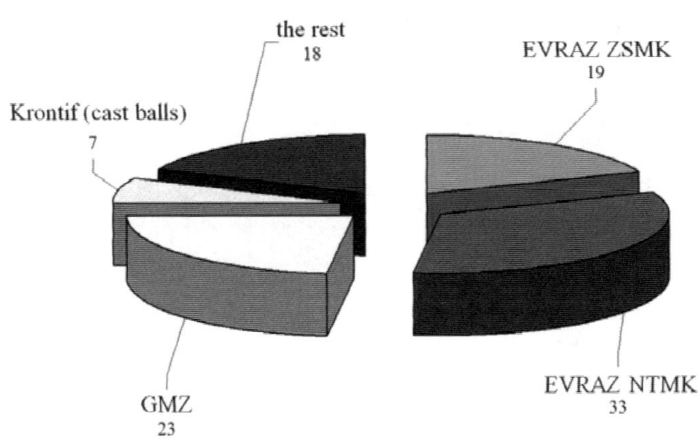

Figure 2. The structure of the grinding balls production in Russia.

In Russia, the bulk of grinding balls is made of medium- and high-carbon unalloyed steels, while in order to achieve their high wear resistance, volumetric hardening of balls from rolling heating with subsequent self-tempering ("EVRAZ ZSMK", "EVRAZ NTMK") or low tempering (Guryev Metallurgical plant) should be performed. At the same time, heat treatment at the Guryevsk Metallurgical Plant is carried out in two stages in two quenching drums with intermediate exposure in the air, which allows the temperature gradient over the cross section of the ball and quenching stresses to be reduced.

The technologies used in the domestic metallurgical enterprises for the production of steel grinding balls make it possible to obtain a surface hardness 55–62 HRC with a hardened layer depth 10–15 mm [5]. It should be noted that the indicated surface hardness is achieved only on balls of small diameter

(40-60 mm), and the surface hardness of grinding balls with a diameter of 80-120 mm does not exceed 50-57 HRC even when using modern quenching devices. To obtain grinding balls of higher hardness groups, specialized steels with an increased content of carbon, manganese and chromium are used. In such steels, the carbon content is 0.60-0.75%; chromium – 0.30-0.75%; Manganese – 0.65-0.85%, in experimental steels – up to 1.05%. Also, a significant share in the total production volume is made up of balls produced from the rejection of rail steels alloyed with vanadium and nitrogen.

Domestic-made steel grinding balls, as a rule, have low wear resistance, which, in addition to low surface and bulk hardness, is also associated with the presence of macrostructure defects (shrinkage defects, flakes) [6, 7]. This fact is due to the fact that until recently this type of product was actually produced from production waste – rolled products of inappropriate quality from various steel grades. For domestic metallurgy, the possibility of using production waste for processing into grinding balls is largely conditioned by the absence in the standard for the production of this type of product (GOST 7524-2015) of strict requirements for the chemical composition (only the carbon content is regulated at a level not less than t 0.40-0, 60% depending on the hardness group of the balls and the carbon equivalent at a level not less than 0.50- 0.85% also depending on the hardness group) – table 1.

Table 1. Requirements for the chemical composition of steel grinding balls in accordance with GOST 7524-2015.

Ball diameter, mm	Hardness group	Carbon content, %	Carbon equivalent
		not less than	
15-55	1. 2	0.40	0.50
	3	0.60	0.70
	4. 5	0.60	0.75
60-120	1. 2	0.50	0.70
	3. 4	0.60	0.75
	5	0.60	0.80

We can also note the complete absence of requirements regarding the macro- and microstructure of the balls produced and the minimum requirements for the quality of the surface of the balls (defects on the surface of the balls, that make the size of the balls beyond the maximum deviations, are not allowed). The main requirements of this standard relate to hardness (on the surface, depth ½ of the radius of the ball and volume), depending on which balls are assigned to one of five groups (table 2).

Table 2. Hardness requirements for steel grinding balls according to GOST 7524-2015.

Ball diameter, mm	Hardness, HRC (not less) by groups						
	1	2	3	4		5	
	surface			surface	at a depth of ½ radius	surface	volumetric
15-45	45	49	55	55	45	61	57
50-70	43	48	53	53	43	60	53
80-100	39	42	52	52	40	58	48

In addition, in the current GOST the absence of requirements for impact resistance of balls, one of the main indicators characterizing the life and specific consumption of these products, should be noted. The standard stipulates only the ability to control shock resistance in agreement with the manufacturer and its methodology for balls of the fourth and fifth hardness groups (the highest hardness).

In recent years, due to increasing requirements for the quality of grinding balls and increasing competition in the market, a significant share of this type of product is supplied not according to GOST, but according to technical specifications (TU) or organization standards (STO), where steel grades with fixed the range of changes in basic chemical elements are provided.

It should be noted that the trend in the production of grinding balls from production wastes, for example, trimmings of rail and beam mills, is to some extent characteristic of foreign manufacturers [8]. However, unlike Russia, in foreign countries, the bulk of steel grinding balls is made from chromium alloyed steels (about 1%). Grinding balls with a diameter of 15-150 mm made of such steels produced by leading Western companies (AGS, Gerdau Ameristeel, Vitkovice) have a hardness on the surface after hardening of 62-65 HRC, while the hardness in the central zone of the balls is at least 60 HRC. The leader in this technological niche is the South African company "Scow Metals". At its ten plants, about 600 thousand tonnes of balls are produced under the common brand "Moly-Sor".

The further prospect in increasing the wear resistance of steel grinding balls is the development and application of specialized spherical hypereutectoid steels. In balls made of such steels, it is possible to achieve two microstructural states that provide higher wear resistance compared to tempering martensite:

- the metal matrix of tempered martensite with evenly distributed granular inclusions of secondary cementite;
- the austenitic-martensitic carbide-free or low carbide structure in which austenite has an adjustable tendency to martensitic transformation during wear. This type of microstructure is preferable because it provides the balls with increased resistance to cracking during hardening and splits during operation due to the presence of viscous austenite.

Regarding the improvement of the heat treatment technology, a promising direction is the use of technology that provides for withholding of quenching cooling and subsequent heating in order to redistribute carbon from martensite to austenite with subsequent tempering [9].

According to the study [10], as a result of Q-n-P treatment, a heterogeneous microstructure is formed in the ball, consisting of sections of tempered martensite and austenite-martensite sections. The redistribution of carbon between martensite and austenite during tempering led to an increase in the fraction of residual austenite to 25–30% at a carbon concentration in Ares of 1.12%.

The resulting structure combines micro-sites with different strengths and ductility, which together provides an increased set of steel properties in connection with the implementation of the effect of the natural composite. Residual austenite lies between martensitic crystals in the form of film inclusions or interlayers. Studies [29] found that austenite in the form of films has a higher stability to deformation martensitic transformation than austenite in the form of blocks.

Although it should be noted that despite the positive results obtained, experience in the industrial use of Q-n-P technology for steel grinding balls is currently lacking.

Characterizing the production of grinding cast iron balls, it can be stated that the share of such balls in the total volume of production and consumption of Russia and the CIS countries is significantly inferior to steel balls (about 20%), although it has some upward trend. The main manufacturer of cast iron grinding balls in Russia and the CIS is the company "Krontif". Abroad, the proportion of grinding balls made of cast iron is much higher and amounts to approximately 35%, also having a tendency to increase. At the same time, gray cast irons are used in Russia and the CIS countries for the production of cast balls, and leading western companies use high-alloyed white cast irons (with a chromium content of 13-23%).

Due to the presence of a significant amount of carbides in the structure, pig-iron grinding balls have a higher operational stability compared to steel balls. In addition to less weight loss, the so-called "toothed" effect is realized in cast iron balls: the directional arrangement of carbide crystals perpendicular to the body surface and their alternation with softer interlayers of the matrix phase, which creates a specific surface roughness and improves the grinding effect by reducing slippage between the balls and grains ground material.

Modern brands of white wear-resistant cast irons are complex alloyed multicomponent alloys with a wide variety of structures and a wide range of physical and mechanical properties. The use of highly alloyed cast irons for casting balls with subsequent heat treatment provides a martensitic-carbide structure with high impact resistance over the entire cross-section of the grinding body. In the production of cast iron balls with a chromium content of 13-23%, a through hardness of 62-64 HRC is achieved. The world leader in this technological segment is the Belgian company "Magotteaux",

IOP Conf. Series: Materials Science and Engineering **866** (2020) 012032 doi:10.1088/1757-899X/866/1/012032

which produces about 320 thousand tonnes of such balls per year at 16 plants. For comparison, the surface hardness of ordinary cast iron balls can reach 50-52 HRC and at the same time they are characterized by low impact resistance.

Thus, the use of alloyed cast irons for the production of grinding balls is a promising direction.

3. Conclusion

The analysis established the distinctive features of the technology for the grinding balls production in Russia and foreign countries, causing lower quality characteristics of domestic grinding balls relative to foreign counterparts.

To improve the quality and competitiveness of grinding balls of domestic production, the following directions are promising:

- increase in the share with the prospect of a complete transition to the production of rolled balls from alloy steels and cast balls from alloy cast irons;
- improvement in the technology of balls heat treatment;
- development of new specialized steel grades for the production of balls.

References

[1] Rakhutin M G and Boyko P F 2017 *Ugol'* **12** 49–52
[2] Aldrich C 2013 *Minerals Engineering* **49** 77–91
[3] Umucu Y and Deniz V 2015 **16(1)** 197–203
[4] Efremenko V G 2002 *Metallurgicheskaya i Gornorudnaya Promyshlennost* **4** 51–54
[5] Lam M M, Serov A I, Smyrnov Y N et al 2017 *Steel in Translation* **47(5)** 325–329
[6] Umanskii A A, Golovatenko A V, Temlyantsev M V et al 2019 *Metallurgist* **63(5-6)** 578–584
[7] Umanskii A A, Golovatenko A V et al *Izvestiya Ferrous Metallurgy* **62(4)** 283–289
[8] Pater Z, Tomczak J, Bulzak T, Cyganek Z, Andrietti S and Barbelet M 2018 *International Journal of Advanced Manufacturing Technology* **97(1-4)** 893–901
[9] Bai X and Jin Y 2017 *Heat Treatment of Metals* **42(5)** 193–196
[10] Efremenko V G, Popov E S et al 2014 *Metallurgist* **57(9-10)** 849–854

IOP Conf. Series: Materials Science and Engineering **866** (2020) 012033 doi:10.1088/1757-899X/866/1/012033

Thermocyclic deformation, annealing, and physical properties of low-carbon steel 10

A N Prudnikov[1], V A Prudnikov[2], A R Fastykovskii[1] and A A Umansky[1]

[1]Siberian State Industrial University, 42 Kirova Street, Novokuznetsk, 654007, Russia
[2]LLC "MSB Engineering", 11-A Pavlovskogo Street, 654007, Russia

E- mail: a.prudnikov@mail.ru

Abstract. The results of the influence of thermocyclic forging and annealing in the range of 100 ÷ 900 °C on the microstructure, coercive force and linear expansion of steel 10. Hot cyclic forging was carried out on slabs 900×700×500 mm in size on the hydraulic press with a force of 20 MN under industrial conditions of ZSMK OJSC (Novokuznetsk). Subsequently, the billets were rolled on a 3 mm thick sheet at OAO NMZ (Novosibirsk). The use of preliminary thermocyclic deformation increases the coercive force by no more than 8%. The possibility of reducing the coercive force of sheet steel made using thermocyclic forging was shown to be almost 3 times lower than the initial untreated state due to subsequent annealing at 900 °C for 10 hours. In addition, this annealing mode reduces the average linear expansion temperature coefficient of sheet steel by 6% in the temperature range of the test 50-450 °C.

1. Introduction

In most industries, the vast majority of blanks, parts and structural elements of products from various materials are used after hardening technologies – deformation, thermal or surface chemical-thermal treatments. However, increasingly stricter requirements for materials used in modern technology have led to the emergence of integrated technologies, including the combined use of various methods and techniques for the formation and management of the structure and, therefore, the properties of these materials. Such technologies include thermo-mechanical, mechano-thermal and more complex type – deformation thermocyclic treatment (DTT).

The technological modes of these treatments combine various types of cold and hot deformation with heating, holding at fixed temperatures and cooling over a wide range of speeds. In the literature, DTT is widely covered as a technology used for hardening ferrous and non-ferrous metals and alloys, including steel, cast iron and aluminum alloys [1-7]. Much less attention is paid to the problem of improving the electrical, magnetic, thermal and other physical properties of various materials. Such publications include the work of the authors [1.8-12].

One of the promising areas for the use of DTT can be the structural low-carbon steel properties that are close to the properties of some groups of soft magnetic materials, in particular commercially pure iron and electrical steel, combined with enhanced mechanical and technological properties. In addition, subsequent heat treatment, which affects its structure and the most important properties, can serve as an additional reserve for improving the properties of low-carbon steel.

For magnetically soft materials, such properties can be electrical, magnetic, and thermal (electrical resistivity, electrical conductivity, magnetization reversal loss at different frequencies, coercive force, coefficient of thermal expansion, etc.). Therefore, the aim of the work was to study the effects of

IOP Conf. Series: Materials Science and Engineering **866** (2020) 012033 doi:10.1088/1757-899X/866/1/012033

preliminary thermocyclic forging and subsequent annealing on physical properties, in particular, coercive force and linear expansion of hot-rolled steel sheet 10.

2. Materials and methods of research

Low-carbon high-quality steel 10 was taken as a research material. Steel was smelted at NKMK OJSC (Novokuznetsk). The chemical composition of the experimental steel are shown in table 1.

Table 1. The chemical composition of the experimental sheet steel 10.

Steel grade	Components, weight. %									
	C	Si	Mn	P	S	Cr	Ni	Cu	As	Fe
10sp	0.13	0.22	0.42	0.014	0.018	0.05	0.04	0.20	0.06	res.

A slab of experimental steel 900×700×500 mm in size was subjected to hot cyclic forging. The forging scheme is a single-pass by platens with flat strikers with a blank setting-up. The slabs were deformed in the forging and thermal workshop of ZSMK OJSC (Novokuznetsk) on the hydraulic forging press with a force of 20 MN. The technology of thermocyclic forging is described in [10], and its main parameters and mode are given below. The heating temperature for forging was 1250 °C; the slab holding time in the furnace before forging was 2 hours (without taking into account the heating time). To heat the workpieces for deformation, a continuous gas furnace was used. The forgings were cooled in air to 200-300 °C. The number of forging cycles was 10, the degree of deformation in each cycle was 6–8%. The total degree of deformation was 65–68%, the total forging coefficient ~ 1.90. Subsequently, the blanks were rolled at OJSC "NMZ" n.a. A.N. Kuzmin (Novosibirsk) per sheet 3 mm thick at the rolling mill 810 according to industrial technology. The heating temperature of the blanks for rolling was 1120÷1250 °C, the holding time was 2÷2.5 hours. The technological mode of sheet manufacturing from steel 10 is described in more detail in [2].

Samples cut from the sheet were annealed in resistance furnaces of SNOL type. To study the steel microstructure the OLYMPUS GX-51F optical microscope was used. To determine the temperature coefficient of linear expansion (TCLE) of steel samples, a high-temperature dilatometer DIL 402C with a measurement error of $00{,}1 \cdot 10^{-6}$ K^{-1} was used. Determination of the coercive force was carried out on the KIFM-1 coercimeter with a flux probe on rectangular sheet samples with a thickness of 3 mm and a size of 90×120 mm. The measurement error was 4 A/m.

3. Results and discussion

One of the most important properties of soft magnetic materials is the coercive force, which determines the energy loss due to magnetization reversal of magnetic circuit elements. First of all, this applies to the cores of magnetic cores, transformers and other structural elements [13]. In this regard, the value of the coercive force was determined for hot-rolled steel 10, manufactured according to the industrial rolling regime and using preliminary DTT. In the first case, it amounted to 214 A/m, and in the case of using thermocyclic forging 232 A/m. Such an increase in coercive force due to the use of preliminary thermocyclic forging modes in the manufacture of hot-rolled sheet steel 10 can be explained by the changes occurring in its microstructure.

This is, first of all, grinding of ferrite grains and pearlite colonies oriented along the rolling direction, which is consistent with the results of metallographic studies carried out in [2]. Such a refinement of the structural components and, consequently, an increase in the length of the interphase grain boundaries, which are places of accumulation of crystal structure defects (dislocations, vacancies, etc.) in the structure of sheet steel 10 made using DTT, is the reason for the increase in its coercive force.

Quite often, in special fields of industry, magnetic elements operate at elevated and lowered temperatures, including the widespread plate-type stacked cores with an electrically insulating coating, sealed magnetic contacts (reed switches), and other similar products. Therefore, for the soft magnetic materials from which such products are made, one of the important characteristics is the characteristic of thermal expansion – the temperature coefficient of linear expansion (TCLE). Studies on the effect of thermocyclic forging on the linear expansion of steel 10 showed that the use of DTT practically does not affect the true thermal expansion coefficient. The magnitude of its change after cyclic forging and rolling on a sheet does not exceed 5% (figure 1). However, one can note a tendency for the coefficient to exceed the coefficient of low test temperatures (up to 300 °C) by 5% compared with the coefficient of hot-rolled steel of industrial manufacture and the opposite change in the thermal expansion coefficient for higher test temperatures (300-450 °C).

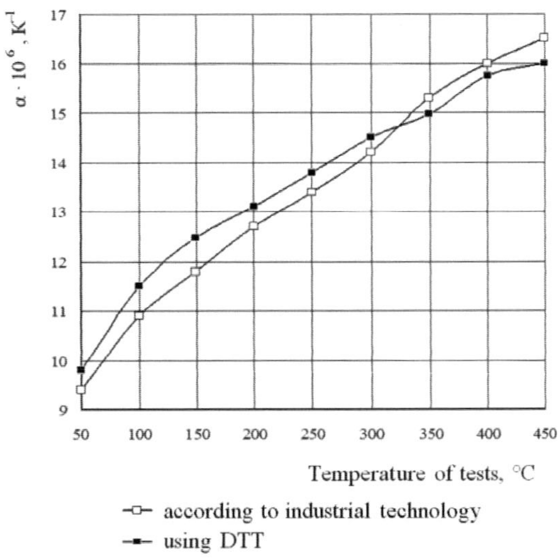

Figure 1. The effect of thermocyclic forging on the linear expansion of hot-rolled steel sheet 10 (sheet thickness 3 mm).

Further, the effect of subsequent annealing for 10 h on the physical properties of sheet steel 10 (thickness 3 mm) subjected to DTT was studied. It was found that an increase in the annealing temperature from 100 to 900 °C with an increments of 100 °C leads to a consistent decrease in the coercive force of steel (figure 2). Its intense decrease was noted at higher annealing temperatures starting from 600 °C. The minimum value of the coercive force corresponds to annealing at a temperature of 900 °C and is 83 A/m, which is almost 3 times lower than that of samples without heat treatment. Apparently, the decrease in the coercive force of sheet steel fabricated using DTT with an increase in the annealing temperature is explained by the general decrease in the level of stresses and defects in the crystal structure (vacancies, dislocations, etc.) formed in the metal as a result of the deformation and cooling modes used as well as a significant increase in ferrite grain and a slight decrease in the volume fraction of pearlite colonies.

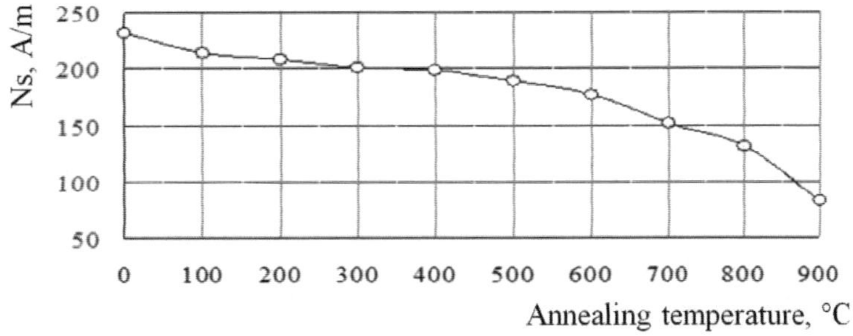

Figure 2. The effect of the heating temperature with cooling in the furnace on the coercive force of steel 10 made using DTT.

IOP Conf. Series: Materials Science and Engineering **866** (2020) 012033 doi:10.1088/1757-899X/866/1/012033

The curves of the temperature dependence of the linear expansion of steel 10 made using thermocyclic forging on the test temperature after annealing for 10 h at 600, 700, 800, and 900 °C and without heat treatment are shown in figure 3.

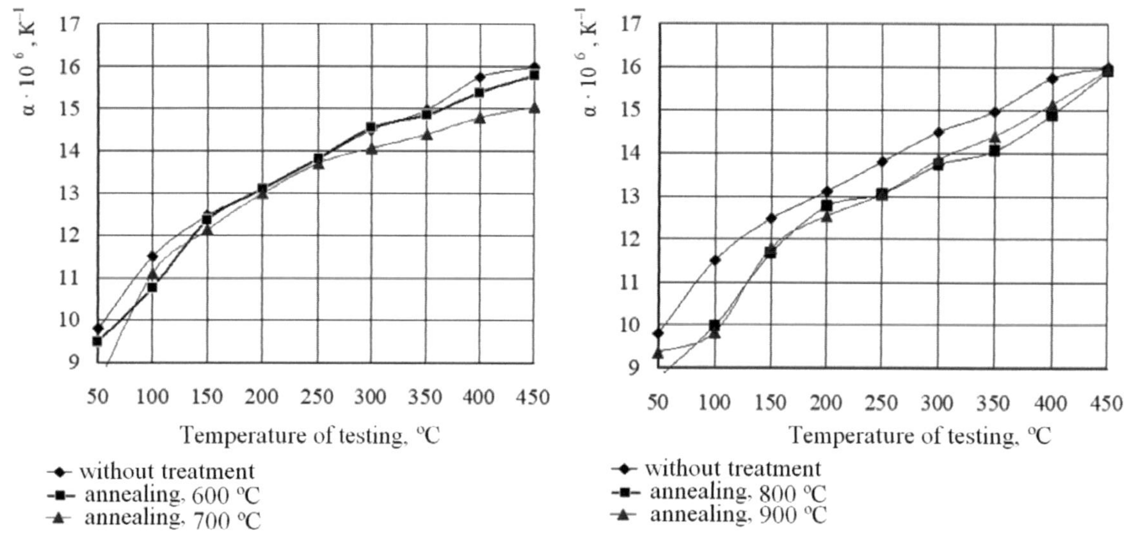

(a) (b)

Figure 3. The effect of annealing temperatures at 600, 700 °C (a) and 800, 900 °C (b) with a holding time of 10 h on the linear expansion of hot-rolled sheet steel 10 made using DTT.

Based on the analysis of the temperature dependence curves of the true thermal expansion coefficient of steel 10 made using preliminary thermocyclic forging, after annealing for 10 h at 600, 700, 800, and 900 °C, graphical dependences of the average coefficient over the test temperature intervals on the annealing temperature are constructed (figure 4).

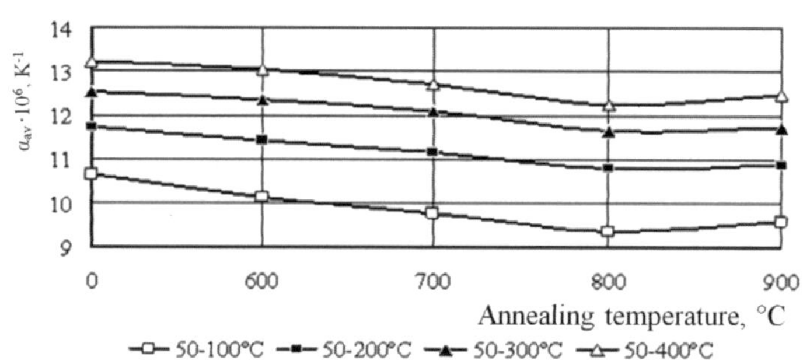

Figure 4. The effect of annealing temperature on the average TCLE in various temperature ranges for testing hot-rolled sheet steel 10 made using DTT.

The obtained dependences $\alpha - T_{test}$ and $\alpha_{av} - T_{test}$ makes it possible to conclude that the tendency of steel 10 to thermal expansion decreases after annealing at all studied temperatures. Moreover, annealing of steel at 800 and 900 °C reduces the temperature coefficient of linear expansion more significantly. This decrease averages over 6% in the entire temperature range of the testings (up to 450 °C). In addition, the most significant annealing at 800 and 900 °C reduces the average TCLE in the low temperature range of 50-100 and 50-200 °C. So, after annealing at 800 °C this decrease is 8 and 12%, and after annealing at 900 ° C – 7 and 12%, respectively, in comparison with thermally untreated samples.

4. Conclusions

1. The use of thermocyclic forging for the manufacture of steel sheet 10sp (thickness 3 mm) increases the value of its coercive force by no more than 8% compared with the industrial mode and does not significantly affect thermal expansion.

2. The subsequent high annealing (900 °C) for 10 h of sheet steel 10 made using thermocyclic forging, allows its coercive force by almost 3 times to be reduced compared to thermally untreated steel, and also to reduce the true values of thermal expansion coefficient on average by more than 6% in the range up to 450 °C.

3. The largest decrease in the average coefficient of thermal expansion of steel 10 after this annealing mode for various test temperatures is observed in the low-temperature zone (from 50 to 200 °C) and amounts to 7-10% compared to its initial state.

References

[1] Fedyukin V K and Smagorinsky M E 1989 *Thermocyclic Processing of Metals and Machine Parts* (Leningrad: Mechanical Engineering) p 255

[2] Prudnikov A N and Prudnikov V A 2015 *Applied Mechaniecs and Materials* **788** 187–193

[3] Prudnikov A N 2013 *Structural and Technological Basis for the Development of Precision Silumin with Regulated Hydrogen Content* (Novosibirsk: NSTU) 40

[4] Afanasyev V K, Chibryakov N V et al 1998 *Pat. of the Russian Federation* No 20130084 appl. 07/07/1998, publ. 05/10/1999

[5] Prudnikov A N 2009 *Steel in Translations* **39** 391–3

[6] Prudnikov A N 2014 *Deformation and Destruction of Materials* **2** 14–20

[7] Prudnikov A N and Prudnikov V A *2015 Metallurgy: Technologies, Innovations, Quality* vol 2 (Novokuznetsk: SibSIU) pp 15–18

[8] Furuya Y and Park Y C 1992 *Nondestructive Testing and Evaluation* **8(1)** 541–554

[9] Bellavin A D, Smagorinsky M E and Shilov I F 1986 *New Materials and Hardening Technologies Based on Progressive Methods of Thermal and Chemical-Thermal Treatment in Automotive Engineering* (Moscow, MADI) pp 86–87

[10] Prudnikov A N, Prudnikov V A and Bogonosov E V 2015 *Metallurgy: Technology, Innovation, Quality* part 2 (Novokuznetsk: SibSIU) pp 35–39

[11] Prudnikov A N and Prudnikov V A 2016 *Actual Problems in Mechanical Engineering* **3** 451–6

[12] Prudnikov V A 2016 *Proc. of the All-Russian Sci. Conf. of Students, Graduate Students and Young Scientists in Natural and Technical Sciences* (Novokuznetsk: SibSIU) pp 188–191

[13] Kekalo I B and Samarin B A 1989 *Physical Metallurgy of Precision Alloys. Alloys with Special Magnetic Properties* (Moscow: Metallurgy) p 496

IOP Conf. Series: Materials Science and Engineering **866** (2020) 012034 doi:10.1088/1757-899X/866/1/012034

Software and hardware for integrated ACS of casting quality

S V Knyazev, A A Usoltsev, D V Skopich, B M Sokolov and N V Oznobikhina

Siberian State Industrial University, 42 Kirova Street, Novokuznetsk, 654007, Russia

E-mail: krookia@mail.ru

Abstract. The implementation of the "Automated system for the operational control of castings production (AS OCCP)" is the first stage and the basis of the integrated automated production control system. It performs three main tasks: control and accounting (the course of production, products, materials, etc.), improvement of the casting quality and the operational control of technological processes. The second stage is based on mathematical processing and analysis of data coming from AS OCCP, which allows the optimal ranges of parameters of technological processes – "Automated system for optimization and analysis of production progress (AS OAPP)" to be determined. AS OAPP consists of two subsystems: quality analysis (QAS) and technology Control (TCS). The first one solves the problems of data analysis and modeling, the second one – real-time calculation of optimal process parameters and forecasting. The tasks of the first and second stages compete for access to different hardware resources. The most critical parameter for the AS OCCP is the performance of the server disk arrays, for the AS OAPP is the processor performance. In both cases, the scaling of the system is effectively solved by parallelizing operations on different servers that make up the cluster, and on different processors (cores) on the same server.

1. Introduction

To collect data on surface defects of castings, a formalized technique has been developed that significantly simplifies the process of technical control and allows the input of results into a computer to be organized in dialogue mode. At the same time, the problem is being solved that virtually eliminates the influence of the subjective perception of the operator-controller on the assessment of signs of casting-technological defects, both at the qualitative and quantitative levels [1, 3-9].

For non-destructive testing of internal casting-technological defects, the most informative in our opinion is the radiation scanning introscope – tomograph [2]. The mode and geometry of transmission were selected through preliminary experiments (figure 1, 2). The control of a complex casting with the help of scanning-type radiation introscopes is easier to automate than others. Scientists from Tomsk Polytechnic University have experience in developing similar equipment and providing it with software.

The implementation of the "Automated system for the operational control of casting production (AS OCCP)" in Rubtsovsk branch of Altayvagon JSC showed that this system allowed casting rejects and the cost of large car casting production to be reduced by an order. Basically, this was due to the full operational control of the production process and the observance of technological instructions by the personnel, as well as a decrease in the influence of the human factor [1,4,8,9].

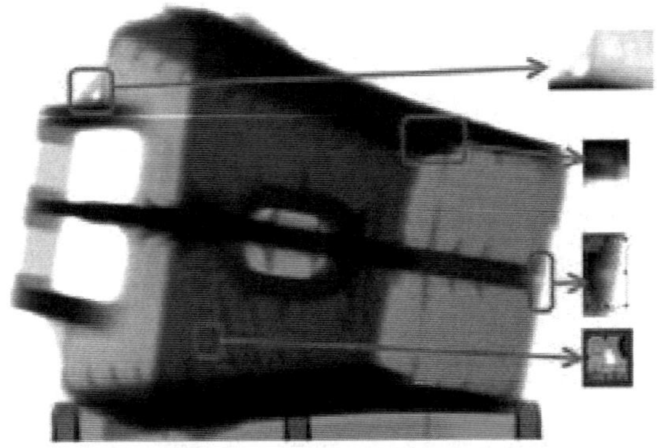

Figure 1. The fragment of a bolster.

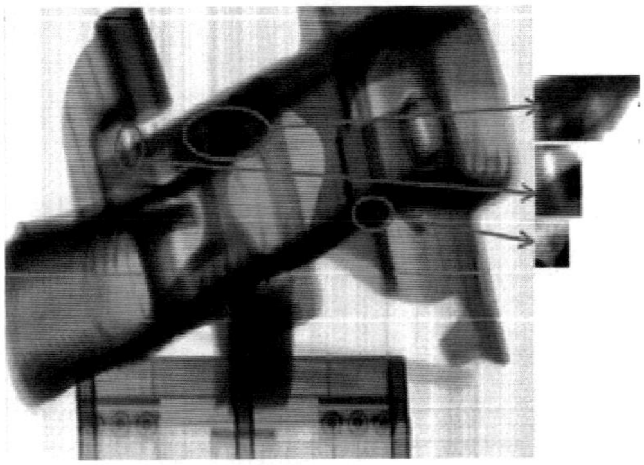

Figure 2. Fragment of the side frame.

2. Results and discussions

AS OCCP is the first stage and the basis of an integrated automated production control system (APCS). The second stage in building an effective automated process control system solves the problems of stabilizing product quality when the external conditions change, for example, the quality of materials, and production optimization (change in the technology in order to reduce costs with constant or higher quality products). The second stage is based on mathematical processing and data analysis, coming from AS OCCP, allows the optimal ranges of parameters of technological processes – "Automated system for optimization and analysis of production progress (AS OAPP)" to be determined.

AS OAPP consists of two subsystems: quality analysis (QAS) and technology control (TCS). The first one solves the problems of data analysis and modeling, the second one – real-time calculation of optimal process parameters and forecasting. The tasks of the first and second stages compete for access to different hardware resources. The most critical parameter for the AS OCCP is the performance of the server disk arrays, for the AS OAPP is the processor performance. In both cases, the scaling of the system is effectively solved by parallelizing operations on different servers that make up the cluster, and on different processors (cores) on the same server.

The construction of cluster systems and parallel computations require appropriate software and hardware, as well as special approaches to the processing and storage of distributed data.

To process images of defects and obtain causal characteristics, the OpenCV software package can be used, which is a computer vision library with an open-source. The example of processing of such image is shown below (figures 3, 4, 5, 6). During processing, the Sobel operator, a Gaussian filter and binarization were used.

IOP Conf. Series: Materials Science and Engineering **866** (2020) 012034 doi:10.1088/1757-899X/866/1/012034

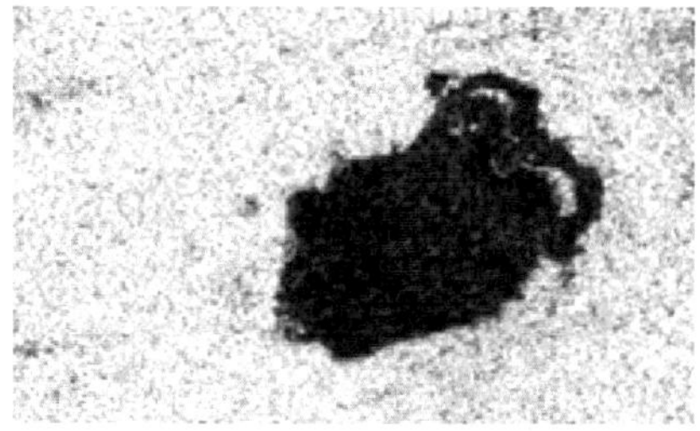

Figure 3. Processing by the Sobel operator.

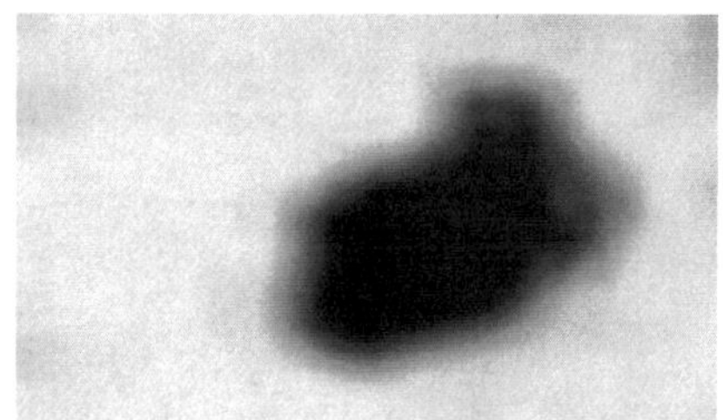

Figure 4. Application of a Gaussian filter.

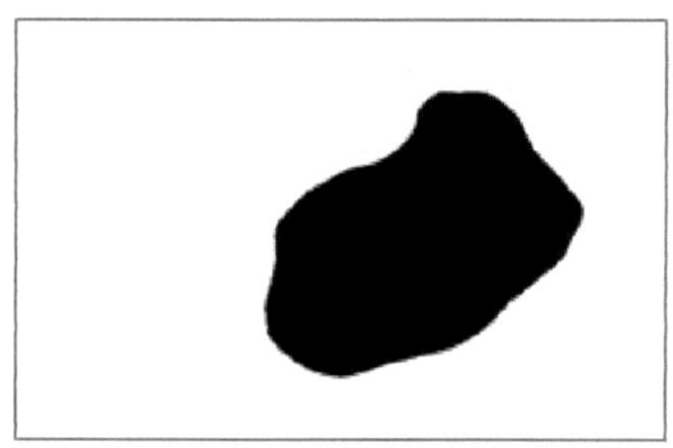

Figure 5. Binarization of the image.

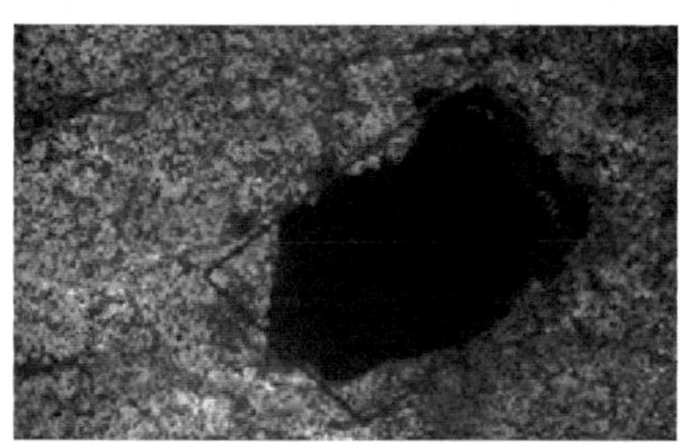

Figure 6. Overlaying a contour on the defect.

According to the found contour, it is possible to calculate the necessary defect parameters. For example, the length of the faces and diagonals of a rectangular region, by which the shape and size of the gas shell can be determined. Further, having a decisive tree of cause-and-effect relationships the cause of the defect can be determined in a first approximation.

The presence of defects in the structure of castings can be influenced by many different factors that are identified on the basis of data statistical analysis. The AS OCCP keeps the history, the value of such factors and defectiveness for each product and group of products of the same smelting. One of the ways to determine the cause of a defect is the clusterization of AS OCCP data and reference of a defect to a particular cluster based on image recognition algorithms using, for example, Euclidean distances.

The task of clusterization is reduced to the determination by an expert method or using various mathematical algorithms of the defects belonging to a particular cluster (data block) according to the total values of dependent factors. Thus, data blocks are formed according to the criterion of the defect cause. Calculating the data block, to which the product defect belongs, can be a very time-consuming operation. It makes sense to place the data clusters on different servers in order to increase the efficiency of pattern recognition and parallelization of search operations.

Modern communication networks are based on asynchronous queries, so, it remains to choose between consistency and data availability, to a large extent it is a choice between relational and NoSQL DBMS. The popularity of NoSQL solutions is growing. The main advantage of such systems is the higher speed and flexibility of working with data due to bypassing the consistency mechanism and SQL syntax.

3. Conclusion

ApacheCassandra DBMS allows a strategy for distributing data among cluster nodes based on keys to be defined. The first distribution strategy is a random markup, distributes data based on a key hash value; the second is an ordinal markup, distributes data over ranges of the key bit values. Thus, it is possible to flexibly distribute various data blocks among the nodes of the system cluster ensuring their high availability. Development of AS OAPP based on a multi-node cluster with installed ApacheCassandra DBMS and usage of Nvidea video cards supporting CUDA technology on each node will be the cheapest and most effective solution. Video cards are selected based on the required number of graphics processors per node.

References

[1] Knyazev S V, Usoltsev A A, Skopich D V, Fatyanova E A and Dolgopolov A E 2016 *IOP Conference Series: Materials Science and Engineering* **150** 012039

[2] Cheprasov A I, Knyazev S V, Usoltsev A A, Dolgopolov A E and Mamedov R O 2016 *IOP Conference Series: Materials Science and Engineering* **150** 012026

[3] Antipenko V I and Knyazev S V 1987 *Soviet Castings Technology* **7** 34–37

[4] Knyazev S V, Skopich D V, Fat'Yanova E A, Usol'Tsev A A and Kutsenko A I 2017 *Izvestiya Vuzov. Ferrous Metallurgy* **60(2)** 128–132

[5] Knyazev S V, Skopich D V, Fat'yanova E A, Usol'tsev A A and Kutsenko A I 2019 *Izvestiya Vuzov. Ferrous Metallurgy* **62(2)** 134–140

[6] Knyazev S V, Skopich D V, Usoltsev A A and Fatyanova E A 2015 *Bulletin of the Siberian State Industrial University* **2(12)** 31–33

[7] Knyazev S V, Antipenko V I, Marchukov V A and Usoltsev A A 1992 *Litejnoe Proizvodstvo* **4** 26–27

[8] Knyazev S V, Usoltsev A A, Skopich D V and Fatyanova E A 2013 *Modern Technologies in Mechanical Engineering* (Penza) pp 111–117

[9] Knyazev S V and Khazan G L 1992 *Litejnoe Proizvodstvo* **12** 24–25

IOP Conf. Series: Materials Science and Engineering **866** (2020) 012035 doi:10.1088/1757-899X/866/1/012035

Features of structure formation and thermal expansion of high alloys of the Al–Si–Cu system

V K Afanasyev, M V Popova, M A Malyukh and A N Prudnikov

Siberian State Industrial University, 42 Kirova Street, Novokuznetsk, 654007, Russia

E-mail: m.popova@rdtc.ru

Abstract: The study results of the copper alloying effect on the microstructure and thermal coefficient of linear expansion (TCLE) of high aluminum alloys for special purposes based on the Al – (20÷30)% Si system are presented. It was found that alloy building of Al – 20% Si with copper in large quantities eliminates the linear expansion anomaly characteristic to the binary alloy and leads to a significant decrease in the thermal expansion coefficient over the entire temperature range of the tests. It was shown that Al–30%Si–20÷40% Cu alloys have stable TCLE in the low-temperature test interval. Metallographic analysis of high alloys showed that copper introduced in amounts of more than 20% contributes to a change in the morphology of the silicon phase and triple eutectic.

1. Introduction

The development and implementation of "breakthrough" technical projects in modern instrument-making industries is determined, first of all, by the materials properties. To date, the best achievement in the field of light alloys with low values of the thermal coefficient of linear expansion (TCLE) is sintered aluminum alloys (SAA) containing 25-30% of alloying elements such as Si, Ni, Fe. The production of alloys of the SAA type is a very expensive process, involving the use of ultrahigh crystallization rates ($\upsilon_{crist.} \geq 10^4$ °C/s), briquetting of powders and hot pressing of blanks [1].

High alloys of the Al– (20÷30)% Si system are special-purpose alloys, because have low TCLE values. To increase the mechanical properties and further more significantly decrease the thermal expansion coefficient in the Al–Si binary alloys, additional alloying elements are introduced, usually having lower TCLE than aluminum [2, 3].

Among alloying elements that reduce the TCLE of Al – Si alloys, copper occupies an important place. Studies of the thermal expansion features of Al–Si and Al–Cu binary alloys show that they can have a wide range of thermal expansion coefficients. The value of the thermal expansion coefficient is primarily determined by the content of alloying elements, as well as technological factors such as melt processing and crystallization conditions [4-7]. Silicon and copper reduce the thermal expansion coefficient of aluminum alloys when their content significantly exceeds the ultimate solubility in aluminum [8, 9].

According to the equilibrium state diagram, in the Al–Si–Cu system, up to 5% of copper can be in solid solution. In industrial alloys based on aluminum, the copper content does not exceed 6% for deformed and 10% for cast alloys. But in the literature there is very little information about the effect of a high copper content on the thermophysical properties of Al–Si alloys. It is known that the increase in the copper content up to 8–10% increases the strength at high temperatures and reduces the creep rate of aluminum, although there is still no consensus on when copper is more effective: if it is in a solid solution or as part of the $CuAl_2$ phase [10, 11]. Therefore, it is of interest to study the thermal expansion of high alloys Al – Si – Cu with a copper content significantly above the solubility limit.

As for the phase composition of Al–Si–Cu alloys, ternary compounds are not formed in this system. The $CuAl_2$ phase and the silicon phase are in equilibrium with the aluminum solid solution [12]. The $CuAl_2$ phase crystallizes directly from the melt at a copper content of 53.3% at a temperature of 591 °C. According to the phase diagram of the Al–Cu state, it has a limited existence interval of 52.5–53.9% Cu, which does not reach the stoichiometric copper content (54.2%), which is due to the defect of the crystal lattice. The $CuAl_2$ phase has a tetragonal lattice; its TCLE is $\alpha_{27\div127}$ = $15.9\cdot10^{-6}$ deg^{-1} and $_{\alpha27\div527}$ = $17.2\cdot10^{-6}$ deg^{-1} [13]. In the Al–Si–Cu ternary system in equilibrium at a temperature of 525 °C and concentrations of alloying elements of 27% Cu and 5% Si, the triple eutectic reaction $L\rightarrow(\alpha)+Si+CuAl_2$ occurs [14].

In the case of nonequilibrium crystallization, the Si phase and the $CuAl_2$ phase are detected at much lower concentrations of copper and silicon compared to the state diagram. Usually, in the molten state, as a result of degeneration of the eutectic, the $CuAl_2$ phase has the form of veins; colonies of triple eutectics ($\alpha+Si+CuAl_2$) are rarely found [15]. During the heat treatment, the silicon phase is more prone to spheroidization than the $CuAl_2$ phase.

2. Materials and methods of research

As an object of study, ternary alloys of the Al–Si–Cu system were chosen at different ratios of Si and Cu. Alloys were smelted in laboratory conditions in furnaces with heaters made of silite, subject to all rules for preparing the charge and conducting melting. The aluminum charge was melted, silicon was introduced into it in the amount of 20 and 30%, as well as copper in the amount of 20, 30 and 40%. Alloys were poured into an aluminum chill mold (t_m = 20 °C). The pouring temperature was 1100 – 1200 °C.

Samples for dilatometric and metallographic studies were made from the obtained ingots. TCLE was determined using a differential optical photo-recording dilatometer of the Schevenar system, the determination error was ± $0.1\cdot10^{-6}$ deg^{-1}. Metallographic analysis was performed on an OLYMPUS GX-5 optical microscope, at magnifications from 200 to 1000 times. Electron microscopy and X-ray microanalysis were carried out using a scanning electron microscope Carl Zeiss AG - EVO 50 Series.

3. Results and discussion

A metallographic analysis of ternary Al – 20% Si – Cu alloys showed that the combined introduction of silicon and copper prompts the formation of smaller primary crystals of the silicon phase than that of a binary alloy (figure 1, a, b). It is seen that between the crystals of the silicon phase there is a dispersed eutectic and solid solution dendrites. It should be noted that in the ternary alloy the size of dendritic branches increases and the distance between them decreases.

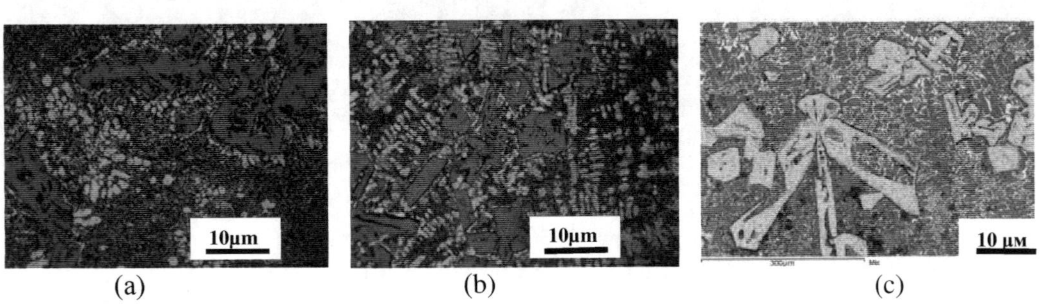

(a) (b) (c)

Figure 1. Microstructure of the Al – 20% Si alloy (*a*); Al-20% Si-% 20 Cu (*b*); Al-20% Si-% 20 Cu (*c*).

The microstructure of the Al–20% Si–20% Cu alloy obtained with a high spatial resolution using a scanning electron microscope allows us to study in more detail the fine structure of the triple eutectic located between the primary crystals of the silicon phase (figure 1, c). It can be seen that the eutectic has a fine needle structure with an average particle size of eutectic silicon of not more than 10 μm. Primary silicon crystals have a predominantly polyhedral shape, but acquire a complex configuration with cavities inside. Sections of α-solid solution are in the form of dendrites.

The results of X-ray microanalysis of the Al–20 % Si–% 20 Cu alloy showed that in some parts of the eutectic a high percentage ratio of Cu and Al is observed (37% and 52%, as well as 41% and 47%, respectively), which confirms the presence of particles in them $CuAl_2$ phases. Thus, silicon is present in the composition of the triple eutectic (α+Si+$CuAl_2$) in the form of needle-shaped particles, and copper is mainly included in the eutectic in the form of the $CuAl_2$ phase (figure 2).

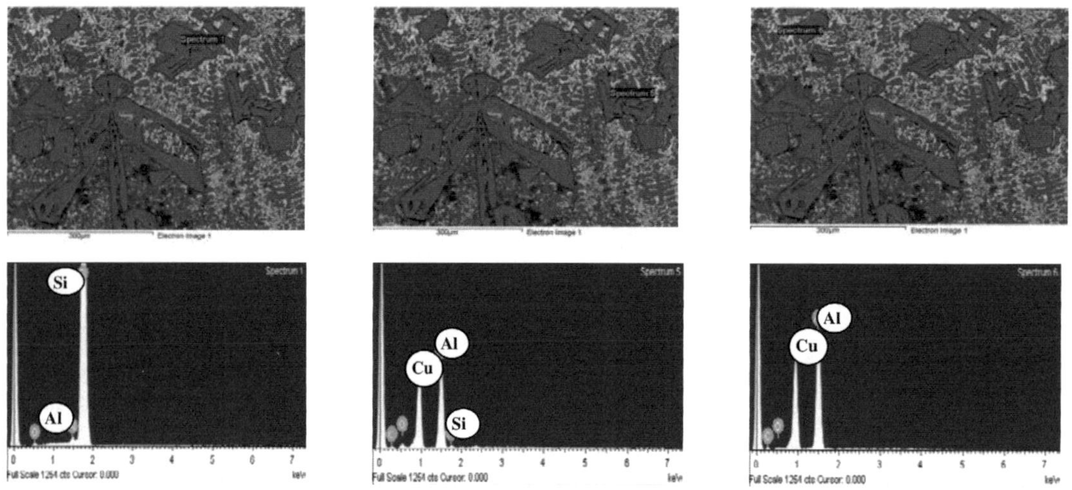

(a) (b) (c)

Figure 2. Microstructure and elemental composition of the Al–20 % Si–20 % Cu alloy in typical sample zones: siliceous phase (a); $CuAl_2$ phase (b, c).

The effect of copper on the microstructure of Al–30% Si–Cu alloys was studied. The results of metallographic analysis are presented in figure 3. It was found that when from 20 to 30% of Cu is introduced into the alloy, the crystals of the siliceous phase mainly have the form of plates, between which a large needle eutectic and dendrites of the α-solid solution are observed. With the introduction of 40% Cu, the crystals of the siliceous phase acquire an irregular shape with many sites of increased etchability. Small sections of the eutectic are visible between the crystals of the siliceous phase, but dendrites of the α-solid solution are completely absent.

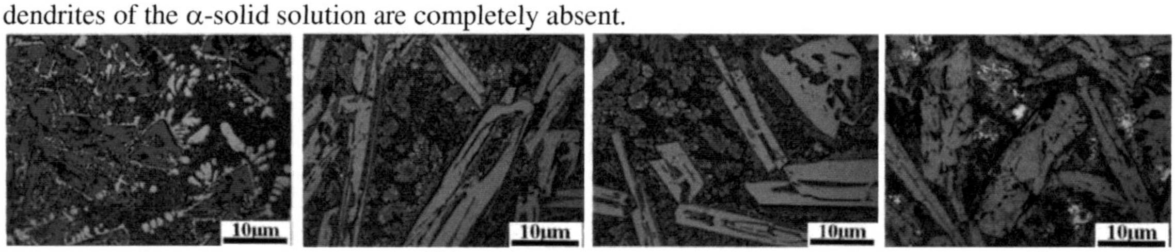

(a) (b) (c) (d)

Figure 3. Microstructure of alloys based on Al–30% Si with different copper contents: without Cu (a), 20% Cu (b), 30% Cu (c), 40% Cu (d).

To study the structural features of high alloys, a metallographic analysis was performed using SEM (figure 4). It was established that the introduction of copper into the Al–30% Si alloy, in the microstructure of which, in addition to the eutectic, large particles of the silicon phase are initially present, contributes to an increase in the size of these particles. It was found that alloying of such alloys with copper in large quantities (20–40% Cu) leads to a decrease in the solubility limit of silicon in aluminum, i.e. the conditions for the formation of alloys change.

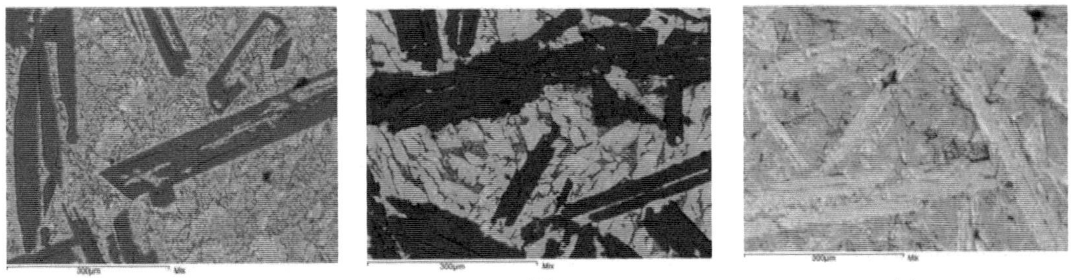

(a) (b) (c)

Figure 4. The microstructure of alloys based on Al–30% Si with different Cu contents: 20% Cu (a), 30% Cu (b), 40% Cu (c).

The results of X-ray microanalysis of Al–30% Si–20–40% Cu alloys showed that a high percentage ratio of Cu and Al is observed in certain zones of the eutectic. Figure 5 shows that between the crystals of the silicon phase there are sections of a Cu solid solution in Al and zones with a high copper content (40% Cu and 49% Al) or (38% Cu and 53% Al), which are the $CuAl_2$ phase. Such changes in the structure lead to a significant decrease in TCLE from $12.4\cdot10^{-6}$ to $6.0\cdot10^{-6}$ deg^{-1} in the range of 50–150 °C (figure 6).

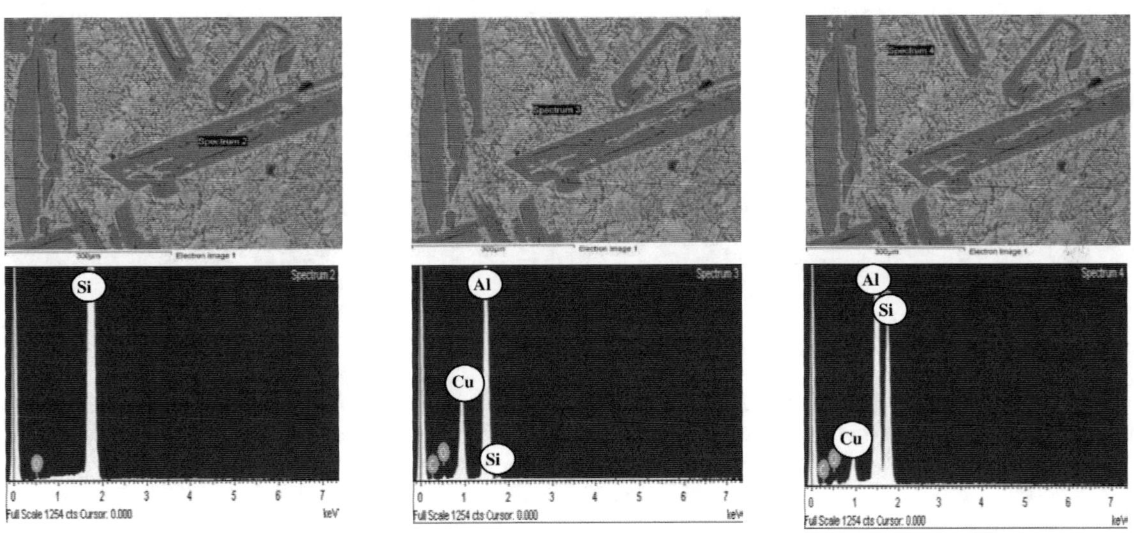

(a) (б) (в)

Figure 5. Microstructure and elemental composition of Al–30% Si–20% Cu alloys in typical sample zones: siliceous phase (a); $CuAl_2$ phase (b); α-solid solution (c).

The effect of copper on the thermal expansion of Al–20% Si–Cu and Al–30% Si–Cu alloys was studied. It was established that the Al–20% Si–20% Cu alloy has stable TCLE in the low-temperature test interval. An increase in the copper content from 20 to 50% contributes to a smooth decrease in the thermal expansion coefficient of ternary alloys over the entire temperature range of the tests, as can be seen in figure 6, a.

It should be noted that alloying Al–20% Si alloy with copper in large quantities eliminates the linear expansion anomaly characteristic of the double alloy and leads to a significant decrease in the thermal expansion coefficient over the entire temperature range of the tests. So, the average TCLE of the Al–20% Si–20% Cu alloy in the range of 50–200 °C is $14.4 \cdot 10^{-6}$ deg^{-1}, while the TCLE of the Al–20% Si–30% Cu alloy is 13, $1 \cdot 10^{-6}$ deg^{-1}. For the Al–20% Si–40% Cu alloy, a decrease in the thermal expansion coefficient to $9.6 \cdot 10^{-6}$ deg^{-1} was established. The greatest decrease in the thermal expansion coefficient is observed for the Al–20% Si–50% Cu alloy; here, the thermal expansion coefficient decreases to $6.0 \cdot 10^{-6}$ deg^{-1}.

Most significantly, copper stabilizes the high-silicon alloys TCLE with a content of 30–50% Si. Figure 6b shows that alloying of the Al – Si binary alloy with copper in the amount of 20% allows us to obtain slightly varying TCLE in the range 50–400 °C: $\alpha = 12.0 \div 12.5 \cdot 10^{-6}$ deg^{-1}. An increase in copper content to 30% makes it possible to reduce the TCLE of the ternary Al–30% Si–30% Cu ternary alloy over the entire test temperature range: from $\alpha_{50} = 6.0 \cdot 10^{-6}$ deg^{-1} to $\alpha^{450} = 6.8 \cdot 10^{-6}$ deg^{-1}.

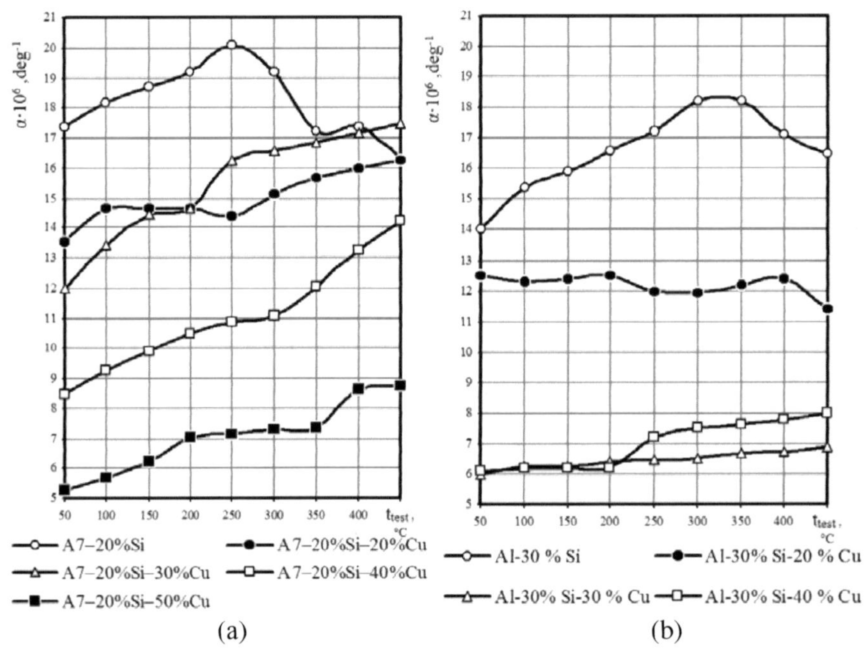

Figure 6. Thermal expansion of Al–Si–Cu alloys: Al–20% Si–Cu (*a*), Al–30% Si–Cu (*b*).

It should be noted that for the Al–30% Si alloy, the introduction of copper of more than 30% is inappropriate since the increase in copper content up to 40% does not lead to a further decrease in the TCLE values, but its increase is observed in the high-temperature test interval.

4. Conclusion

With the introduction of large amounts of copper, the conditions for the structure formation of hypereutectic alloys Al–(20÷30)% Si change. Copper introduced in amounts of more than 20% contributes to an increase in size and a change in the shape of the silicon phase, as well as a change in the structure of the triple eutectic.

Building of alloys Al– (20÷30)% Si with copper in the amount of more than 20% leads to a significant decrease in the thermal expansion coefficient of ternary alloys over the entire temperature range of the tests. It was established that the Al–30% Si–20% Cu alloy has practically constant thermal expansion coefficients over the entire test temperature range of $12.5 \div 12.0 \cdot 10^{-6}$ deg^{-1}, which is typical for invar alloys [16]. This allows such alloys to be used when paired with steel products.

Thus, the joint alloying of aluminum with silicon and copper allows low and stable TCLE in the low and medium temperature test ranges to be obtained: Al–30% Si–20% Cu alloy has $\alpha = 12.0 \div 12.5 \cdot 10^{-6}$ deg^{-1} in the range of 50–400 °C; Al–30% Si–30% Cu alloy has $\alpha = 6.0 \cdot 10^{-6}$ deg^{-1}.

Acknowledgments
All electron microscopic studies presented in this article were carried out in the collective use center "Structure, Mechanical and Physical Properties of Materials" (TsKP SSM) of Novosibirsk State Technical University (Head – Professor Bataev Vladimir Andreevich).

References

[1] Gopienko V G, Smagorinsky M E and Gregory A A 1993 *Sintered Materials from Aluminium Powders* (Moscow: Metallurgy) p 320
[2] Novikova S I 1974 *Thermal Expansion of Solid Bodies* (Moscow: Nauka) p 292
[3] Afanasyev V K, Popova M V, Ruzhilo A A and Frolov V F 2002 *Russian Metallurgy (Metally)* **6** 539–44
[4] Afanasyev V K, Gorshenin A V, Popova M V, Prudnikov A N and Starostina M A 2010 *Metallurgy of Machine Building* **6** 23–6
[5] Popova M V, Frolov V F and Lyubushkina A N 2003 *Izvestiya Ferrous Metallurgy* **2** 38–40.
[6] Afanasyev V K, Popova M V, Malyukh M A et al 2018 *IOP Conf. Series: Materials Sci. and Eng.* **411** 012010
[7] Afanasyev V K, Popova M V, Dolgova S V, Gorshenin A V and Malyukh M A 2019 *Metallurgist* **63(1–2)** 87–95
[8] Afanasyev V K, Ruzhilo A A, Lyubushkina A N and Popova M V 2003 *Izv.Vuzov. Ferrous Metallurgy* **10** 16–7
[9] Popova M V 2004 *Processing of Metals* **3(24)** 16–9
[10] Zolotarevskiy V S and Belov N A 2005 *Metal Science of Cast Aluminium Alloys* (Moscow: MISiS Press) p 376
[11] Stroganov G B *High-Strength Casting Aluminium Alloys* (Moscow: Metallurgiya) p 216
[12] Mondolfo L F 1979 *Structure and Properties of Aluminium Alloys* (Moscow: Metallurgiya) p 639
[13] Kikoin I K 1976 *Tables of Physical Quantities: Reference book* (Moscow: Atomizdat) p 1006
[14] Hatch J E 1989 *Aluminum: Properties and Physical Metallurgy: Reference book* (Moscow: Metallurgiya) p 422
[15] Prigunova A G, Belov N A et al 1996 *Silumines. Atlas of Microstructures and Fractograms of Industrial Alloys: Reference book* (Moscow: MISiS Press) p 175
[16] Afanasyev V K, Popova M V et al 2006 *Invars* (Novokuznetsk: SibSIU) p 126

IOP Conf. Series: Materials Science and Engineering **866** (2020) 012036 doi:10.1088/1757-899X/866/1/012036

Modeling of the chromium compounds synthesis in the nanocrystalline state

L S Shiryaeva and G V Galevskiy

Siberian State Industrial University, 42 Kirova Street, Novokuznetsk, 654007, Russia

E-mail: kafcmet@sibsiu.ru

Abstract. In this work, the technology of nanocrystalline chromium compounds by carbidization of chromium-containing raw materials with hydrocarbons in a plasma flow generated in a three-jet direct-flow reactor was selected as the object of study. The computer program was developed that implements a comprehensive multifactor model of the plasma synthesis process and allows multivariate engineering and research calculations of the effective carbidization parameters of chromium-containing raw materials.

1. Introduction

Computer modeling is one of the most effective tools of cognition, analysis and design, which are available to specialists responsible for the development and operation of complex chemical and metallurgical industries. The idea of computer modeling is simple and at the same time attractive, it allows the researcher to experiment with objects in cases where it is almost impossible or inexpedient to do this on a real object. Plasma synthesis is a complex metallurgical process for research. This is due to its short duration, high temperatures and the small volume of the carbide formation zone. Under such conditions, a model-mathematical approach seems to be very promising, involving the thermodynamic modeling of synthesis processes, the simulation of the interaction of streams of chromium-containing raw materials and heat carrier gas, and an experimental study with the method of the planned experiment.

Comparison of different technologies for producing powders of refractory compounds indicates that, to achieve the nanoscale, technologies based on the use of highly concentrated energy flows for gasification of raw materials and the formation of the target product during volumetric condensation from the gas phase are primarily demanded. Among such technologies, the plasma method is characterized by comparative ease of implementation and is the most studied and competitive [1-6]. The processes of plasma synthesis can conditionally be divided into three stages: 1) the evolution of the feedstock, including its movement in the flow of heat carrier gas, heating and phase transitions; 2) chemical reactions, i.e. synthesis itself; 3) the formation of a dispersed product (condensation, coalescence, crystallization, coagulation).

High speeds of the reaction media, which are 30-60 m/s in the channels of three-jet direct-flow reactors, depending on the operating parameters of the plasma torches and thereby contributing to a reduction in the residence time of raw materials at temperatures of its transition to the vapor phase (10-4 - 10-5 s), impose constraints on the completeness of the first stage, which, in the case of using dispersed raw materials, limits and determines the quantitative yield of the target product in this regard. In this regard, of all the factors affecting the degree of feedstock conversion, namely: the thermophysical and thermodynamic properties of the plasma and feedstock, the ratio of the mass flow rates of the plasma forming gas and feedstock, the initial plasma temperature, the design features of the reactor, specific energy consumption, the organization of the process of mixing dispersed feedstock with plasma, etc., – the main limiting factor is the heat exchange of a high-temperature gas flow with the particles of the processed material moving in it.

Therefore, high degrees of conversion of raw materials at the first stage of synthesis can be achieved with such hydrodynamic and energy parameters of the reactors that provide for a very limited time the supply of sufficient energy from the heat carrier to the dispersed raw materials for the required thermophysical and physicochemical transformations [7-9]. However, an experimental study of this stage of synthesis is difficult due to its short duration and the lack of diagnostic and monitoring tools that work reliably under such conditions and, as a rule, comes down to a model-mathematical one, which makes it possible, with a certain "idealization" of the process, to determine the effect on the transition degree of raw materials to the gas phase of the temperature and velocity of the gas flow in the raw materials input zone, consumption of reagents, input conditions and particle sizes of the raw material [1].

In the present work, as the object of study we selected the modeling technology for the production of chromium carbide by carbidization of chromium-containing raw materials with hydrocarbons in the plasma flow generated in a three-jet direct-flow reactor.

2. Modeling results

Technological options, parameters and indicators of chromium carbide synthesis (forecast) are presented in table 1. The calculation took into account possible losses of raw materials in the reactor in the amount of 4% and losses of synthesis products in the capture system in the amount of 5%. An analysis of the data in the table shows that the most promising is the technological option for producing carbide by carbidization of chromium powder, the technical and economic indicators of which are significantly higher than expected in other options.

Table 1. Technological options, parameters and indicators for the synthesis of chromium carbide (forecast).

Technological parameters and indicators	Technological options		
	chromium carbidization	chromium oxide reduction Cr_2O_3	chromium chloride reduction $CrCl_3$
Reactor power, kW	150	150	150
Consumption of plasma-forming gas (nitrogen), kg/s	$9 \cdot 10^{-3}$	$9 \cdot 10^{-3}$	$9 \cdot 10^{-3}$
Initial temperature of the plasma flow, K	5400	5400	5400
Quenching temperatures, K	2000	2000	2000
Mass consumption concentration kg of raw material / kg of nitrogen	0.10	0.12	0.15
The degree of raw materials conversion to carbide	0.96	0.96	0.96
The yield of carbide, % of the mass	91.4	90.0	91.8
Productivity for chromium carbide, kg / h	3.42	2.79	1.69
The intensity of production, $kg/h \cdot m^3$	2011	1641	994
Consumption of chromium-containing raw materials, kg/kg	0.90	1.30	2.76
Methane consumption kg/kg	0.84	2.74	0.89
Gas consumption – heat carrier, kg/kg	9.47	11.6	19.70
Specific consumption of energy, kW·h/kg	43.85	53.76	88.70
Cooling water consumption, m^3/kg	1.08	1.32	2.19

3. Computer program for the model implementation

Based on the results of theoretical and experimental studies, the computer program "Generalized Model of Carbide Formation in Plasma Synthesis" [10] has been developed, designed to conduct multivariate engineering and research calculations of the content of chromium carbonitride and basic impurities in the synthesis products, their level of dispersion depending on the main technological parameters. Variable synthesis parameters are: the initial temperature of the plasma stream, the temperature of the quenching of the synthesis products, the amount of hydrocarbon for chromium carbidization, the concentration of hydrogen, atomic nitrogen in the plasma-forming gas. The calculation results are analyzed. The report is generated for each calculation option. The structural composition of the program includes: an internal DBMS, an interface for entering data, report output in Microsoft Excel.

The usage of software product makes it possible to carry out multivariate engineering and research calculations determining the content of chromium carbonitride and basic impurities, the level of dispersion and identifying the dependence of the content on the main parameters. To create a software product, the necessary information was collected on the development and development of the technology of plasma-metallurgical production of chromium carbonitride. A database has been created to store the necessary process data. Data from the database is transferred to the program, where it is used and processed.

The software makes it possible to calculate and present graphically the results of a study of the dependence of the content of chromium carbonitride in the synthesis products on the initial temperature of the plasma flow, the quenching temperature of the synthesis products, the amount of hydrocarbon for chromium carbidization, the concentration of hydrogen and atomic nitrogen in the plasma-forming gas.

The software product is based on the calculation of the content of chromium carbonitride in the synthesis products, carried out using field model data. The field model data obtained during the implementation of the plasma metallurgical production technology of chromium carbonitride are the values of the following synthesis parameters: the initial temperature of the plasma flow, the temperature of the quenching of the synthesis products, the amount of hydrocarbon for chromium carbidization, the concentration of hydrogen and atomic nitrogen in the plasma-forming gas.

When conducting the computational process, the listed synthesis parameters are input data. Data entry is carried out by changing the current value within the studied range under the supervision of a teacher, to whom the corresponding disciplines are assigned, which excludes the entry of incorrect data.

The content of chromium carbonitride in the synthesis products is described by the first-order equations. The result of the calculations is a graph that determines the nature of the dependence of chromium carbonitride content in the synthesis products and basic impurities on its conditions. Then an analysis of the results is carried out. Using a software product allows to be analyzed: the yield of chromium carbonitride and its associated phases; the dispersion level of chromium carbonitride; the effect of the initial temperature of the plasma flow; the effect of the amount of hydrocarbon for chromium carbidization; the effect of the concentration of hydrogen in the plasma-forming gas; the effect of the concentration of atomic nitrogen in the plasma-forming gas.

Based on the analysis, the following conclusions are made:

- on predicting the content of chromium carbonitride and impurities in the synthesis products depending on the main parameters (initial temperature of the plasma stream; temperature of quenching of the synthesis products; amount of hydrocarbon for chromium carbidization; concentrations of hydrogen in and atomic nitrogen in the plasma-forming gas);
- about the optimality of synthesis conditions and obtaining the required content of chromium carbonitride in its products;
- on the quantitative prediction of the chromium carbonitride dispersion.

A report is generated for each option.

The purpose of conducting multivariate engineering and research calculations is to optimize the formation of chromium carbonitride by determining the content of chromium carbonitride and basic impurities, the level of dispersion, and identifying the dependence of the content of synthesis products on its conditions.

The main tasks are: ensuring maximum yield of the target product; determination of factors affecting the composition of the product; search for effective control actions to obtain the desired content of chromium carbonitride in the synthesis products; development of user skills with the most common applications; independent performance by the user (student) of the work and the ability to control it by the teacher with mandatory user registration.

The program uses elements of an object-oriented approach and technology of traditional structural design. The program algorithm should provide calculation and graphical representation of the results of the study of the dependence of chromium carbonitride content in the synthesis products on its parameters. This software is installed on the computer using the Setup.EXE installation file. The main working window contains all the main bookmarks necessary for the user's further work, with the help of which he selects chromium-containing raw materials and calculation options (figure 1).

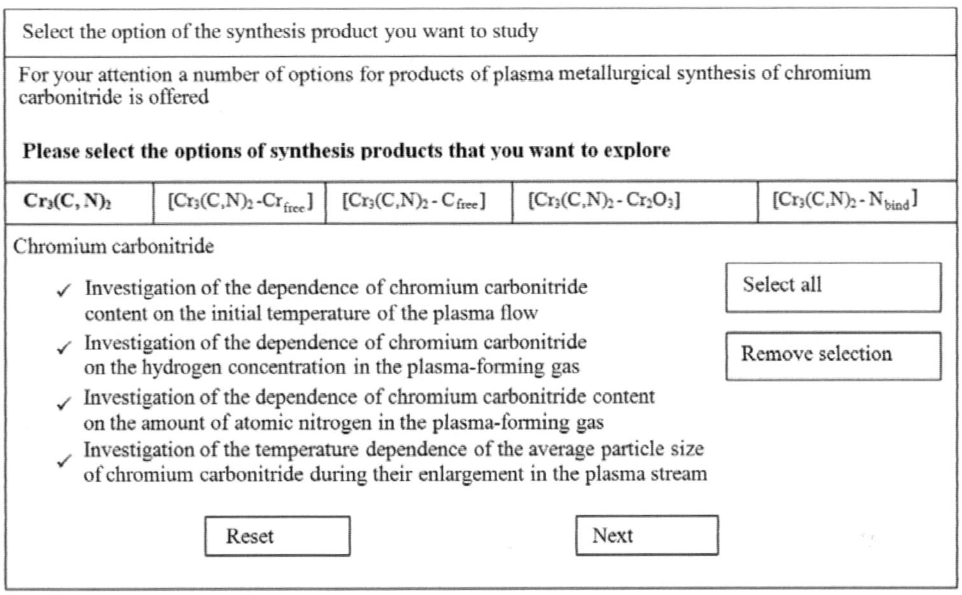

Figure 1. The main working window of the program.

The software product allows a report template in Microsoft Excel to be created, which presents the calculation results and the resulting graphical dependencies. An alternative output of the received information to a screen, printer or file located on an external medium is possible. The following is information on the content in products of plasma metallurgical synthesis of chromium carbonitride Cr_{free}, C_{free}, Cr_2O_3, N_2.

4. Conclusion

The developed computer program is a complete software product that helps to carry out multivariate engineering and research calculations to optimize the parameters of the plasma metallurgical production of chromium carbide, as well as research on the creation of new highly efficient processes using refractory nanocarbides, nanoborides and their compositions.

References

[1] Nozdrin I V, Rudneva V V, Shiryaeva L S and Terentyeva M A 2012 *Izv. Vuzov. Ferrous Metallurgy* **2** 13–18
[2] Nozdrin I V 2011 *Izv. Vuzov. Ferrous Metallurgy* **10** 3–7
[3] Shiryaeva L S, Galevsky G V, Nozdrin I V and Rudneva V V 2017 *Bulletin of Irkutsk State Technical University* **10** 177–188
[4] Tsymbal V P 2006 *Mathematical Modeling of Complex Systems in Metallurgy* (Kemerovo,

Moscow: Rossiyskiye Universitety, Kuzbassvuzizdat) p 431
[5] Tseluikin V N 2017 *Composite Electrochemical Coatings: Deposition, Structure and Operational Properties. Advances in Nanotechnology* 67–90
[6] Shiryaeva L S, Nozdrin I V and Galevsky G V 2015 *IOP Conference Series: MSE* 012004
[7] Karpov I V, Ushakov A V et al 2017 *Tech. Physics. The Russian J. of Applied Physics* **62(1)** 168–173
[8] Mossa A L and Burov I S 1980 *Processing of Dispersed Materials in Plasma Reactors* (Minsk: Nauka i Tekhnika) p 208
[9] Rudneva V V, Galevskii G V and Yurkova E K 2007 *Steel in Translation* **37(2)** 115–118
[10] Shiryaeva L S, Nozdrin I V et al 2012 *Cert. No. 18396 on Registration of the Electronic Resource "Program "Generalized Model of Carbide Formation in Plasma Synthesis" Version 1.0.2" in the joint fund of electronic resources "Science and Education" RAO* (Moscow: INIPI)

Features of treatment of metal surface layers by pulsed plasma jets formed during electrical explosion of conductors

E A Budovskikh, L P Bashchenko and S V Raykov

Siberian State Industrial University, 42 Kirova Street, Novokuznetsk, 654007, Russia

E-mail: budovskikh@mail.ru

Abstract. The results analysis of the experimental studies obtained by the methods of modern physical materials science of metals and alloys surface layers after electroexplosive alloying and coating is presented.

1. Introduction

One of the promising methods of surface treatment that have been developed in recent years is electric explosive alloying (EEA) and electric explosive coating (EEC) [1 – 3].

2. Technological processes of surface alloying and coating

The diagram of the realized technological processes of surface treatment on EVU 60/10 installation is shown in figure 1 (here q is the power density absorbed by the surface during processing and determined by the charge energy; m is the mass of the exploding conductor and powder sample placed in the explosion zone). Coating with powder particles is possible, coatings with condensed explosion products, application of thin films from explosion products, and hardening from a solid state are possible in the first zone, when the absorbed power density is 10^3 W/cm^2, depending on the mass of the exploding conductor. The mass of the exploding conductor of approximately 100 mg provides coating by powder particles.

Reduction in the mass up to values 10 mg provides coating with condensed explosion products. With a further decrease in the mass of the exploding conductor, the content and dispersion of the condensed particles of the explosion products in the jet decreases. In this case, it is possible to deposit thin films on the irradiated surface. For small values of the mass of the exploded conductor, a positive treatment effect is also achieved due to the pulsed hardening of the surface layers from a solid state. In the second zone, where the absorbed power density is approximately 10^5 W/cm^2, treatment with powder coatings makes it possible to conduct reinforcement of the melted surface layers with particles of various substances (carbides, oxides, borides, etc.) placed in the explosion zone and transferred to the irradiated surface.

In the case when powder samples are not used, it is possible to carry out such type of EEA in which the melt formed on the surface is saturated with the products of the explosion of conductors, the material of which can be, for example, thin foils of metals and alloys, carbon and other fibers.

Prospects for further scientific research and practical development in the field of EEA and EEC are associated with the electric explosive formation of new structural phase states on the surface of metals and alloys using heat-sensitive components. For example, in the case of electroexplosive titanium and nickel plating of the aluminum substrate, despite the pulsed character of processing with a duration of 100 μm, surface layers 150–200 μm thick are formed.

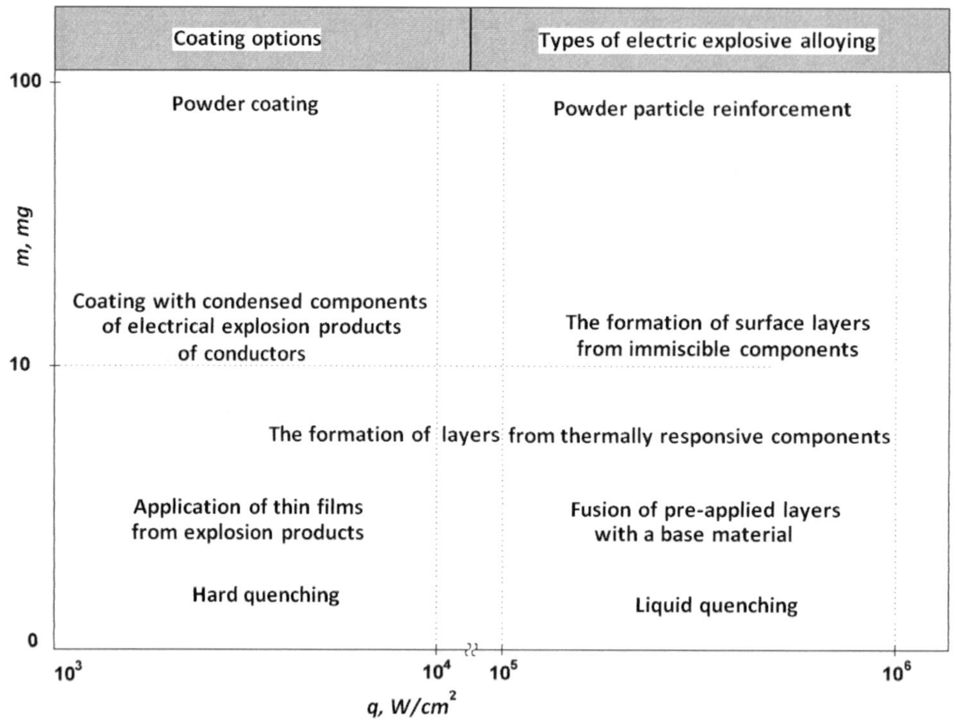

Figure 1. Diagram of technological processes of surface alloying and protective coatings application using multiphase plasma jets implemented on an EVU 60/10 installation.

This is due to the fact that during the interaction of the components, in addition to heat release, their active convective mixing occurs. It is in particular proved by the structural features near the interface between the substrate and the surface layer formed during EEA (figure 2). The acceleration of the interaction processes of alloying elements with the substrate material due to heat generation and convective mixing is accompanied by the synthesis of hardening phases (figure 3).

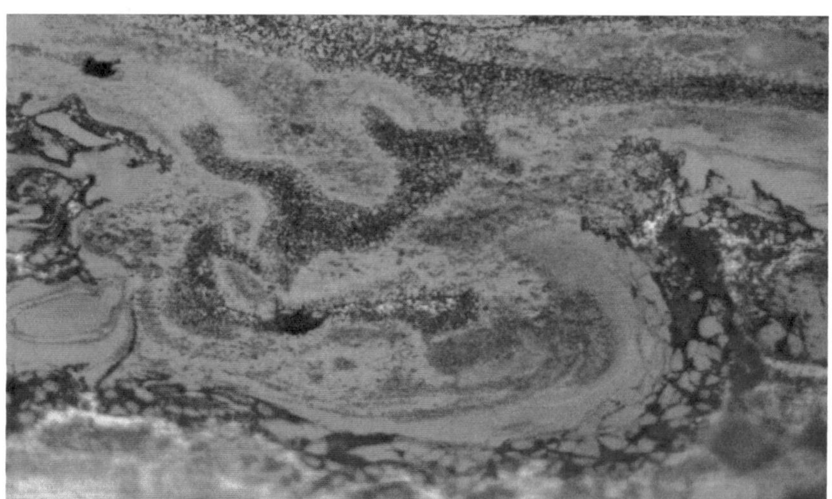

Figure 2. The image of the interface structural features of electric explosive titanium-plating zone of aluminum substrate. Light microscopy, × 1000.

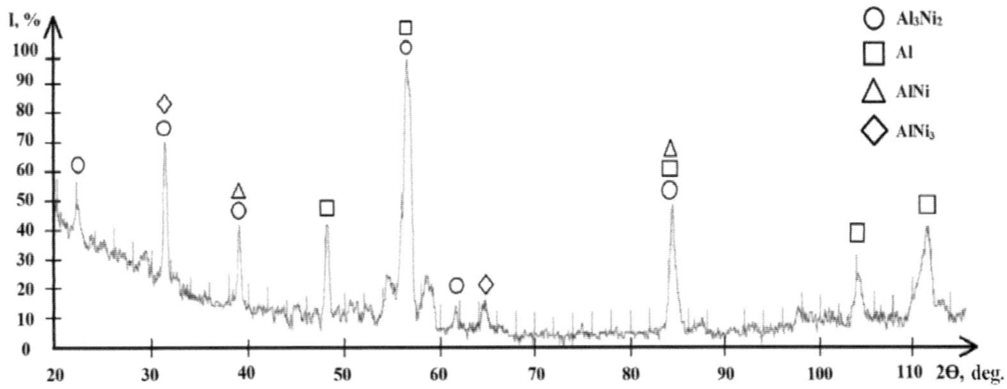

Figure 3. The surface diffraction pattern of electroexplosive nickel-plating zone of aluminum substrate.

3. The results and consideration of issues on electroexplosive carburization and titanium carboboronation

Electroexplosive carburization and titanium carboboronation make it possible to obtain modified surface layers with a thickness up to 30 μm [4]. During EEC, coatings with a thickness of 200–250 μm are formed on the surface of a titanium substrate using a composite electric conductor consisting of titanium foil, carbon fibers, and amorphous boron powder, containing titanium carbide particles synthesized during the spraying process and uniformly distributed over the layer depth. In this case, the interface between the coating and the substrate is even, while the coating has a metallurgical bond with the substrate. It can be seen (figure 4) that the particles of the main phase have a dendritic morphology. The first-order axes reach a length of 10–20 μm. Axes of the second order are weakly expressed. Fine particles with characteristic sizes of several micrometers are observed between the axes.

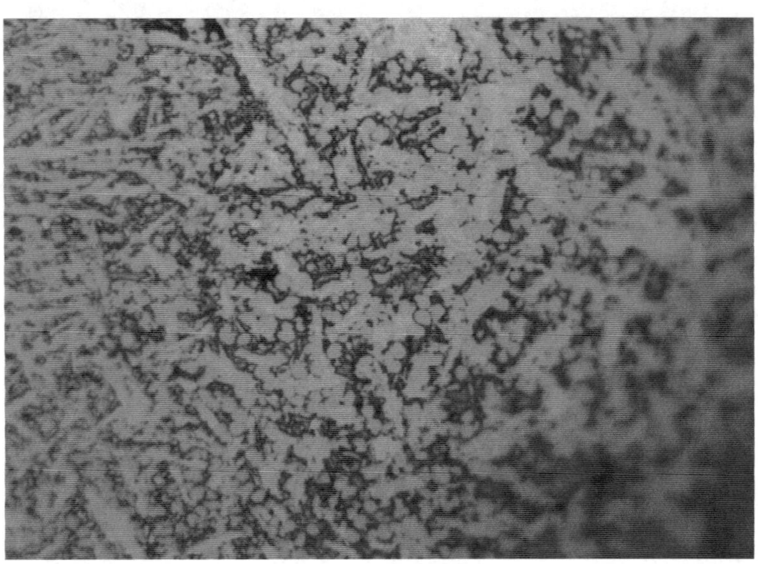

Figure 4. Features of the crystallization structure of the Ti–C–B coating systems. Light microscopy, × 200.

The phase composition of the coating is formed mainly by two phases – titanium carbide TiC and titanium diboride TiB_2 (figure 5).

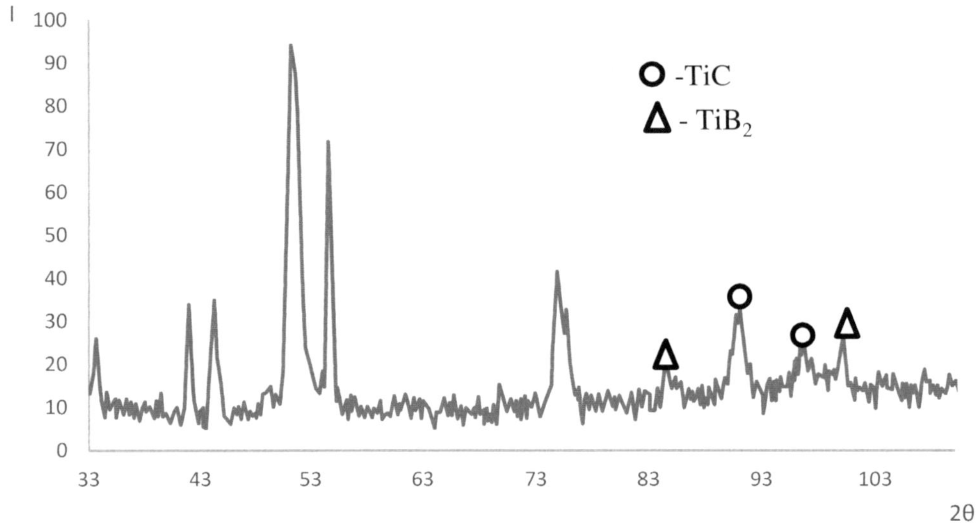

Figure 5. Diffraction pattern of the Ti–C–B coating system.

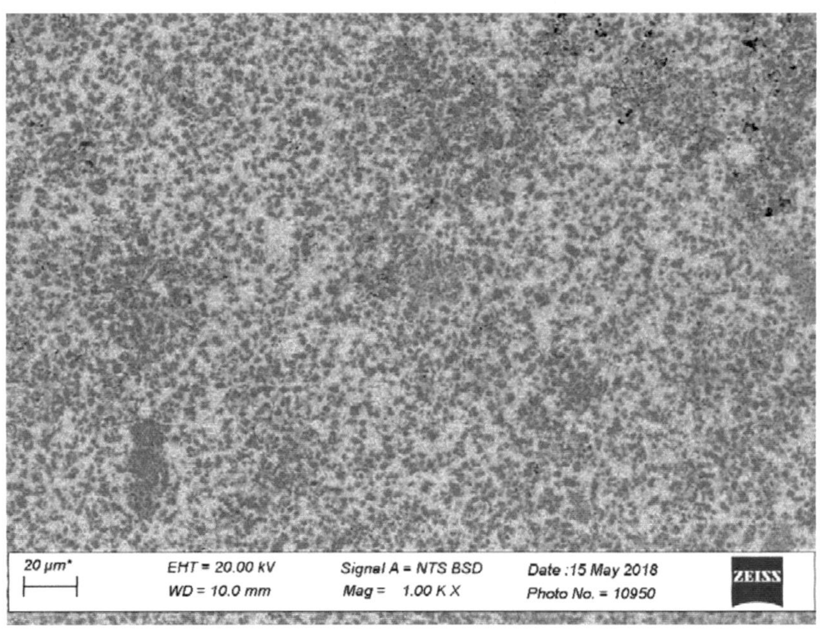

Figure 6. Structural features of electroexplosive coatings of the Ti – B system. Scanning electron microscopy.

During EEC on the titanium substrate using a composite electric conductor consisting of titanium foil and amorphous boron powder, coatings are formed mainly hardened by particles of titanium diboride TiB_2 (figures 6 and 7). In this case, the particles of the strengthening phases located in the viscous metal matrix have predominantly equiaxial morphology with characteristic sizes of 1-3 µm.

IOP Conf. Series: Materials Science and Engineering **866** (2020) 012037 doi:10.1088/1757-899X/866/1/012037

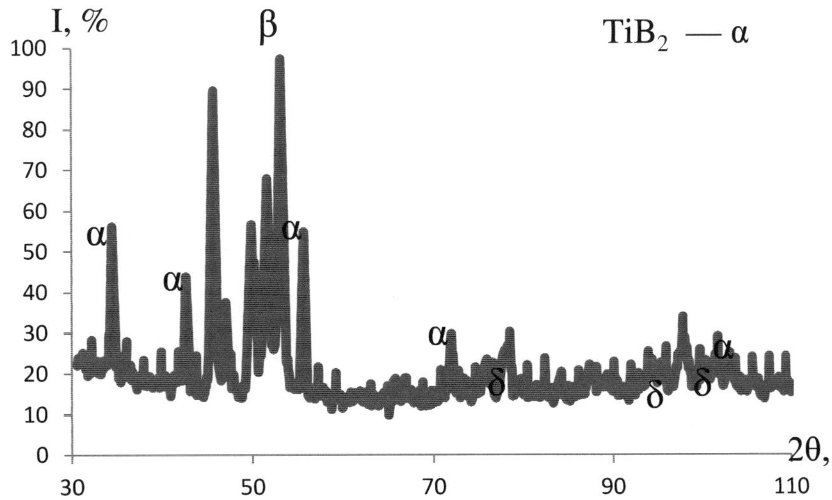

Figure 7. Diffraction pattern of the Ti – B system coating.

4. Conclusion

The most promising directions for the practical use of EEA and EEC are the following:

1. Hardening of cutting and stamping metalworking tools (drills, cutters, punches, dies, etc.).
2. Hardening of drawing, drilling and mining cutting tools from tungsten carbide hard alloys.
3. Surface protection of the copper electrical contacts from electroerosive destruction. Currently, more than a dozen types of coatings have been developed for the W – Cu, Mo – Cu, TiB_2 – Cu, W – C – Cu and Mo – C – Cu systems. The maximum surface area to be treated is 30 cm^2.
4. Improvement of the wear and heat resistance of titanium alloys with various types of alloying. When choosing the nomenclature of parts and tools for treatment, it is necessary to take into account, first of all, the size and shape of the hardened surface. Best results are achieved on flat surfaces with a diameter up to 30 mm.

References

[1] Bagautdinov A Y, Budovskikh E A, Ivanov Y F and Gromov V E 2007 *Physical Bases of Electroexplosive Alloying of Metals and Alloys* (Novokuznetsk: SibSIU) p 301

[2] Romanov D A, Budovskikh E A, Gromov V E and Ivanov Y F 2014 *Electroexplosive Spraying of Wear-and Electroerosion-resistant Coatings* (Novokuznetsk: Poligrafist) p 203

[3] Raikov S V, Budovskikh E A, Gromov V E et al 2015 *Formation of the Structure, Phase Composition, and Surface Properties of Titanium Alloys During Electroexplosive Alloying and Subsequent Electron Beam Processing* (Novokuznetsk: Poligrafis) p 290

[4] Bashchenko L P, Soskova N A, Budovskikh E A et al. 2012. *Russian Metallurgy (Metally)* **3** 15–19

IOP Conf. Series: Materials Science and Engineering **866** (2020) 012038 doi:10.1088/1757-899X/866/1/012038

Use of waste sludges from chemical enterprises as secondary metallurgical raw materials

S V Krivosheev[1], O I Gordievskiy[1], I V Nozdrin[2] and R G Garaev[1]

[1]Novokuznetsk Aluminum Plant, 7 Ferrosplavny Proyezd, Novokuznetsk, 654084, Russia
[2]Siberian State Industrial University, 42 Kirova Street, Novokuznetsk, 654007, Russia

E-mail: kafcmet@sibsiu.ru

Abstract. The characteristics of dump zinc-containing sludges from chemical enterprises producing artificial fibers are given. The analysis of quantitative, chemical and phase compositions of sludge fields is carried out. The possibility of pyro- and gyrometallurgical processing of these materials was established. The need for preliminary preparation of sludge for processing is shown. Technological recommendations for obtaining marketable sulfate and zinc oxide are formulated.

1. Introduction

Zinc belongs to the group of major non-ferrous metals in terms of production and consumption. In 2018, its production amounted to more than 13 million tonnes, second only to aluminum and copper. At the same time, zinc consumption is growing steadily from year to year. In this regard, the amount of zinc-containing waste is proportionally increasing and inevitably there are questions about their disposal. The domestic zinc industry has now mastered the processes of processing rich wastes with zinc content of more than 30%, and poorer materials are practically not used due to the high costs of their delivery and processing [1].

However, the increase in the amount of accumulated waste, the aggravation of environmental issues related to warehousing and storage, and toughening of environmental legislation brings the task of recycling to the fore. Rather high prices for basic non-ferrous metals in the world market make solution of this problem economically attractive. The initiator of active work in this direction was the flagship of the Russian zinc industry – Chelyabinsk Zinc Plant. Issues of using secondary raw materials are on the agenda due to an acute shortage of zinc concentrates conditioned by the inaccessibility of domestic raw materials associated with the redistribution of property in the extractive industries, the increased cost of zinc in the world market and logistics costs because of the rise of transportation tariffs.

As part of the "Raw Material" plant program, efforts were made to expand the ore base of zinc concentrates, the characteristics and reserves of secondary zinc represented by the waste from ferrous and non-ferrous metallurgy, as well as a number of chemical plants were attested. Research and design work was carried out to introduce new and modernize existing technological processes for recycling. Employees and students of Siberian State Industrial University were involved in these works. The studies were carried out jointly with the engineering and technical center of Chelyabinsk Zinc Plant.

IOP Conf. Series: Materials Science and Engineering **866** (2020) 012038 doi:10.1088/1757-899X/866/1/012038

2. Attestation of dump zinc-containing sludges from enterprises producing artificial fibers

To develop an effective technology for extracting zinc from waste, first of all, it is necessary to attest these materials in terms of quantity and composition. The paper summarizes information on the types, amount and chemical composition of zinc-containing sludge from wastewater treatment and waste technological solutions at domestic enterprises for the production of artificial fibers.

A comprehensive attestation of reserves, chemical and mineralogical composition of sludge from enterprises producing artificial fibers in the West Siberian region (LLC "Sibvolokno", LLC "Krasnoyarskiye Volokna") were carried out. The total amount of accumulated zinc-containing materials is estimated at about 1.5 million tonnes (up to 200 thousand tonnes of zinc). Table 1 summarizes information from available literature on the amount of zinc-containing sludges from various chemical plants with similar chemical and material composition [2]. The amount of zinc in the sludge is comparable to the annual programs of large zinc plants.

Table 1. Characteristics of sludges from chemical plants.

Enterprises	Amount of zinc in the sludges, thousand tonnes	The zinc content in the sludge,%
LLC "Sibvolokno", Zelenogorsk, Krasnoyarsk Territory	25.0	18.0-42.0
LLC "Krasnoyarskiye Volokna", Krasnoyarsk	20.0	10.0-22.0
LLC "Balakovo Fiber Materials Plant", Balakovo, Saratov Region	103.0	13.0-24.0
OJSC "SvetlogorskKhimvolokno", Svetlogorsk, Republic of Belarus	24.0	34.0-42.0
PO "Khimvolokno", Kalinin	15.0	9.0-12.0
NPO "Khimvolokno, Barnaul	2.0	24.0-26.0
AO "Azot", Kemerovo	0.8	24.0-34.0

Zinc-containing sludge is formed during the production of artificial fibers using technology of zinc sulfate to form fibers through spinning jets. Precipitation bath is an aqueous solution containing sulfuric acid, zinc sulfate and sodium sulfate. The jets leaving the die are deposited (coagulated) due to the formation of zinc – xanthate bonds and the desolvating effect of electrolytes.

The waste solution is neutralized with soda ash or lime milk. As a result, a suspension of zinc carbonate is formed, which is sent to the sump. The precipitate is a pasty, slightly sticky powder of dirty gray color, insoluble in water [3].

According to the technical reports of the enterprises, it was found that the total amount of zinc in the sludge damps of "Krasnoyarskiye Volokna" and "Sibvolokna" is about 20,000 and 25,000 tonnes of zinc, respectively, with an average zinc content of 16 to 30%. To certify the chemical, particle size and phase composition of the sludge, samples were taken from various places of the dumps using steel rods: along the perimeter of the dumps in the warm season, from the surface in winter. The depth of sampling ranged from 0.3 to 4 m.

It was found that the color of the samples depends on the content of the main impurities and varies from light gray to light brown or brown. In all samples, cellulose fiber residues are present that significantly increase the water-holding capacity of the sludge. The material is a clay-like mass with a moisture content of 40 to 80%. The particle size of the material is in the range of 0.1 – 2.5 mm. The bulk density of the samples is in the range of 0.87 – 1.25 g/cm^3, and the density (true) is 1.60 – 1.68 g/cm^3 [4].

The results of the samples chemical analysis are shown in table 2. For comparison, the compositions of zinc-containing sludges from the production of artificial fibers at Balakovo Fiber Materials Plant and SvetlogorskKhimvolokno (Republic of Belarus) attested in the chemical laboratory of the Chelyabinsk Zinc Plant are shown.

Table 2. The chemical composition of zinc-containing sludges from enterprises of artificial fibers.

Component	The chemical composition of the sludge samples, % mass.			
	Krasnoyarsk	Zelenogorsk	Balakovo	Svetlogorsk
Zinc total	10.8 - 22.3	14.2 - 29.6	21.2	46.7
Zinc soluble	n/f	18.6	13.6	45.7
Zinc water-soluble	n/f	0.33	0.12	0.36
Iron	0.80 - 1.65	1.92 – 2.70	1.5	1.55
Lead	0.029	0.21	0.041	0.13
Cadmium	n/f	0.0004	0.005	0.0033
Copper	0.022	0.009	0.011	0.011
Nickel	n/f	0.012	0.009	0.011
Cobalt	n/f	0.004	0.0031	0.0023
Tin	n/f	0.023	0.041	N/f
Chlorine	n/f	0.071	n/f	0.028
Fluorine	n/f	0.044	n/f	0.005
Aluminum	0.95	3.05	n/f	n/f
Arsenic	n/f	0.007	0.0026	0.001
Antimony	n/f	0.0025	0.0018	0.0048
Silica	0.5	7.21	11.2	10.97
Calcium oxide	n/f	1.73	10.3	0.19
Sulfur total	n/f	16.27	16.9	1.76
Magnesium oxide	n/f	N/f	0.79	0.24

The analysis of the chemical and phase composition of the sludge shows that zinc in them is represented by carbonates, sulfates and sulfides. The main ballast impurities are calcium and silicon oxides. There is a large scatter of data on the chemical composition of sludge even within the same enterprise, which is determined both by changes in the parameters of the technological regimes of enterprises, and by the conditions of waste storage.

The total zinc content in the materials in question varies between 11–47%, calcium oxide – 2–10%, silicon oxide – 1–11%. The presence of sulfide sulfur in the samples of sludge, reaching, in some cases, 2%, was confirmed. It should be noted that there are residues of cellulose fiber in the sludge and, in some cases, traces of organic impurities that negatively affect the processes of direct leaching of sludge and filtering products. The moisture content of sludge in sludge collectors is 70-90%, after natural drying in air it decreases to 30-40%.

Despite the similarity of the technologies for producing synthetic fibers and wastewater treatment at these enterprises, the zinc content and mineralogical composition of the samples taken are significantly different. So, in the samples of sludge from LLC "Krasnoyarskiye Volokna", the zinc content is significantly lower than in samples taken at LLC "Sibvolokno". In addition, the content of impurities in the first case is much higher – most likely this is determined by the fact that wastewater from all adjacent industries, including household ones, was directed to the sludge collectors. The high zinc content in the sludge of the OJSC "SvetlogorskKhimvolokno" is explained by the use of caustic soda solution to neutralize acidic effluents instead of lime milk at the other compared enterprises.

3. Features of zinc-containing sludge processing

A sufficiently high zinc content, phase composition and insignificant content of "critical" impurities make it possible to process these materials at existing zinc production facilities. Experience in the processing of sludge confirms the effectiveness of the Waelz process for these purposes [5]. The cost-effective zinc content in the sludge, in this case, should exceed 14%.

Due to the existing geographical structure of existing zinc plants, the distance to the nearest enterprises is more than 2000 km, which does not allow for a positive profitability of sludge delivery and zinc extraction because of the high logistics costs even at relatively high prices for zinc, currently exceeding $ 2000 per tonne. This fact necessitates the development of mobile sludge processing technologies in place.

IOP Conf. Series: Materials Science and Engineering **866** (2020) 012038 doi:10.1088/1757-899X/866/1/012038

Worthy of attention is the proposal for the processing of materials similar in chemical composition according to the standard scheme of carbothermal sludge recovery with coke or coal in a rotary tube furnace for expanded clay firing [6]. A significant reduction in capital costs is achieved through the use of standard equipment, and a high concentration of zinc in the resulting Waelz oxyde (more than 50%) makes the processing process cost-effective. The disadvantages of the technology under consideration include the need for a bulky exhaust gas cooling system and capture of Waelz oxyde, as well as high professional requirements for the personnel serving the process.

Figure 2 shows one of the possible instrumentation-technological schemes of hydrometallurgical processing of viscose slimes [7]. Direct leaching of sludge shows a low extraction of zinc, accompanied by a breakdown in the regular process. It is connected with the presence of carbonates and organic impurities in these materials, due to the presence of residues of viscose fiber and waste products of the existing biocenosis in the area where sludge collectors are located. Thus, to remove the organic component in the materials under study, it is recommended that the technological stage of oxidative burning (calcination) of sludge be introduced.

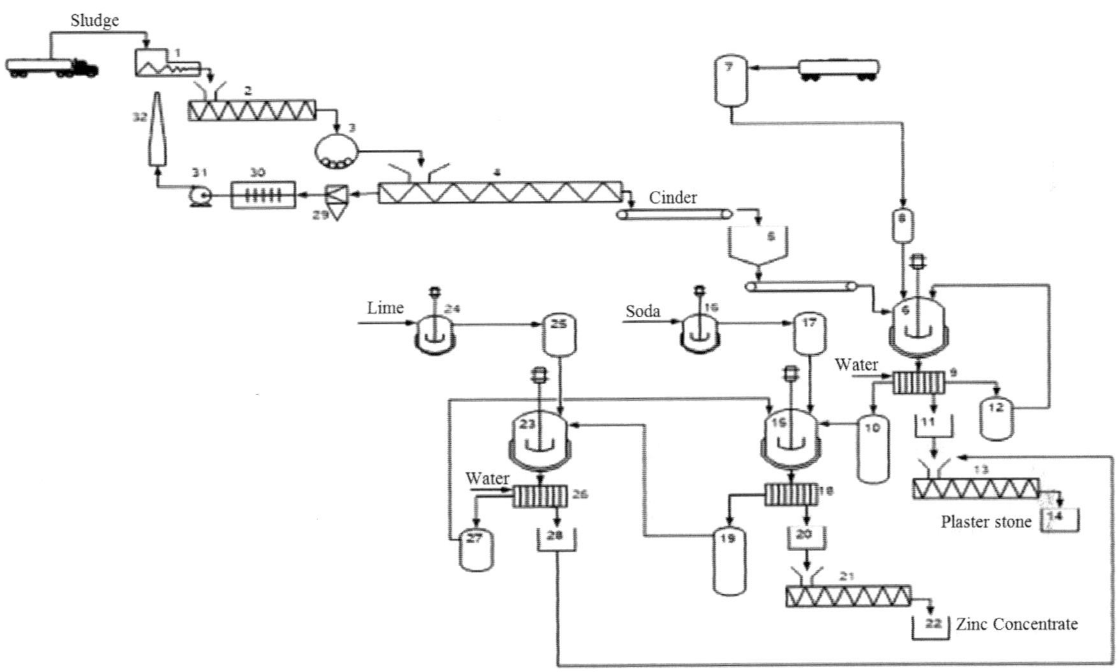

Figure 1. The hardware setup for processing viscose slurries: 1 – screw wet sludge feeder, 2 – dryer drum, 3 – ball mill, 4 – rotary kiln, 5 – chilled cinder, 6 – leach tank, 7 – tank of sulfuric acid, 8 – dispenser of sulfuric acid, 9,18,26 – filter presses, 10,19 – storage tank for solution, 11,14,22,28 – containers for cakes, 12,27 – tank for industrial water, 13 – calcining drum of plaster stone, 15 – precipitator of zinc concentrate, 16 – reactor for the preparation of soda solution, 17 – dispenser of soda solution, 22 – drying drum for zinc concentrate 23 – alkali regeneration reactor, 24 – reactor preparing lime milk, 25 – doser of milk of lime, 29 – cyclone, 30 – electric precipitator, 31 – induced draft fan, 32 – pipe.

4. Conclusion

Taking into account the specifics of the location of considered chemical enterprises, the existing on-site infrastructure, human resources and the consumption environment of possible products as the basic technology for the processing of zinc-containing sludge, in our opinion, it is currently most appropriate to consider the hydrometallurgical scheme for the extraction of zinc in the form of zinc sulfate or zinc oxide. Application of such processes makes it possible to use high-performance standard equipment that provides high technological performance and cheap, non-deficient leaching reagents, such as sulfuric acid. High mobility, adjustability and adjustment of leaching processes, cleaning solutions and isolation of finished products provide high quality materials at low cost.

References

[1] Tarasov A V, Besser A D and Maltsev V I 2004 *Metallurgical Processing of Secondary Zinc Raw Materials* (Moscow: Gintsvetmet) p 219
[2] Pavlovskaya E D, Chistyukhin E A, Dzhalolov Kh O and Nozdrin I V 2017 *Comprehensive Attestation of Zinc-containing Sludges from Enterprises Producing Artificial Fibers in the West Siberian Region* (Dusseldorf: Aluminum – Verlag) p 698
[3] Serkov A T 1981 *Viscose Fibers* (Moscow: Khimiya) p 296
[4] Tarasov A V 2008 *Secondary Metallurgy of Heavy Non-ferrous Metals* vol 3 (Moscow: Academkniga) p 448
[5] Kozlov P A 2002 *Welz-process* (Moscow: Ruda i Metally) p 197
[6] 2011 *Research into the Possibility of Utilization of Zinc-containing Sludge From "Sivinit" OJSC: Report on Research Work* (Krasnoyarsk: Aeratsionniye sistemy) p 48
[7] 2010 *Technological Regulations for the Production of Zinc Sulfate* (Ust-Kamenogorsk: Litera) p 57

IOP Conf. Series: Materials Science and Engineering **866** (2020) 012039 doi:10.1088/1757-899X/866/1/012039

Assessment of the required profitability of metallurgical companies based on a subject-oriented approach

S G Galevskiy

St. Petersburg Mining University, 2, 21st Line, St Petersburg, 199106, Russia

E-mail: sgalevskii@gmail.com

Abstract. In this paper a developed subject-oriented approach to determining the required return is presented. A subjective risk measure is proposed that takes into account the peculiarities of risk perception by the investor and its awareness. Based on the concept of matching risk and profitability, an equation was developed that reflects the relationship between the subjective measure of risk and the level of profitability. All this allows the determination of the required return to the full individually for each investor to be approached.

1. Introduction

The key to sustainable development of any – including metallurgical – company is the fulfillment of all its obligations and meeting the expectations of investors. This is what gives the company the opportunity to attract financial resources on favorable terms and, thus, implement projects that ensure the growth of volumes and quality of products, optimize resource consumption, compliance with environmental standards, etc. In turn, any projects at the stage of substantiating their feasibility need criteria that can separate effective initiatives from ineffective ones. The required rate of return, showing the minimum acceptable risk-adjusted return on invested capital that meets the expectations of the investor, is one of the most important indicators, without which it is virtually impossible to assess the effectiveness of any project and, accordingly, make an informed management decision.

2. Problem statement

There is no single universal methodology for determining the required return. To calculate this parameter, the following are used: capital asset pricing model (CAPM), incl. with a variety of adjustments; discounted dividend model; data on current or historical dividend return adjusted for estimated growth rate; leverage adjusted for the risk premium; risk-adjusted market returns; multiplier "price/profit" and many other methods. According to studies [1-4], both in the Anglo-Saxon countries (USA, UK, Canada) and in countries of Continental Europe (Germany, France, the Netherlands), the most popular way to determine the required return is the CAPM. Similar conclusions are confirmed on the basis of local studies devoted to studying the situation in individual countries – for example, Australia [5] or the Czech Republic [6].

However, according to the above studies, often the required return is determined precisely on the basis of investor expectations: in France and the Netherlands this method turned out to be the second most popular, while in Germany the determination of the required return based on investor

IOP Conf. Series: Materials Science and Engineering **866** (2020) 012039 doi:10.1088/1757-899X/866/1/012039

expectations was even more common than the CAPM. Thus, the feasibility of determining the required return solely on the basis of certain objective indicators has to be questioned.

It seems, however, that it is necessary to distinguish risk as a kind of objective entity and subjective perception of risk by a person (including an investor). Obviously, due to differences between people, the same amount of risk will be perceived differently, ranging from completely ignoring this risk (risk-neutral investor) to recognizing such a risk level as unacceptable (investor-absolute risk phobia). Differences in risk perception will entail a different assessment of the required return: if in the first case the investor is ready to invest in a risky asset without any additional risk premium, then in the second case, no return will encourage the investor to risk his money.

In view of the above, it is necessary to propose a subject-oriented approach to determining the required return and supplement the basic algorithm for all considered methods: "assessing the objective level of risk – calculating the required return – choosing an asset to invest according to the ratio of objective risk and profitability" with one more link: assessing the subjective level of risk by specific investor.

3. Methodology and research results

As an objective measure of risk, the author proposes the expected amount of losses, determined as follows:

$$R=\sum_{i=1}^{n}\left(k_f-k_i\right)*p_i*c_i \tag{1}$$

k_f – return of risk-free asset;
k_i – the yield of this asset at the i-th outcome;
p_i – the probability of the i-th outcome;
c_i – the selection parameter.

The parameter c_i is necessary to give this measure of risk a one-sided nature and is determined as follows:

$$c_i=\begin{cases}0, k_f-k_i<0\\1, k_f-k_i>0\end{cases} \tag{2}$$

In the framework of this work, a hypothesis is put forward about the similar nature of risk perception and perception of physical stimuli (light, sound, weight, etc.). This will allow the subjective measure of risk to be determine using psychophysical laws of perception, long ago derived and experimentally confirmed. It seems appropriate to use the generalizing law of Zabrodin, which is a solution to the following differential equation:

$$\frac{\mathrm{d}s}{s^z}=K*\frac{\mathrm{d}R}{R}, \tag{3}$$

R – the stimulus intensity (expected loss size);
S – a sense of risk;
K – a constant designed to ensure comparability of stimulus intensity and sensations;
z – a constant characterizing awareness (0 – complete ignorance, 1 – absolute awareness).

The solution of this equation with respect to S gives the following expression:

$$\frac{\mathrm{d}s}{s^z}=K*\frac{\mathrm{d}R}{R}, \tag{4}$$

C is a constant that occurs during integration. The solution of this equation with respect to S gives the following expression:

IOP Conf. Series: Materials Science and Engineering **866** (2020) 012039 doi:10.1088/1757-899X/866/1/012039

Since the perception of risk is determined depending on the expected losses, it can be argued that S shows the expected losses on the asset, taking into account the peculiarities of the perception of the initial data by the investor. In other words, a sense of risk is a kind of "depreciation" of a risky asset in the eyes of the investor. Parameter S shows how much this asset is less valuable compared to a risk-free asset that requires similar investments and provides the same return. Accordingly, the perception of risk should range from 0 to 1.

In this regard, it is necessary to identify three types of risk (risk area) depending on the investor's perception of risk: zero risk (S = 0); acceptable risk (0 <S <1); unacceptable risk (S = 1).

At first glance, a zero sense of risk is possible only with a zero value of an objective risk measure, i.e. in the absence of risk as such. However, in practice, sensations arise only when a stimulus overcomes a certain threshold. From the point of view of investments, an example of this is the recognition of certain assets (for example, government bonds) as conditionally risk-free, although any investment is obviously associated with risk, albeit small.

Let us denote the threshold value of the expected losses at which the perception of risk still does not arise, like R_0. Then the following expression is true:

$$0 = \sqrt[1-z]{(1-z) * K * \ln(C * R_0)} \qquad (5)$$

Transforming this expression, we get:

$$c = \frac{1}{R_0}, \qquad (6)$$

Accordingly, then the following is also true:

$$s = \sqrt[1-z]{(1-z) * K * \ln(R / R_0)} \qquad (7)$$

We denote the threshold value of the expected losses at which a sensation of unacceptable risk arises as R_1. Then the following expression is true:

$$1 = \sqrt[1-z]{(1-z) * K * \ln(R_1 / R_0)} \qquad (8)$$

Transforming this expression, we get:

$$(1-z) * k = \frac{1}{\ln(R_1 / R_0)} \qquad (9)$$

Accordingly, then the following is also true:

$$s = \sqrt[1-z]{\frac{\ln(R / R_0)}{\ln(R_1 / R_0)}} = \sqrt[1-z]{\log_{\frac{R_1}{R_0}} \frac{R}{R_0}} \qquad (10)$$

It should be noted that this expression is applicable only for $R_0<R<R_1$, i.e. in the area of acceptable risk. In the area of zero risk, S = 0, and in the area of unacceptable S = 1.

The parameter z, denoting investor awareness, significantly affects the perception of risk by the investor. The feeling of risk reaches its maximum value with complete ignorance, while with absolute awareness the feeling of risk tends to zero. Thus, an investor who considers himself completely knowledgeable does not feel the risk – no matter how objectively the asset is risky.

Next, it is necessary to establish the nature of the relationship between the subjective sense of risk and the minimum required return.

IOP Conf. Series: Materials Science and Engineering **866** (2020) 012039 doi:10.1088/1757-899X/866/1/012039

A subjective measure of risk shows how much of its value in the eyes of an investor loses an asset, taking into account expected losses refracted through the prism of perception of that investor. However, if a risky asset loses part of its value (as opposed to risk-free), then there should be an incentive that will force an investor to invest in a risky asset if there is a risk-free alternative. Obviously, only high returns can be such an incentive. Accordingly, equality must be respected:

$$(1+k)*(1-S)=(1+k_f)*(1-S_f), \tag{11}$$

k – the required return on the asset;
S – a subjective measure of the risk of the asset;
S_f – a subjective measure of the risk of a risk-free asset.

Given that S_f is zero, we can express the required return:

$$k=\frac{k_f+S}{1-S}, \tag{12}$$

The return on a risk-free asset is the return on Russian government bonds (federal loan bond OFZ) for the period under review (ten years, 2009-2018), which averaged 7.74%. At the same time, it is logical to evaluate the lower perception threshold R_0 as the value of the expected losses on the asset, which the investor considers as risk-free – 0.96%.

The author of the present study conducted a study in which respondents answered the question: “At what level of expected losses is investing in this asset not considered?”. According to its results, the average unacceptable level of losses is 42%, but it must be remembered that in our case, losses are understood not only as a loss of invested funds, but a loss of funds relative to capital that was invested in a risk-free asset. Therefore, to determine the average value of the threshold of perception, the profitability of a risk-free asset should be added to the indicated 42%. Therefore, the upper threshold of perception is 49.74%.

Awareness as well as the upper threshold of perception characterizes a particular investor and should be determined by him. However, it seems appropriate to calculate the average value of this quantity to obtain a certain reference point, which the investor can focus on. Using the averaged parameters obtained earlier and calculating the risk of investing in the MICEX market index by years for the period 2009-2018, it was determined that the average investor awareness in the Russian stock market during this period was 0.591.

We illustrate the proposed methodology by the example of the largest Russian metallurgical companies, whose shares belong to the so-called Blue chips. Table 1 presents the values of the objective risk measure (expected loss, R), the subjective risk measure (S), the required and expected (actual) profitability calculated on the basis of a subject-oriented approach and stock price data for these companies for the analyzed period.

As it can be seen from table 1, the MMK company showed the highest profitability, but it also has the highest risk value, which leads to an extremely high required profitability. In general, all the largest metallurgical companies demonstrate returns below the required value, which means insufficient asset management and can lead to difficulties in attracting financial resources. To the least extent, this is characteristic of Severstal and NorNickel companies, which actual profitability is close to the required one, and, taking into account the payment of dividends, may exceed it.

Table 1. Required and expected returns of the largest Russian metallurgical companies for the period 2009-2018.

	Objective measure of risk	Subjective risk measure	Required return	Expected return
MMK	15.2%	0.417	84.9%	50.4%

NLMK Group	13.9%	0.385	75.3%	30.0%
Polyus	11.1%	0.311	56.4%	32.1%
Severstal	10.3%	0.288	51.2%	46.8%
Nornickel	8.4%	0.232	40.2%	38.5%

4. Conclusion

The proposed methodology allows taking into account not only the objective risks associated with investing, but also the investor's attitude to risk, as well as the degree of his awareness (taking into account information asymmetry). This is an important advantage of this approach compared to traditional methods, which involve assessing the required return only on the basis of the characteristics of the asset (investee), without taking into account the characteristics of the investor (investor). It seems that the introduction and development of the proposed approach will make it possible to approach more flexible to determination of the required return and solve the issue of investing funds in full individually for each investor.

Reference

[1] Graham J R and Harvey C R 2001 *Journal of Financial Economics* **60** 187–243
[2] Brounen D, De Jong A and Koedijk K C G 2004 *Corporate Finance in Europe Confronting Theory with Practice. ERIM Report Series Research in Management, Erasmus Research Institute of Management*
[3] Gitman L and Vandenberg P 2000 *Cost of Capital Techniques Used by Major US Firms: 1997 vs. 1980. Financial Practice and Education, Fall/Winter* pp 53–68
[4] Bruner R, Li W, Kritzman M, Myrgren S and Page S 2008 *Market integration in developed and emerging markets: Evidence from the CAPM. Emerging Markets Review* **9** pp 89–103
[5] Truong G, Graham P and Peat M 2008 *Australian Journal of Management* 95–121
[6] Kolouchová P and Novák J 2010 *Cost of Equity Estimation Techniques Used by Valuation Experts. IES Working Paper* (IES FSV, Charles University)

IOP Conf. Series: Materials Science and Engineering **866** (2020) 012040 doi:10.1088/1757-899X/866/1/012040

Correct accounting of risks in assessing the effectiveness of investment projects of metallurgical companies

S G Galevskiy

St. Petersburg Mining University, 2, 21st Line, St Petersburg, 199106, Russia

E-mail: sgalevskii@gmail.com

Abstract. This article proposes the methodology that modifies the capital asset pricing model (CAPM). The proposed methodology allows for separate accounting of risks of the first and second kind and, accordingly, to correctly reflect them in the calculations using the discounted cash flow method. This makes it possible to adequately assess the effectiveness of investment projects and make informed management decisions.

1. Introduction

The implementation of investment projects (in the broad sense of the word) is the key to the continuous development and improvement of any – including metallurgical – companies. However, for metallurgical companies, a correct assessment of the effectiveness of projects is extremely important, since many projects are very large-scale, and the implementation of inefficient solutions leads to the inefficient use of a large amount of resources. Moreover, the feasibility assessment of project implementation is based, as a rule, on the method of discounted cash flows, the application of which is based on the forecast of future cash flows. However, in accordance with the concept of the time value of money, not only the total value of the generated cash flows is important, but also their distribution over time. Therefore, the projected cash flows must be discounted to determine their current value, and this means the need to use a discount rate in the calculations, which is the amount of the required return for the owner of the capital.

2. The problem statement

In addition to mediating the temporary difference between the cash flows of different periods, in modern financial management practice, the discount rate is usually assigned one more function – taking into account the risks inherent in the estimated asset. However, the impossibility of an accurate and objective assessment of the risk magnitude leads to many diverse approaches to solving this problem.

Thus, the author of this article recommends to determine the required return on equity on the basis of a subject-oriented approach described in detail in [1–2], and, according to studies [3-8], the main method for determining the discount rate is the capital asset pricing model (CAPM).

However, regardless of the model used, the calculation method using the discounted cash flow method leads to a distorted accounting of risks if they are taken into account in the discount rate. Traditionally, the value (value) of an asset under the discounted cash flow method is calculated using the following formula:

$$V = \sum_{i=1}^{n} \frac{FCF_i}{(1+k)^i}, \tag{1}$$

FCF_i – net cash flow of the i-th period;

k – discount rate.

Net cash flow, in turn, represents the difference between the inflows and outflows generated by the asset being valued. In this regard, all types of risk that affect the value of this asset can be grouped into two categories:

1) the risk that inflows will be less than expected (risk of the first kind);

2) the risk that outflows will be greater than the expected value (risk of the second kind).

It is these two types of risk that can lead to a decrease in net cash flow and, as a result, to the fact that the value (value) of the asset will be less than the expected value. Accordingly, it is these two types of risk that should be correctly taken into account in the calculations. Consider whether this is true. To do this, we will consider formula (1), taking into account that the net cash flow consists of inflows and outflows, and the discount rate consists of a risk-free component and a risk premium:

$$V = \sum_{i=1}^{n} \frac{FCF_i}{(1+k)^i} = \sum_{i=1}^{n} \frac{CFP_i}{(1+k_f+k_r)^i} - \sum_{i=1}^{n} \frac{CFN_i}{(1+k_f+k_r)^i}, \tag{2}$$

FCF_i – net cash flow of the i-th period;

k – discount rate;

CFP_i – inflows of the i-th period;

CFN_i – outflows of the i-th period;

k_f – risk-free rate;

k_r – risk premium.

We see that this approach only correctly takes into account the risk of the first kind – an increase in the risk premium leads to a decrease in the discounted value of the inflows, but at the same time also reduces the discounted value of the outflows. Thus, the risk of the second kind is not taken into account correctly. It turns out that such an approach is not applicable for evaluating assets that generate not only inflows, but also outflows, and the risk accounting methodology in the discount rate requires the introduction of new tools that would correctly take into account risks of both the first and second kind.

3. Methodology and research results

Since the most popular and frequently used model for determining the discount rate based on risks is CAPM, this work proposes a methodology for adapting this particular model to correct account of risks of both the first and second kind when discounting cash flows. The basic formula of this model, which allows the required profitability to be determined, is as follows:

$$k = k_f + \beta * (k_m - k_f), \tag{3}$$

k_f – risk-free rate;

k_m – return on market portfolio;

β – beta ratio.

Theoretically, the beta coefficient is defined as the ratio of the covariance of stock returns with market portfolio returns and the variance of market portfolio returns:

$$\beta = \frac{COV(\text{k};\text{k}_m)}{\sigma_m^2}, \tag{4}$$

$COV(\text{k};\text{k}_m)$ – covariance of stock returns and market portfolio;

σ_m^2 –variance of return on market portfolio.

In practice, the beta coefficient is usually defined as a regression coefficient, in which the risk premium for the market portfolio is the explanatory factor, and the explained risk premium for the asset being assessed is:

$$k - k_f = \beta * (k_m - k_f), \tag{5}$$

This approach to determining the required return, as was shown above, distorts the results of applying the discounted cash flow method in the case of assets that generate both inflows and outflows. For adequate discounting of cash flows of an investment project, business, etc. it is necessary to divide the risk of the asset, assessed by the beta coefficient, by the risk of the first (decrease in inflows) and second (increase in outflows) kind. The usage of the CAPM model to estimate the required return when discounting cash flows generated by real assets suggests that the risk of the company shares (estimated by the beta coefficient of its shares) is comparable to the risk of cash flows generated, for example, by an investment project.

Thus, the risk of fluctuations in the market value of company shares is equal to the risk of fluctuations in the value of cash flows of a real asset. Leaving the validity of this assumption outside the scope of this study, we can assume that in this case, fluctuations in stock returns and the beta coefficient are quite capable of adequate reflection of not only the risk of a business or investment project, but also the risk of both the first and second kind. For this, it is necessary to separately calculate two beta coefficients: a beta coefficient for positive cash flows (inflows) and a beta coefficient for negative cash flows (outflows). In this case, the beta coefficient for the inflows should be determined on the basis of observations with a positive stock return, and the beta coefficient for the outflows should be determined on the basis of observations with a negative stock return. Thus, to find two beta coefficients, it is necessary to build two regression dependencies. Consider the algorithm for finding discount rates on the example of one of the largest Russian metallurgical companies: PJSC MMC Norilsk Nickel.

To calculate the traditional beta coefficient, we used data on the weekly risk premium for the shares of this company (the difference between the yield on shares and government bonds) and the weekly market risk premium for the risk, calculated as the difference between the yield on the MICEX market index and government bonds (OFZ). To build the regression dependence, we used data from the last five years (07/01/2014 – 06/30/2019), which provided 260 observations, the beta coefficient was 0.99. It should be noted that the return on the risk-free asset (federal loan bonds) was adopted at the level of 7.01% per annum, the yield on the market portfolio (calculated by the IMOEX index) was 14.24% per annum. Using the above data, it is possible to determine the value of the required yield by the traditional CAPM model:

$$k = k_f + \beta * (k_m - k_f) = 7.01\% + 0.99 * (14.24\% - 7.01\%) = 14.17\%, \tag{6}$$

The same observations were used to determine two beta coefficients reflecting two different types of risk, but only the observations with positive values (138 observations) were used to determine the beta coefficient of inflows, and for the outflow beta coefficient – with negative values (122 observation) risk premium on shares of Norilsk Nickel. Based on the constructed regression dependencies, the beta coefficient of inflows was 0.564, and the beta coefficient of outflows was 0.432. Accordingly, the discount rate for inflows can be determined by the formula:

$$k_p = k_f + \beta_p * (k_m - k_f) = 7.01\% + 0.564 * (14.24\% - 7.01\%) = 11.09\%, \tag{7}$$

k_p – discount rate for inflows;

β_p – beta coefficient of inflows.

When determining the discount rate for outflows, the risk premium should not be added to risk-free profitability, but, on the contrary, subtracted from it.

$$k_n = k_f - \beta_n * (k_m - k_f) = 7.01\% - 0.432 * (14.24\% - 7.01\%) = 3.89\%, \tag{8}$$

k_n – discount rate for outflows;

β_n – beta coefficient of inflows.

Let us consider the mechanism of applying two discount rates on the example of a conditional project. Suppose PJSC MMC Norilsk Nickel intends to implement a three-year project that requires an initial investment of 100 million rubles. Project inflows over the years will amount to 150, 250 and 200 million rubles, outflows – 100, 150 and 120 million rubles. Calculate the net present value of the project (NPV).

In the case of applying the traditional CAPM model, it is necessary to discount the project's net cash flows at a rate determined by formula (6) – 14.17%. The calculations are presented in table 1.

Table 1. The calculation of the value of discounted cash flows using the traditional CAPM model.

	0	1	2	3
Initial investments	100	-	-	-
Inflows	-	150	250	200
Outflows	-	100	150	120
Net cash flow	-100	50	100	80
Discounted Net Cash Flow	-100	43.8	76.7	53.8

The amount of the discounted cash flows (NPV of the project) when calculated according to the traditional CAPM model is 74.3 million rubles.

When applying two discount rates, it is necessary to separately discount the inflows and outflows, and then find the difference between them to determine the discounted net cash flow. The calculations are presented in table 2.

Table 2. Calculation of the value of discounted cash flows using two discount rates.

	0	1	2	3
Initial investments	100	-	-	-
Inflows	-	150	250	200
Outflows	-	100	150	120
Discounted inflows	-	135,0	202,6	145,9
Discounted outflows	-	96,3	139,0	107,0
Discounted net cash flow	-100	38,8	63,6	38,9

The amount of discounted cash flows (NPV of the project) when calculating using two discount rates is 41.2 million rubles.

As can be seen from the calculations, when applying two discount rates, the NPV of the project is lower by 33.1 million rubles, or almost twice. Obviously, the incorrect accounting of risks of the second kind when using a single rate leads to an overestimation of project indicators, as well as an overestimation of the value of any other valued asset that generates not only inflows, but also outflows.

4. Conclusion

Thus, as shown above, the application of the proposed methodology avoids overstatement of project performance indicators and, as a result, the implementation of inappropriate management decisions.

IOP Conf. Series: Materials Science and Engineering **866** (2020) 012040 doi:10.1088/1757-899X/866/1/012040

And this, in turn, makes it possible to manage the available resources more effectively and direct them to the projects implementation that are most beneficial for the metallurgical company, taking into account all risks.

References

[1] Galevsky S G 2017 *Scientific and Eechnical Statements of SPbSPU. Economic Sciences* **10(3)** 197208
[2] Galevsky S G 2018 *SPbSPU Scientific and Technical Reports. Economic Sciences* **11(3)** 128139
[3] Graham J R and Harvey C R 2001 *Journal of Financial Economics* **60** 187–243
[4] Brounen D, De Jong A and Koedijk K C G 2004 *Corporate Finance in Europe Confronting Theory with Practice. ERIM Report Series Research in Management, Erasmus Research Institute of Management*
[5] Gitman L and Vandenberg P 2000 *Financial Practice and Education* Fall/Winter 53–68
[6] Bruner R, Li W, Kritzman M, Myrgren S and Page S 2008 *Emerging Markets Review* **9** 89–103
[7] Truong G, Graham P and Peat M 2008 *Australian Journal of Management* 95–121
[8] Kolouchová P and Novák J 2010 *Cost of Equity Estimation Techniques Used by Valuation Experts. IES Working Paper* (IES FSV, Charles University)

IOP Conf. Series: Materials Science and Engineering **866** (2020) 012041 doi:10.1088/1757-899X/866/1/012041

Mathematical modeling of radiant heat transfer in the melting furnace working space

V S Shvydkii, V V Lavrov and E A Devyatykh

Ural Federal University n.a. the first President of Russia B.N. Yeltsin, 28 Mira Street, Ekaterinburg, 620002, Russia

E-mail: v.s.shvydkiy@urfu.ru

Abstract. The paper provides mathematical modelling results of radiation flow formation in the work space of the smelting (glassmaking, open-hearth) furnace with different heating patterns and flow distribution over the surface of the bath and enclosing structures. A significant influence of the flame length upon technical and economic parameters of furnace operation was found out.

1. Introduction

For assessment of the relation between the flame length and flow density we have built a three-dimensional mathematical model of the external problem in the furnace with a liquid bath [1-3]. For solution of the mathematical model, we used the zonal method [4-5]. In zonal method problems, separate flows have different meanings and, therefore, their relation with the flame length has its own specific features [6-8]. In terms of the balance, resulting flows have the biggest relation with heat consumptions (including fuel consumption). Here, first of all, differences in zone types and heat-exchange patterns become evident. The surface of melt is a zone of class II. For these zones heat consumptions are set as values of specific resulting flows (heat losses through the basin walls and bottom are also considered). For this reason, the flame length has an impact only on local heat-exchange characteristics – at average, the situation shall not depend on the flame length.

2. Density of Thermal Flows on the Surface of Melt and Furnace Roof

Table 1 presents data confirming this statement. The figures in the last column of table 1 characterize the spread in values of resulting flows (their root-mean-square deviation). It is determined by the formula:

$$\sigma_q = \sqrt{\frac{1}{L \cdot B} \int_0^L \int_0^B \left[q(x,y) - q_{mean} \right]^2 dx\, dy},$$

where L is a furnace length, m and B is a furnace width, m.

Table 1. Parameters of resulting flows on the surface of glass melt.

Flame Length, m	q_{max}, kW	x_{max}, m	y_{max}, m	Q_{mean}, kW	σ_q, kW
3.178	1906.513	6.339	1.773	546.624	441.634
6.300	1913.720	6.350	1.722	548.651	442.946

9.534	1917.190	6.347	1.723	548.945	443.322
11.580	1918.799	6.349	1.723	548.945	443.322
13.620	1916.399	6.349	1.727	547.699	442.840

Figure 1 shows a typical pattern of flow distribution over the surface of melt (in all figures the numbers on the curves are flow values, kW).

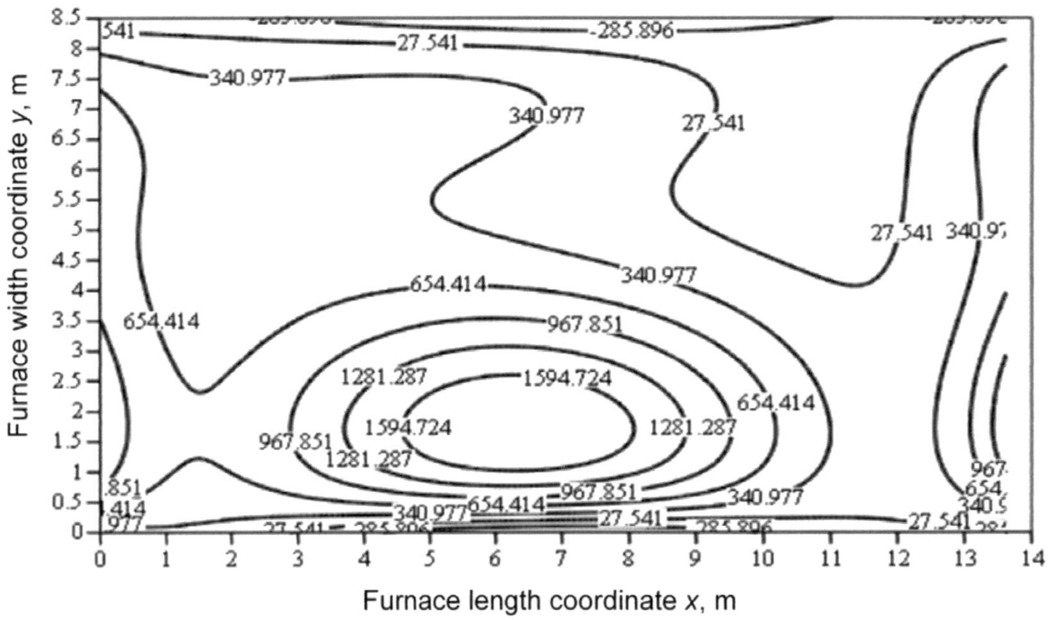

Figure 1. Resulting flows on the bath surface.

The zones of the furnace roof refer to zones of class III. For these zones, the relation between the zone temperature and resulting flow density is set. When the flame length changes, distribution of heat liberations also changes, i.e. the pattern of resulting flows depends on the flame length. The resulting flows reflect the summary total of radiant and convective heat exchange and, therefore, this dependence will not be too high but detectable anyway.

Table 2 shows parameters of resulting flow distribution over the surface of the furnace roof. As it can be seen from the data of Table 2, generally, when the flame length increases, the conditions for operation of roof refractory are getting worse as the flow values grow both at the maximum point and in the mean.

The pattern of resulting flow distribution on the furnace roof is quite typical and differs only in details at different flame length. Figure 2 shows flow distribution for a basic flame length (9.534 m) as an example.

Table 2. Parameters of resulting flows on the surface of the furnace roof .

Flame length, m	q_{max}, kW	x_{max}, m	y_{max}, m	$q_{\text{сред}}$, kW	σ_q, kW
3.178	13.266	6.087	6.758	6.936	3.277
6.300	13.624	6.563	6.529	7.613	2.997
9.534	12.998	6.985	6.236	7.997	2.827
11.58	13.466	7.202	6.268	8.395	3.020
13.62	14.089	7.113	6.313	8.412	3.093

Unlike the resulting flows that implicitly allow for heat consumptions, the incident heat flows in terms of the balance are "a half-finished product" as they characterize the heat input without tracing its further use. In this case, the difference between the glass melt surface and furnace roof surface as radiation detectors is shown only through radiative characteristics and distance to heat liberation (to the flame). As the flame is geometrically closer to the bath surface, the value of the incident flows on the bath will be somewhat higher, which is clearly seen from the comparison of the data given in table 3 and table 4.

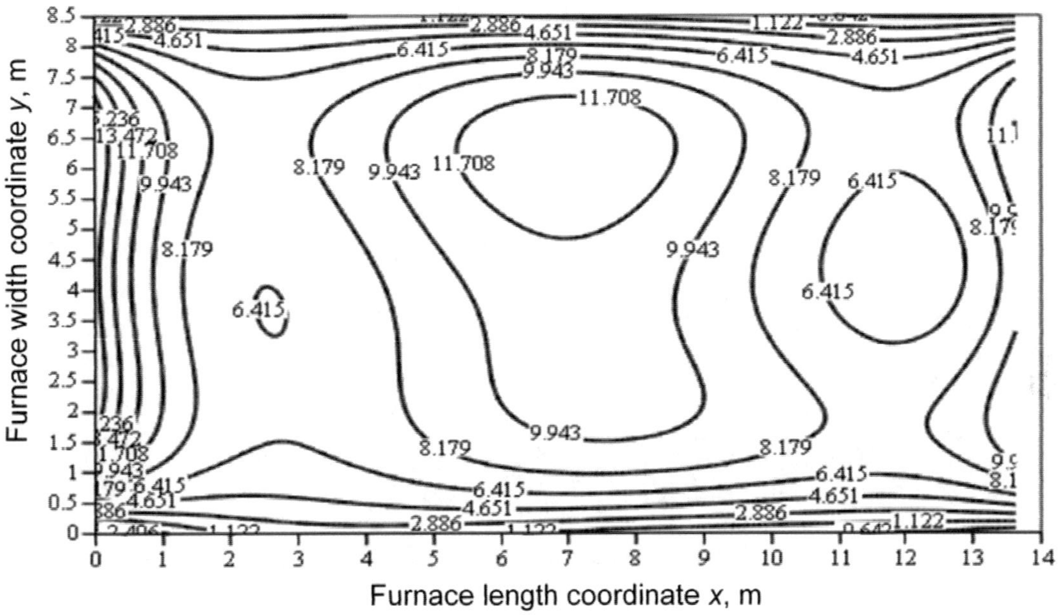

Figure 2. Distribution of resulting heat flows over the surface of the furnace roof at the basic flame length.

As it can be seen, in both cases the shortest flame (3.178 m) is out of approximately the same patterns. Flow distribution for this flame is shown in figure 3.

It shall be noted that distribution of incident flows over the surface of glass melt at this flame length is the same and there are only quantitative differences in the flows, which is already seen from the data in table 3 and table 4.

At any other flame length, distribution of incident flows both over the glass melt surface and furnace roof surface is of the same type in qualitative terms. An example of this distribution is shown in Fig.4 for the bath surface at the flame length of 11.58 m.

Table 3. Parameters of incident flows on the surface of glass melt.

Flame length, m	q_{max}, kW	x_{max}, m	y_{max}, m	q_{mean}, kW	σ_q, kW
3.178	5622.650	0.000	1.932	1817.145	1267.785
6.300	4976.533	5.655	1.792	2119.416	1221.137
9.534	5205.960	6.854	1.780	2191.285	1197.814
11.580	4975.711	7.123	1.856	2231.086	1199.315
13.620	4485.394	7.098	6.764	2125.690	1165.064

IOP Conf. Series: Materials Science and Engineering **866** (2020) 012041 doi:10.1088/1757-899X/866/1/012041

Table 4. Parameters of incident flows on the surface of the furnace roof.

Flame length, m	q_{max}, kW	x_{max}, m	y_{max}, m	q_{mean}, kW	σ_q, kW
3.178	6045.190	0.000	1.860	1774.035	1126.632
6.300	4316.125	5.335	1.786	2112.410	965.099
9.534	4424.280	6.818	1.777	2228.115	937.690
11.580	4250.628	7.169	1.842	2288.739	971.194
13.620	4035.914	7.263	6.710	2179.024	962.680

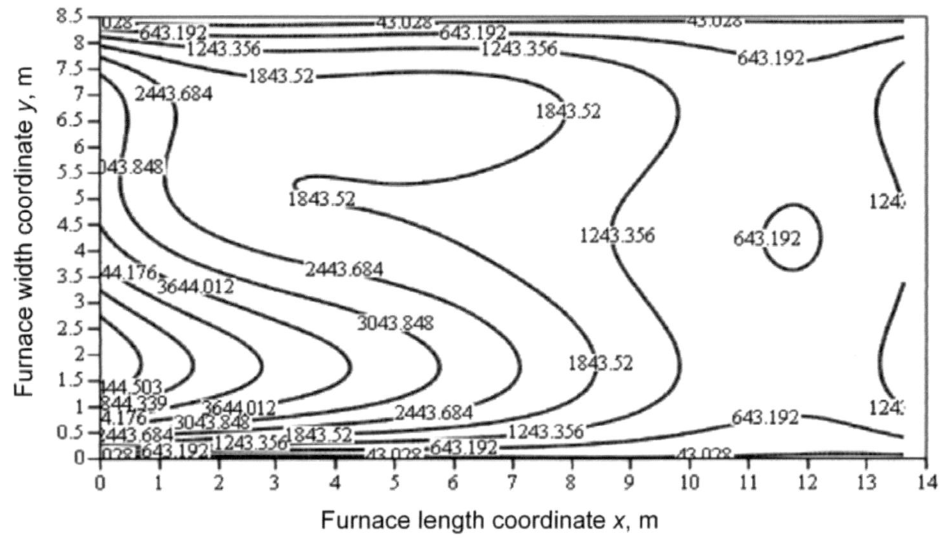

Figure 3. Distribution of incident flows over the surface of the furnace roof (at the flame length of 3.178 m).

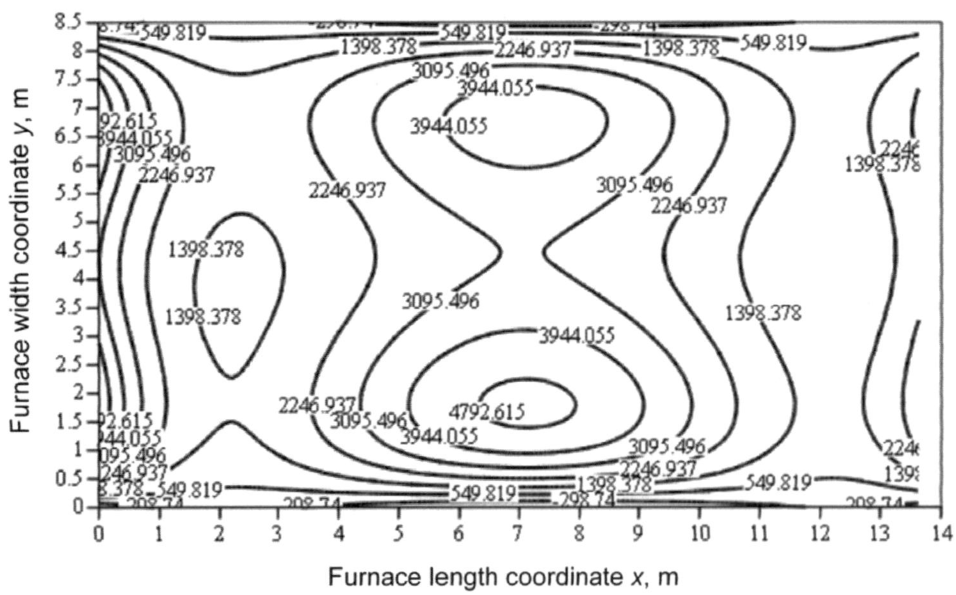

Figure 4. Distribution of incident flows over the bath surface at the flame length of 11.58 m.

3. Conclusion
Mathematic modelling of this type based on calculations within the zonal method of solution of integrodifferential equations for radiative and convective energy transfer enables to make a right choice of refractory materials for enclosing the furnace as well as optimize design and operating parameters of the furnace.

References
[1] Shvydkii V S and Dzyuzer V Ya 2009 *Designing of Energy-efficient Glassmaking Furnaces* (Moscow: Teplotechnik) p 339
[2] Dzyuzer V Ya and Shvydkii V S 2016 *Refractories and Industrial Ceramics* **56(6)** 597–600
[3] Dzyuzer V Ya, Shvydkii V S and Sadykov E B 2013 *Glass and Ceramics* **69(9-10)** 302–305
[4] Shvydkii V S et al 1999 *The Elements of the System Theory and Numerical Methods for Modelling Processes of Heat and Mass Transfer: College Textbook* (Moscow: Intermet-Engineering) p 520
[5] Turchak L I and Plotnikov P V 2002 *The Fundamentals of Numerical Methods: College Textbook* (Moscow: Fizmatlit) p 304
[6] Dzyuzer V Ya 2017 *Refractories and Industrial Ceramics* **58(1)** 25–28
[7] Sokolov B A and Abakin D A 2017 *Glass and Ceramics* **74(7-8)** pp 278–281
[8] Sborshchikov G S and Terekhova A Yu 2017 *Glass and Ceramics* **74(1-2)** 18–19

IOP Conf. Series: Materials Science and Engineering **866** (2020) 012042 doi:10.1088/1757-899X/866/1/012042

Algorithms and software for optimal management of raw materials, fuel and energy resources in blast furnace production

N A Spirin, I A Gurin, V V Lavrov and K A Shchipanov

Ural Federal University n.a. the first President of Russia B.N. Yeltsin, 28 Mira Street, Ekaterinburg, 620002, Russia

E-mail: n.a.spirin@urfu.ru

Abstract. The structure of optimization model of optimal management of raw materials, fuel and energy resources in the blast-furnace shop of iron and steel works is represented. The following blocks are taken as system basis: 1) calculation of the set of parameters that characterize the thermal, gas-dynamic, slag and blasting modes for every blast furnaces of the shop during the base period; 2) calculation of linearized model coefficients (constants of transferring via different exposure pathways) individually for every blast furnace as well as properties of iron ore raw materials, fluxing additions, blasting parameters, parameters of fuel-enriched blast influencing the technical-and-economic indices of separate furnaces performance, their thermal, gas-dynamic and slag operation modes in the course of blast-furnace melting according to UrFU-MMK blast-furnace production model within the base period; 3) solution of tasks that consider the optimal allocation of raw materials, fuel and energy resources for the project period of blast furnaces operation; 4) analysis of obtained results and providing of recommendations on the optimization of blast furnaces parameters. The developed functional model of optimal distribution of raw materials, fuel and energy resources for the engineering and technology personnel of blast-furnace shop is illustrated; the main functions and interconnections between the separate functional blocks are defined. The functions of created "Optimal management of raw materials, fuel and energy resources in the blast-furnace production" software that is realized in the Microsoft Visual Studio 2017 (C# programming language) programming environment in the form of web application are pointed out. The program product provides the engineering and technology personnel of blast furnace shop of iron and steel works with the opportunity to solve the tasks of optimal distribution of fuel and energy resources (natural gas and oxygen consumption) within the group of blast furnaces in the different technological situations.

1. Introduction

To solve the tasks relating to the management of complex difficult and energy-consuming process units in the dry metallurgy, it is necessary to use the modern automated systems of optimal management of raw materials, fuel and energy resources. The blast furnace processing is given the special place in the set of technological operations directed at obtaining the metal products as it is the most resource-intensive, energy-consuming and difficult operation using up to 50% of fuel materials consumed by the iron industry.

The analysis of current state of blast furnaces complex management provides an opportunity to draw the conclusion that it is necessary to be concentrated on the solving of challenging applied research task to create the automated subsystem for the optimal management of raw materials, fuel and energy resources, as well as forecasting of blast furnaces complex operation under the conditions of change in raw materials and fuel market, instability of composition and quality of smelt iron ore raw materials. Such task is to be solved based on development and application of new methods and algorithms [1-8]. The mathematical models of technical processes and production [9-13] represent the intellectual core of such system.

2. Model structure

The structure of optimization model of optimal raw materials, fuel and energy resources management within the blast-furnace shop is represented in Fig. 1 and includes the following main blocks:

1. *Calculation of the set of parameters that characterize the thermal, gas-dynamic, slag and blasting modes of every blast furnace within the shop during the base period.* The parameters of blast furnace melting are evaluated with the help of balance model of blast-furnace process developed within the natural simulation approach that may be conventionally divided into two parts, i.e. model of base state and prediction model [14-17]. The model of base (reference) state enables the evaluation of the state of every blast furnace and influence of input parameters on the thermal, gas-dynamic and slag modes of blast furnace melting with the use of all actually available information on parameters of burden material, fuel-enriched blast, blast-furnace gas and melting products. The prediction model with reference to the results obtained with the help of base (reference) state model provides an opportunity to evaluate the indices of blast furnace melting when the melting conditions are changed. The operating results of blast furnace and blast-furnace shop at all are known according to the data of technical report of blast furnace shop. The UrFU-MMK software and 'Technical report of blast-furnace shop' subsystem are used for this purpose [15-17]. The total number of technical report indices comprises more than 300 units. 'Blast-furnace shop technologist WKS' program unit is used to calculate the set of parameters characterizing the thermal, gas-dynamic, slag and blasting modes of every blast furnace within the shop during the base period (for about 30 parameters for every blast furnace) [15, 17].

2. *Calculation of linearized model coefficients.* The influence coefficients (constants of transferring via different exposure pathways), in particular, the properties of iron ore raw materials, fluxing materials, blasting parameters, parameters of fuel-enriched blast that influence the technical-and-economic indices of work of certain furnaces and their thermal, gas-dynamic and slag modes of blast-furnace melting under the base period conditions, are calculated for every blast furnace with the use of blast-furnace process model [16]. Considering the model linearity, these coefficients for given furnace are further regarded to be the constant ones in relation to the chosen base period when solving the prediction tasks. The probability of linearization of mathematical model of blast-furnace process are evaluated for the operating conditions of separate furnaces; the possible limits of raw materials, blasting and mode parameters change, as well as use of linear mathematical programming methods to solve the optimization tasks are determined. At the same time, the methods evaluating the blast-furnace process linearity that were developed earlier on and represented in the works [15, 17] are also used.

Thus, to improve the rating reliability, the model uses the incremental relations that allow bringing of task to the task of linear mathematical programming as well as excluding the influence of instrumentation and accuracy errors for information channels.

Furthermore, the tasks of optimal raw materials, fuel and energy resources distribution during the project period are solved. This includes the following blocks:

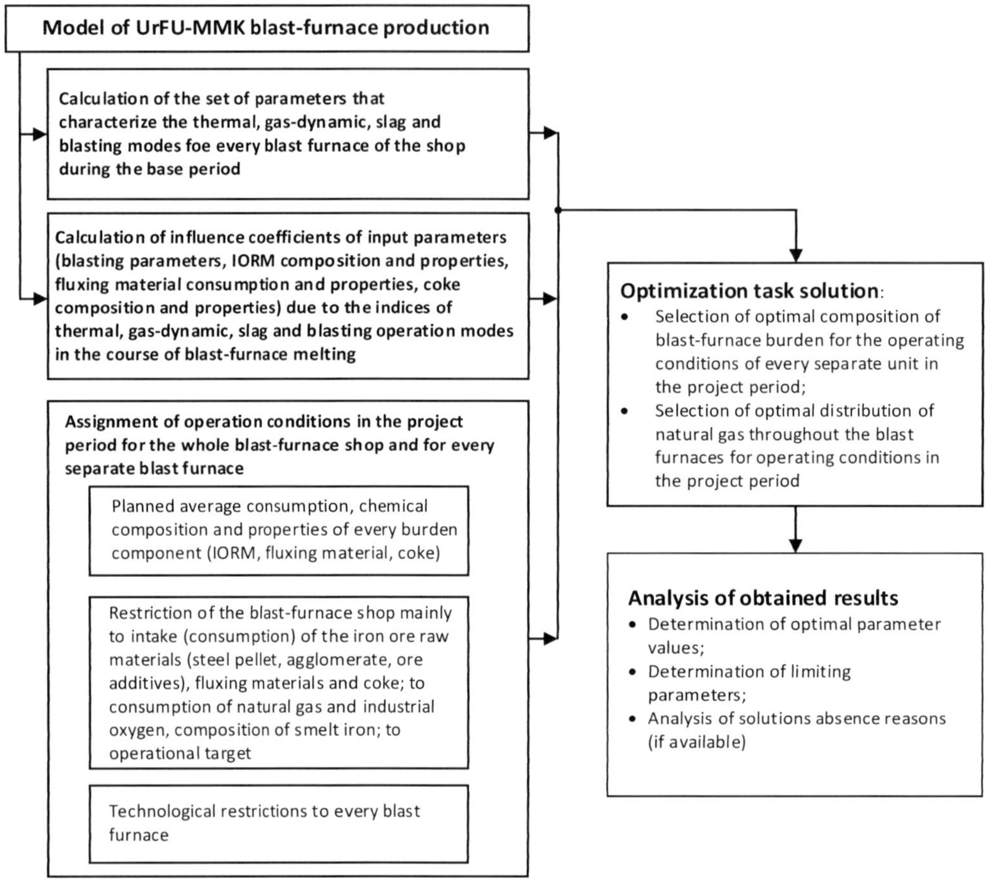

Figure 1. Structure of optimization model of optimal raw materials, fuel and energy resources management in the blast-furnace shop.

1. *Setting of working conditions in the project period for the whole blast-furnace shop and every furnace in particular.*

- The task of planned average consumption, chemical composition and properties of every burden component including the iron ore raw materials, fluxing materials and coke are executed. The general planned shop production is evaluated by means of the conversion model factor and reference data (technological instruction) acceptable for the operating conditions of blast-furnace shop at all; the production is distributed during the project period for every blast furnace. The consideration of technical condition for every blast furnace, their repair schedules and planned furnace shutdown is carried out by means of Delphi approach.
- Setting of restrictions of the blast-furnace shop mainly to intake (consumption) of iron ore raw materials (steel pellet, agglomerate, ore additives), fluxing materials and coke; to consumption of natural gas and industrial oxygen, composition of smelt iron; to operational target.
- Setting of technological restrictions to every blast furnace characterizing the thermal, gas-dynamic and slag modes of blast-furnace melting, composition and properties of found iron.

2. *Solution of task relating to the selection of optimal composition of blast furnace burden material under the operating conditions in the projection period for every separate unit.* Solution of task that considers the optimal composition of blast-furnace burden material for every blast furnace at the set bypass ratio of agglomerate and steel pellet with the possible correcting of fluxing material and agglomerate basicity as well as consideration of limitations on the thermal, slag and gas-dynamic

melting modes. At the same time, the following variants of tasks solution are provided in priority order:

- Correcting of burden material composition by the bypass ratio of agglomerates/steel pellets if the fluxing additives are absent;
- Correcting of burden material composition only by means of fluxing additions consumption;
- Any combination of stated variants is possible.

3. Solution of task relating to the optimal distribution of natural gas throughout the blast furnaces under the operating conditions of project period:

- Formalization of restrictions to the thermal, gas-dynamic and slag operation modes of blast-furnace melting for every shop furnace;
- Formalization of general shop restrictions to the consumption of fuel and energy resources: consumption of natural gas, process oxygen, coke, as well as cast-iron production;
- Bypass ratio of natural gas and process oxygen is set due to the guard condition of maintenance of burning temperature on the injection lances at the scheduled interval.
- The objective function is set as the task to minimize the coke consumption.

4. *Analysis of obtained results.* The peculiarity to solve the complex optimization tasks of mathematical programming relates to the fact that there are possible cases when the limitations laid on the furnace operation modes are contradictory (there is no region of feasibility). Due to this, the solution algorithm provides the analysis stage for the task solution. In case of absent region of feasibility and conditions contradiction, it is required to apply the correcting stage (to evaluate the correct use of burden materials, restrictions, reliability of source information etc.).

3. Functional modeling of optimal raw materials and energy resources distribution

The method used for the development of functional model is based on the idea, notation of IDEF0 (Integrated computer aided manufacturing DEFinition) structural analysis and designing methodology. The use of this methodology provided an opportunity to create the functional structure of program complex, find out the activities it performs as well as connections between these activities, controlling actions and execution mechanisms of every function.

As a result, the functional model of optimal raw materials and energy resources distribution for engineering and technology personnel of blast-furnace shop is developed based on system approach; the main functions and interconnections between the separate functional units are determined. The total number of decomposed units of functional model embraces 115; the fragment is shown in Fig. 2.

'*Prepare the initial data for calculation*' function (A1) ensures the retrieval of necessary actual reporting indicators of blast furnaces operation characterizing the thermal, gas-dynamic, slag and blasting modes of every blast furnace within the shop during the base period from the database of blast-furnace production; the averaging of indicators for the user-defined period; setting of value characteristics of all types of fuel and energy resources; calculation of all coefficients of linear UrFU-MMK model due to the assessment of impact of input parameters (blasting parameters, composition and properties of IORM, fluxing materials consumption and properties, coke composition and properties) on the indices of thermal, gas-dynamic, slag and blasting operation modes of blast-furnace melting.

'Work out restrictions and perform calculations for base period' function (*A2*) ensures for the base period: formation of restrictions of the blast-furnace shop mainly to intake (consumption) of iron ore raw materials (steel pellet, agglomerate, ore additives), fluxing materials and coke, to consumption of natural gas and process oxygen, composition of smelt iron and operational target; formation of technological restrictions to every blast furnace (theoretical burning temperature; composite index of thermal condition of furnace bottom; ratio of heat capacities of shaft flows; degree of burden material compensation; slag properties (slag viscosity at different temperatures, viscosity gradients); sulfur and silicon content in iron; consumption of natural gas by blast furnaces; coke consumption by the furnace); calculation of optimization task for the base period.

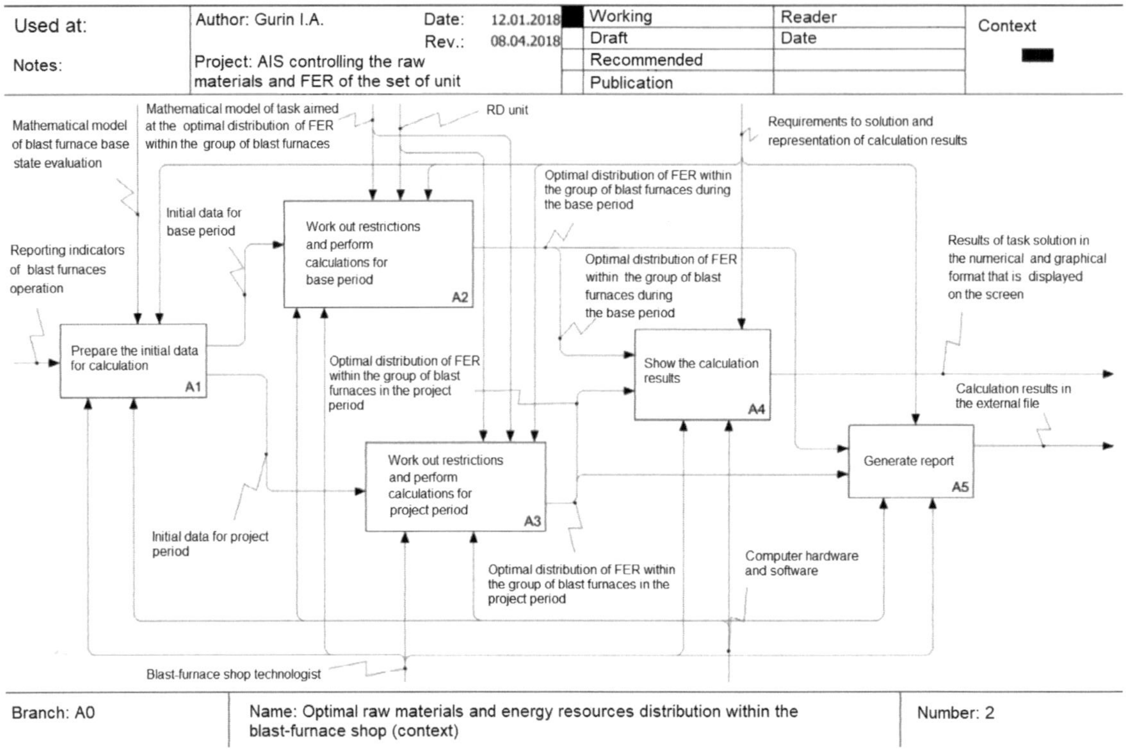

Figure 2. The first level of decomposition of functional model of optimal raw materials and energy resources distribution.

'*Work out restrictions and perform calculations for project period*' function (A3) ensures the following settings in the project period: planned average consumption, chemical composition and properties of every burden component (IORM, fluxing materials, coke); formation of restrictions of the blast-furnace shop mainly to intake (consumption) of iron ore raw materials (steel pellet, agglomerate, ore additives), fluxing materials and coke, to consumption of natural gas and process oxygen, composition of smelt iron and operational target; technological restrictions to every blast furnace; calculation of optimization task for the project period.

'*Show the calculation results*' function (A4) ensures the display of solved results of optimization task on the screen in the numerical and graphical format.

'*Generate report*' function (A5) provides an opportunity to prepare the list of indices for exporting to the external file of commonly used office documents formats (Word, Excel, CSV, PDF and others).

4. Mathematical, algorithmic and information support

The structural system analysis and designing of mathematical model units are performed based on procedure-oriented approach. The basis of this approach consists in the use of DFD (Data Flow Diagrams) of information model the main components of which are represented by data flows that transfer information from one unit to another. The notation of DFD method suggests the partition of mathematical model on the separate functional components (processes) and their representation in the form of network connected with the data flows.

4.1. Blast furnace burden material optimal composition selection. The solution block chart for the task calculating the optimal composition of blast-furnace burden material is represented in figre 3. The

IOP Conf. Series: Materials Science and Engineering **866** (2020) 012042 doi:10.1088/1757-899X/866/1/012042

optimization model includes two main components: objective function and technological restrictions to every blast furnace.

4.2. Calculation of the optimal natural gas distribution. The optimal distribution of fuel and energy resources within the group of blast furnaces and, particularly, injected fuel and process oxygen within the group of blast furnaces is an actual task because the technological parameters of operation of separate furnaces differ considerably. It makes sense to have the strategic methodology to evaluate the efficiency of specified resources use for the blast furnaces as well as carry out their optimal distribution if having the set general consumption of these resources within the blast-furnace shop. Figure 4 shows the flow diagram solving the optimization task of fuel and energy resources distribution within the group of blast furnaces.

The following variants of objective functions are provided:

- Minimal coke rate:

$$Z = K \rightarrow \min_{X_{i,B} \in G_{xB}} ; \tag{1}$$

- Maximal performance:

$$Z = Đ \rightarrow \max_{X_{i,B} \in G_{xB}} . \tag{2}$$

where i – type index of burden materials and fuel-energy resources; $X_{i,B}$ – vector characterizing the types, rate and chemical composition of i materials of blast-furnace burden, fuel and energy resources; $X_{i,B} \in G_{x_B}$ – restriction on burden parameters, i.e. types, chemical composition of fed burden materials, fuel and energy resources; G_{x_B} – region of feasibility of burden, fuel and energy resources parameters.

Formation of technological restrictions presupposes the accounting and mathematical description of restrictions to the thermal, gas-dynamic and slag melting modes.

5. Software for optimal management of raw materials, fuel and energy resources

The 'Optimal management of raw materials, fuel and energy resources during the blast furnace production' software in the form of web application is developed in accordance with the modern principles of application programs development (functionality, expandability, integration with databases, user-friendly interface, safety, information estimation) [18-20]. .NET technology is the basis for program implementation thus giving the additional freedom to choose the platform and programming language. The program product is written in C# language with the use of Microsoft Visual Studio 2017 programming environment.

The software for choosing the blast furnace burden optimal composition provides the solution of the following technological tasks:

- Selection of optimal ratio of agglomerate parts and one of types of imported steel pellets (the material that helps correct the burden to further obtain the necessary burden basicity is chosen from the dropdown list);
- Selection of consumption of one of fluxing materials (the type of fluxing material that helps correct the burden to further obtain the necessary burden basicity is chosen from the dropdown list);
- Selection of agglomerate chemistry (basicity) during the project period at the set consumption as well as portions of iron ore and fluxing materials.

IOP Conf. Series: Materials Science and Engineering **866** (2020) 012042 doi:10.1088/1757-899X/866/1/012042

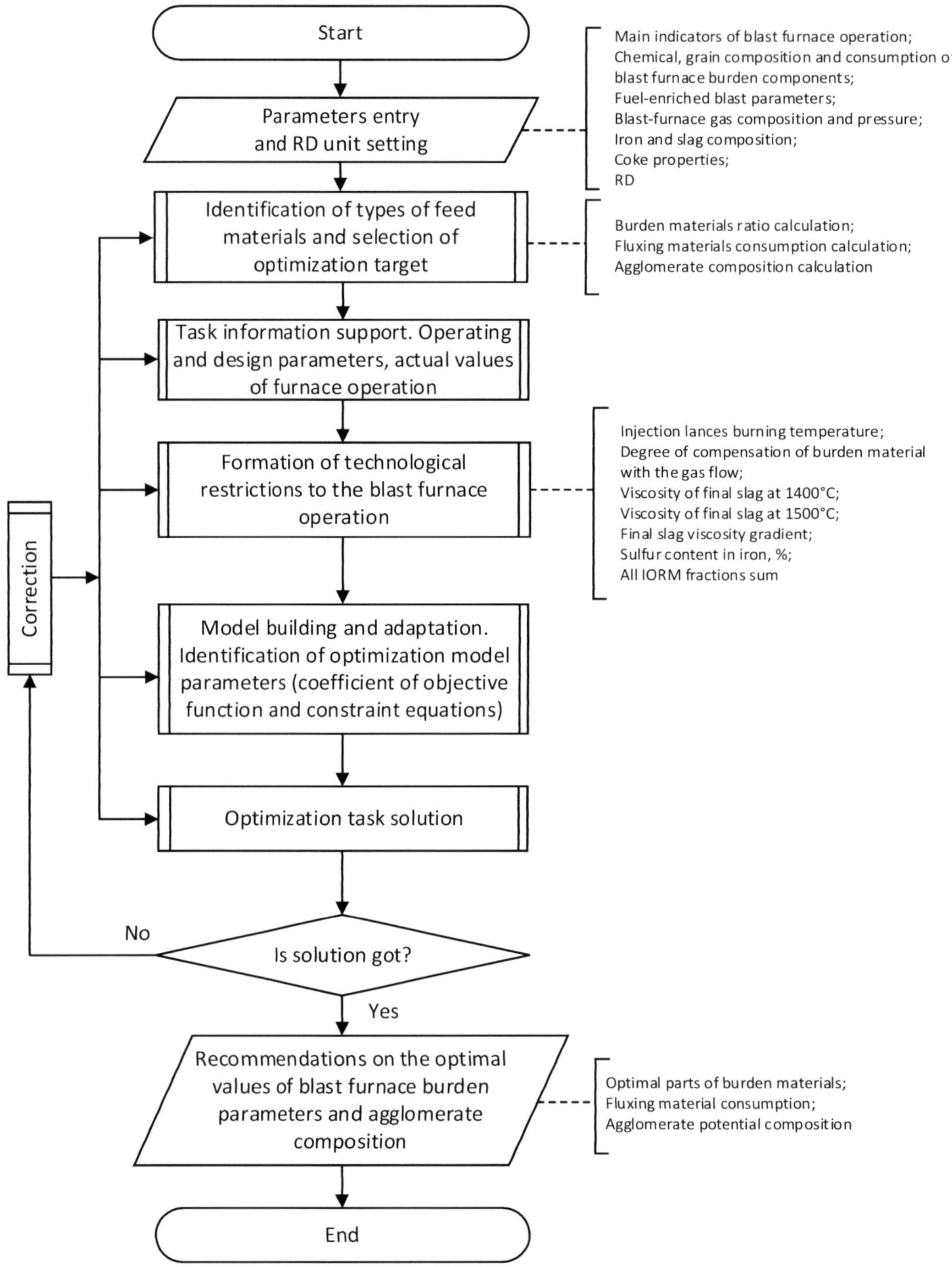

Figure 3. Solution sequence of task on calculating the optimal composition of blast-furnace burden.

IOP Conf. Series: Materials Science and Engineering **866** (2020) 012042 doi:10.1088/1757-899X/866/1/012042

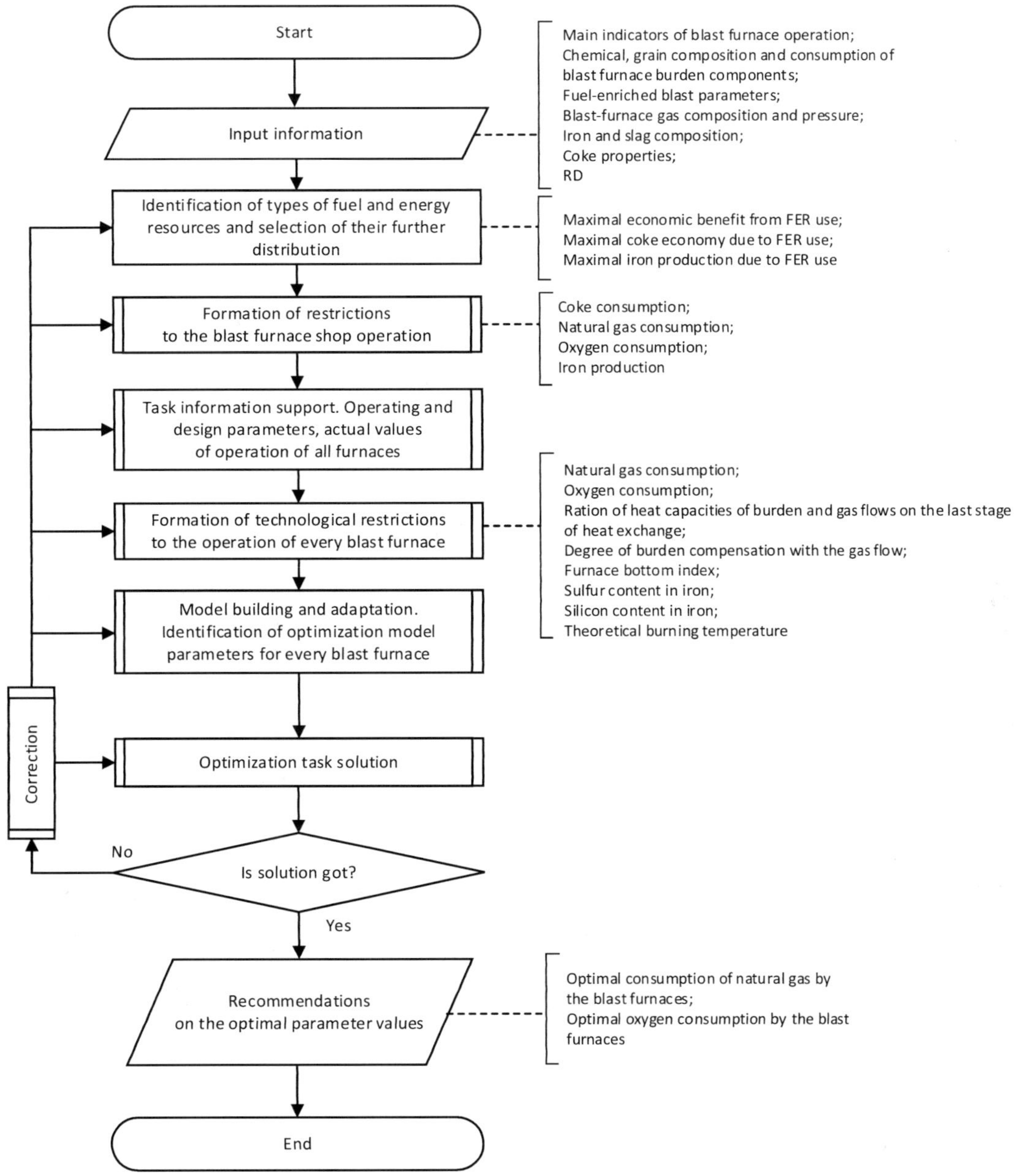

Figure 4. The block chart solving the task of optimization of fuel and energy resources distribution within the group of blast furnaces.

The subsystem software calculating the distribution of fuel and energy resources provides an opportunity to solve the tasks of optimal distribution of consumed natural gas and oxygen in the following technological situations:

- If changing the volume of resources due to the consumption of natural gas and oxygen within the whole blast-furnace shop;
- If determining the maximal volume of resources of fuel-enriched blast that may use the complex of blast furnaces;
- If changing the resource by the volume of coke possessed by the blast-furnace shop;
- If changing the requirements of assignment for the volume of the found iron by the complex of

blast furnaces;

- If changing the market conditions, i.e. requirements ensuring the maximal coke economy, maximal production, economical performance of furnaces complex as well as prices of fuel and energy resources etc.

6. Conclusion

The functional model of optimal management system of raw materials, fuel and energy resources in the blast-furnace production is developed.

The following subsystems structures are realized:

- Selection of optimal composition of blast-furnace burden that allows the determination of necessary ratio of burden materials, fluxing materials consumption and agglomerate composition that ensures its set basicity if considering the set technological restrictions to the slag, gas-dynamic and thermal operation modes of blast furnaces;
- Selection of optimal distribution of fuel and energy resources within the group of blast furnaces in the different technological situations (maintenance or change of general consumption of natural gas, oxygen and coke for the blast-furnace shop at all; correction of operating and design parameters of furnaces included into the concerned group; change of market conditions) differing by the consideration of technological restrictions to the operation of separate furnaces.

The complex of algorithms for optimal distribution of raw materials, fuel and energy resources between the furnaces of blast-furnace shop with consideration of technological restrictions in the conditions of unstable composition and quality of smelt iron ore raw materials and fuel as well as changeable market conditions is developed. The proper software is developed with the help of modern technologies.

References

[1] Porzio G F et al 2014 *Applied Thermal Engineering* **70(2)** 1148–55
[2] Bandyopadhyay S and Sahu G C 2011 *Chemical Engineering Transactions* **25** 279–284
[3] Ganjehkaviri A et al 2014 *Applied Thermal Engineering* **67(1-2)** 566–578
[4] Parshakov V M, Polinov A A et al 2017 *Metallurgist* **61(7-8)** 535–542
[5] Mitra T, Pettersson F et al 2017 *Materials and Manufacturing Proc.* **32(10)** 1179–88
[6] Zablotskii P A, Petrenko V A and Kovshov V N 2017 *Metallurgist* **61(3-4)** 175–178
[7] Wang B-X, Chen W et al 2017 *Ironmaking and Steelmaking* **44(1)** 59–65
[8] Harvey J-P and Gheribi A E 2014 *Metallurgical and Materials Trans. B: Process Metallurgy and Materials Proc. Sci.* **45(1)** 307–327
[9] Kazarinov L, Shnayder D and Barbasova T 2017 *Key Engineering Materials* **743** 363–368
[10] Sibagatullin S K, Kharchenko A S et al 2017 *Chem. Tech. and Metall.* **52(4)** 694–701
[11] Filatov S V, Zagainov S A et al 2017 *Izv. Vuzov. Ferrous Metallurgy* **60(8)** 637–642
[12] Kuang S, Li Z and Yu A 2018 *Steel Research International* **89(1)** 1700071
[13] Ariyama T, Natsui S et al 2014 *ISIJ international* **54(7)** 1457–71
[14] Spirin N A, Lavrov V V et al 2011 *Model Systems of Decision Support in CAPCS of Metallurgical Blast-furnace Melting* (Ekaterinburg: UrFU) p 462
[15] Spirin N A, Lavrov V V et al 2011 *Metallurgist* **54(9-10)** 566–569
[16] Spirin N, Gileva L et al 2015 *AISTech 2015 Iron and Steel Technology Conference and 7th International Conference on the Science and Technology of Ironmaking, ICSTI 2015* (Cleveland, United States) **1** 1225–32
[17] Spirin N A, Lavrov V V et al 2016 *Metallurgist* **60(5-6)** 471–477
[18] Gurin I A, Lavrov V V et al 2017 *Steel in Translation* **47(7)** 463–468
[19] Lavrov V V, Spirin N A et al 2017 *Steel in Translation* **47(8)** 538–543
[20] Lavrov V V, Gurin I A and Spirin N A 2018 *International Russian Automation Conference RusAutoCon* (Sochi, Russia, Sept 9-16)

IOP Conf. Series: Materials Science and Engineering **866** (2020) 012043 doi:10.1088/1757-899X/866/1/012043

Expert system of diagnostics blast furnace process

N A Spirin, I A Gurin and A S Istomin

Ural Federal University n.a. the first President of Russia B.N. Yeltsin, 28 Mira Street, Ekaterinburg, 620002, Russia

E-mail: n.a.spirin@urfu.ru

Abstract. The expert system of diagnostics blast furnace process is presented. It is based on a logical-mathematical model for assessing the progress of blast furnace smelting. The model provides an opportunity to evaluate the normal operation mode of blast furnace and further deviations from this mode such as overdeveloped gas flows (peripheral and central), violation of thermal melting conditions (hot and cold course of melt), violation of smooth descent of burned materials in the furnace (tight furnace operation, higher and lower suspension of burden). The functional capabilities of developed software are represented.

1. Introduction

To improve the technologies of iron smelting and solve the tasks of control of such complex and power-intensive process units as blast furnaces it is necessary to use the intelligent expert control systems [1-9]. The contemporary process of iron smelting in the blast furnace is subject to the influences of numerous and variable controlled and uncontrolled factors which cause the violations of blast-furnace melting operation. One of the methods to diagnose and control the blast-furnace melting operation consists in using the expert systems that include either determined knowledge about the process or formalized practical knowledge of blast furnacemen. The state analysis of question regarding the real practical use of mathematical models of blast furnace management on a real-time basis provides an opportunity to state that these methods and corresponding software are not practically developed today thus determining the actuality of this paper [10-15].

2. Detection of the normal operation mode of blast furnace

The deviation of ΔX_i feature characterizing the furnace operation during the base X_i^B (set values) and project period X_i^P in i modulus shouldn't exceed the accepted ΔX_i^{acc} value that is a model setting.

$$\Delta X_i = \left| X_i^B - X_i^P \right| \le \Delta X_i^{acc} \quad (1)$$

If condition (1) is met («True»), the value '1' will be assigned to i's identifier of P_i indication. In other way ('False'), it will be equal to 0. Therein, all indications are ranked. Each of indications is assigned with the value of its R_i rank that is changed from 0 to 1 and determined by means of the expert evaluation.

The probability of normal operation of blast furnace (P_n) is calculated according to the following correlation:

IOP Conf. Series: Materials Science and Engineering **866** (2020) 012043 doi:10.1088/1757-899X/866/1/012043

$$P_n = \sum_{i=1}^{n} \left(P_i \times \frac{R_i}{\sum_{i=1}^{n} R_i} \right) = \sum_{i=1}^{n} (P_i \times \alpha_i), \qquad (2)$$

where α_i – weighting factor of i's indication identifier changing in the range of 0 to 1; n – number of indications.

To identify the normal operation mode of blast furnace, the values of X_i^B indicators are used for model setting and corrected for operating conditions of the certain blast furnace. There are 15 controlled indications used in the model to identify the normal operation mode of blast furnace. Unlike the known works of other authors, this research integrates the complex of controlled indications and designed parameters of blast-furnace melting operation. The mathematical support and software additionally use the complex of 9 main estimate indicators applied in the model of blast furnace melting operation and adapted from MMK perspective. Thus, the number of indicators is equal to 24.

The main complex design parameters for diagnostics of blast-furnace melting operation embrace the following ones [15-17]:

- *Temperature condition* – generalized parameters characterizing the thermal state of upper (index of the shaft thermal state etc.) and lower part of the blast furnace (theoretical flame temperature; bottom thermal state index, i.e. a specific enthalpy of melting products including the iron melting heat except of slag formation heat (heat input for physical heating of melting products and recovery of oxides of hard-to-recover elements in the iron according to the reactions of direct iron recovery etc.)).
- *Gas-dynamic mode* – degree of compensation of burden material by means of gas in the upper and lower parts of furnace as well as in the separate annular areas etc. The forecasting and evaluation of gas-dynamic parameters of the blast-furnace melting operation.
- *Slag mode* – viscosity and viscosity polytherms of final slag; slag melting temperature; slag viscosity at given temperature; gradients of slag viscosity, basicity of slag; melting temperature, melting temperature range of iron-ore materials and thickness of zone of IOM viscoplastic state.
- *Melting intensity* – volume of melt burden per unit time, m^3/ min. The main complex parameters (some of specified design parameters) were used for further diagnostics of blast-furnace melting operation.

The configuration and location of zone of viscoplastic state of iron-ore materials (melting temperature and temperature range of iron-ore materials melting; height and thickness of zone of IOM viscoplastic state in the different vertical elements of blast furnace) are additionally identified apart from the modes evaluation.

The algorithm scheme for identification of deviations from the normal operation mode of blast furnace is presented in figure 1.

3. Detection of deviations from the normal operation mode of blast furnace

The diagnostics of the following deviations from the normal operation mode of blast furnace is specified:

1. Violation of gas flow stability (peripheral and central gas flows).
2. Violation of thermal melting condition (hot and cold course of melt).
3. Violation of smooth descent of burned materials in the furnace (suspension of burden (upper and lower), tight furnace operation).

In addition, the same methodology as for evaluation of normal operation mode of blast furnace is used. To find out the type of melting deviation from the normal operation mode it was considered expedient to compare two periods. One of such periods is a base one for which the values of melting parameters are a model setting that characterizes its normal operation mode. The second (project period) is chosen to identify the types of deviations from the normal operation mode of blast furnace. The data are collected during this period within 2 hours.

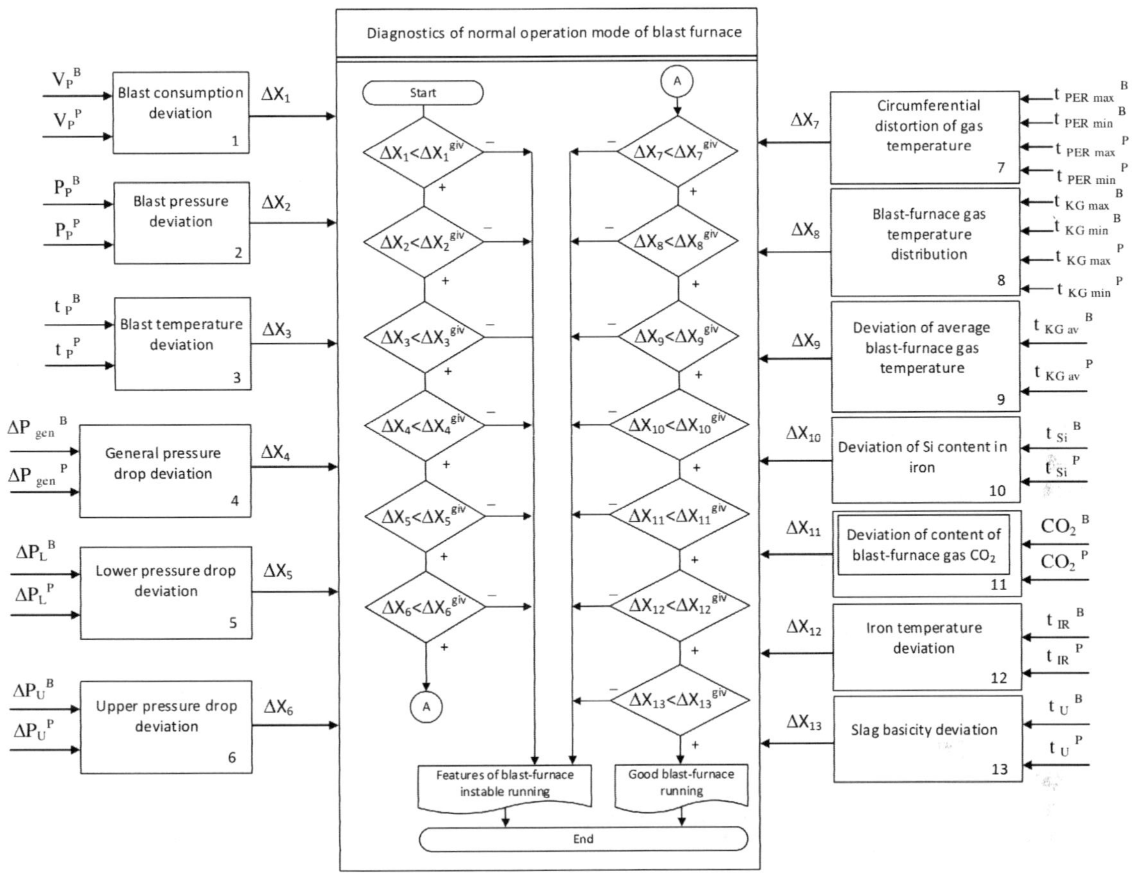

Figure 1. Fragment of algorithm identifying the deviations from the normal operation mode of blast-furnace melting.

The diagnostics of normal operation mode of blast-furnace and types of deviations from the normal operation mode of blast-furnace, the number of controlled and design (due to model) parameters for evaluation of these modes are represented in table 1. Thus, 111 parameters including 40 complex design values are used to evaluate the blast-furnace melting operation.

4. Functional modeling

The formation of functional model was the first stage of program product development. It was developed by means of AllFusion Process Modeler (BPwin) program according to IDEF0 (Integrated computer aided manufacturing DEFinition) standard.

The use of IDEF0 technique provided an opportunity to create the functional structure of program complex, find out its activities and connections between those activities, administrative actions and mechanism to execute every function that finally allowed the prevention of possible errors in early stages of design.

Table 1. Number of controlled and design (due to model) parameters for evaluation of normal operation mode and types of deviations from the normal operation mode of blast-furnace.

Type of deviation from the normal operation mode	Number of controlled parameters	Number of complex design parameters	Total
Normal blast furnace operation mode	15	9	24
Violation of gas flow stability:			
– peripheral gas flow;	9	5	14
– central gas flow.	8	5	13
Violation of thermal melting condition:			
– hot course of melt,	8	4	12
– cold course of melt.	8	4	12
Violation of smooth descent of burned materials in the furnace:			
– upper suspension of burden;	8	4	12
– lower suspension of burden;	7	4	11
– tight furnace operation	8	5	13
Total:	71	40	111

The general number of decomposed blocks of functional model comprises 90 blocks. The fragment of 1st level chart of functional model for subsystem that predicts the occurrence of deviations in the course of blast furnace melting shown in figure 2 embraces the following functions:

- *"Data collection and preprocessing" (A1)* ensures the automatic filling of system with the data from CAPCS and enterprise information system (EIS).
- *"Detection of features of normal blast furnace operation" (A2)* ensures the revaluation and analysis of parameters deviation features.
- *"Detection of features of deviation from the normal blast furnace operation" (A3)* ensures the revaluation and analysis of such deviations as peripheral gas flow; central gas flow; hot course of melt; cold course of melt; tight furnace operation; upper suspension of burden; lower suspension of burden; number of feeds.
- *"Development of recommendations on the maintenance technologies of blast furnace melting operation» (A4)*. The recommendations on the melt process correction are developed based on found deviations.

The results of functional modeling permitted to further proceed to the next stages of system development, i.e. creating the software architecture and implementation.

The "Identification of deviation type of blast-furnace melting operation" software in the form of web application is developed in accordance with the modern principles of development of application programs (functionality, expandability, integration with databases, user-friendly interface, safety, information estimation) [18-20]. .NET technology is the basis for program implementation thus giving the additional freedom to choose the platform and programming language. The program product is written in C# language with the use of Microsoft Visual Studio 2015 programming environment.

The program module is a part of system that optimizes the technological process of blast furnace melting being included into the composition of automated information system of analysis and forecasting of manufacturing situations of 'MMK' PJSC blast-furnace shop and intended for the engineering and technology personnel.

Figure 4, as an example, shows the home screen of program where the blast furnace working state is represented. Three types of violations are considered in the course of analysis: thermal conditions violation, gas flow violation and burden descent violation. The state of every deviation is specified. Moreover, there is a possibility to provide the detail information on the deviation. The form represented in figure 4 as diagram shows the information about parameters influencing the probability of occurrence of the certain deviation.

IOP Conf. Series: Materials Science and Engineering **866** (2020) 012043 doi:10.1088/1757-899X/866/1/012043

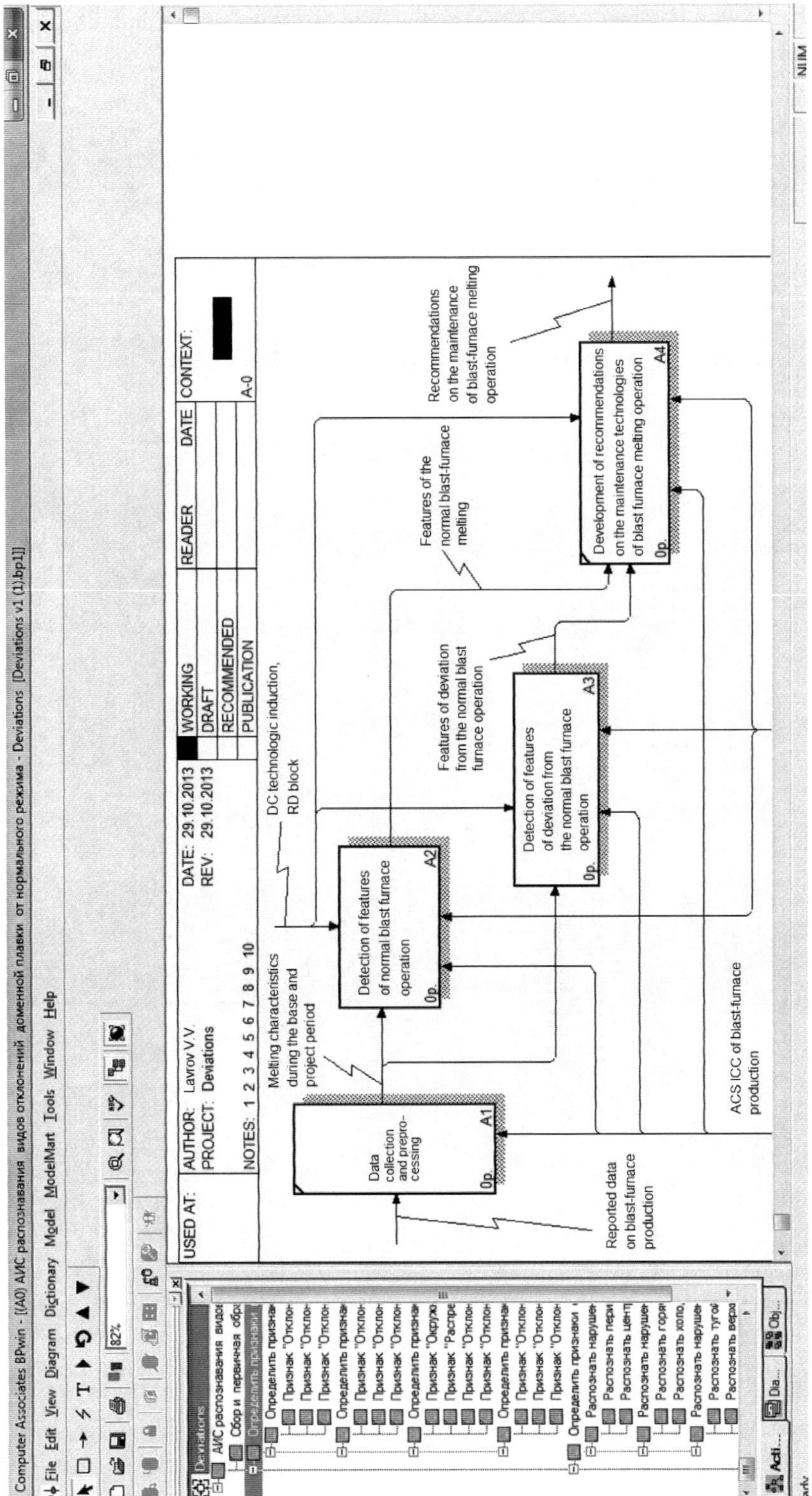

Figure 2. The first level of functional model of information system that identifies the types of blast-furnace melting deviations.

IOP Conf. Series: Materials Science and Engineering **866** (2020) 012043 doi:10.1088/1757-899X/866/1/012043

Figure 3 shows the architecture of information system software in which the basic components of its implementation are pointed out.

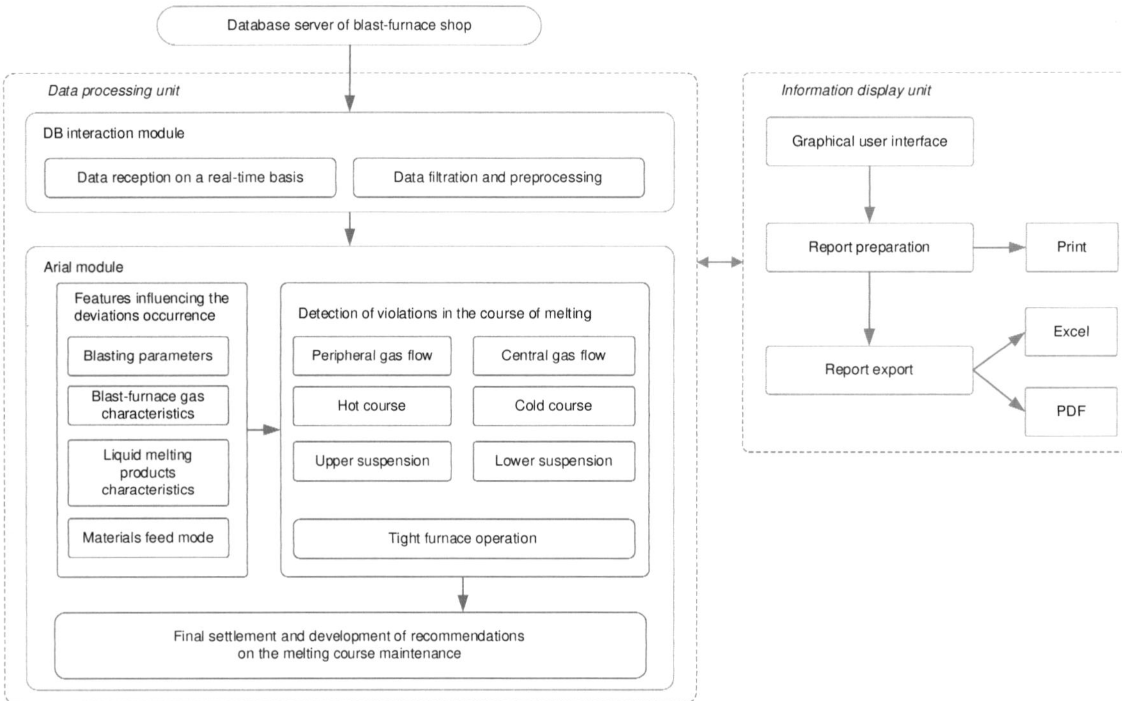

Figure 3. Software architecture.

Figure 4. Fragment of software screen identifying the type of deviation of blast furnace melting from the normal operation mode.

5. Conclusion

The expert system of diagnostics blast furnace process is developed. It is based on a logical-mathematical model for assessing the progress of blast furnace smelting.

IOP Conf. Series: Materials Science and Engineering **866** (2020) 012043 doi:10.1088/1757-899X/866/1/012043

The expert system, logical-mathematical model and program product are intended for the current assessment of blast furnace smelting.

The use of computer information logical system allows the operating personnel to perform the diagnostics of blast-furnace operation on a real-time basis as well as solve the operational tasks relating to the controlling of blast-furnace melting technology.

References

[1] Storm P 1998 *Proceedings of the Int. Conf. of the Manufacturing Value (Chain)* **2** pp 623-630

[2] Frerichs D K 1992 *Tappi Journal* **75(6)** 138–141

[3] Lida O et al 1992 *Kawasaki Steel Techn. Dept* **26** 30–37

[4] Yagi J 1991 *ISIJ International* **31(5)** 387–394

[5] Vapaavuori E 1997 *Expert Systems with Applications* **12(3)** II

[6] Matsuzaki S et al 2006 *Nippon Steel Technical Report* **94** 87–95

[7] Iida M et al 2009 *ISIJ International* **49(8)** 1123–1132

[8] Hera P et al 2011 *U.P.B. Scientific Bulletin, Series B* **73(4)** 171–182

[9] Ueda S et al 2010 *ISIJ International* **50(7)** 914–923

[10] Luo S 2013 *Asian Journal of Control* **15(2)** 553–561

[11] Chu M S et al 2014 *Iron and Steel* **49(11)** 1–8

[12] Ariyama T 2014 *ISIJ International* **54(7)** 1457–1471

[13] Bi X et al 2013 3rd *Proc. of the International Conference on Mechatronics and Intelligent Materials* (MIM 2013) **706-707** 1971

[14] Kuang S et al 2018 *Steel Research Int.* **89(1**) 1700071

[15] Spirin N A et al 2011 *Model Systems of Decisions Support in CAPCS of Metallurgical Blast Furnace Smelting* (Ekaterinburg: UrFU) p 462

[16] Spirin N et al 2015 *AISTech 2015 Iron and Steel Technology Conf. and 7th Int. Conf. on the Science and Technology of Ironmaking, ICSTI 2015* (Cleveland, United States) **1** 1225–32

[17] Spirin N A et al 2016 *Metallurgist* **60(5-6)** 471–477

[18] Gurin I A et al 2017 *Steel in Translation* **47(7)** 463–468

[19] Lavrov V V et al 2017 Steel in Translation **47(8)** 538–543

[20] Lavrov V V et al 2018 *Proc. of the Int. Russian Automation Conf. (RusAutoCon)* (Sochi, Russia, Sept 9-16)

IOP Conf. Series: Materials Science and Engineering **866** (2020) 012044 doi:10.1088/1757-899X/866/1/012044

Resource and energy saving technology for producing high-quality steel castings with heat-time treatment

D A Lubyanoy[1], R O Mamedov[2,3], S V Knyazev[3], B M Sokolov[2,3] and N V Oznobikhina[2,3]

[1]Kuzbass State Technical University, 7 Ordzhonikidze Street, Novokuznetsk, 654000, Russia
[2]"Kuznetskoye Litye" LLC, 7 Ferrosplavniy proezd, build. No. 1, Novokuznetsk, 654034, Russia
[3]Siberian State Industrial University, 42 Kirova Street, Novokuznetsk, 654007, Russia

E-mail: Lubjanoy@yandex.ru

Abstract. Today, the current direction in the development of steel production is resource and energy saving. So, one of the problematic issues in the steelmaking process is to reduce the consumption of ferroalloys. The possibility of improving the quality of castings, due to the heat-time treatment of the melt is considered. The melt is overheated to high temperatures by holding in the furnace. Because of that, the yield of suitable products increases, the metal consumption of products decreases.

1. Introduction
In modern conditions of industrial production, the improvement of the technological and service properties of metal products, the achievement of a world-class level of quality indicators, is possible only if the fundamental scientific research is deepened and new technological solutions are developed on their basis. The production of the vast majority of metallic materials is associated with their smelting. Therefore, the issues of obtaining fundamental data on the structure and properties of liquid metal, their changes as a result of various external influences are of significant importance.

As a rule, the classical technology of steel production from a fresh batch includes an oxidative melting period in which 0.2 – 0.5 % C is oxidized [1]. When iron oxides are added to the bath with a melt, in the form of iron ore concentrate, metal oxidation is also performed at the same time. It was found that during the oxidation period of the smelting during the carbon burnout process, not only harmful gases are removed, but also useful elements such as manganese, chromium and silicon.

2. The problem and its solution
Resource and energy saving is currently the most important direction in the development of steel production. Reducing the consumption of ferroalloys in the process of steelmaking is an urgent task. For this reason, if possible, it is necessary to reduce the oxidation period during smelting. From the data analysis in [2], it can be seen that when burning 0.1-0.15% C from the initial value, the maximum removal of hydrogen is achieved. Then, the hydrogen content decreases less intensely. Its final content depends on the raw lining of furnaces, an undried chute and a casting ladle, charge materials,

ferroalloys and various additives introduced into the metal or into slag during melting and casting, the atmosphere surrounding the liquid metal during the production and casting of metal, etc.

According to the published data [2], the carbon content in the metal by melting should be 0.1 – 0.2% higher than in the finished steel. From the point of view of P.Ts. Peshev and R.P. Todorova high-quality deoxidation of steel is the most important means of preventing sieve shells. To deoxidize one tonne of steel, it is recommended to introduce 1 – 1.3 kg of aluminum. The residual amount of aluminum in steel should be more than 0.03% [3].

This means that the deoxidation of steel and the presence of residual aluminum in it have a decisive influence on the most important indicators of steel quality such as mechanical properties and sieve-like shells [4]. For this reason, one of the main tasks of the oxidation period to remove hydrogen in the metal becomes less significant [5]. As a result, oxidation of a high carbon content from the melt, which is accompanied by the oxidation of useful impurities such as silicon, manganese, and chromium, is not rational in meltings for castings.

In the conditions of modern production, it is often not possible to obtain castings of the required quality without additional processing. This is due to many factors, for example, metal heredity, crystallization rate, etc. Many years of production experience suggests that one of the most important reserves for reducing structural and chemical heterogeneity of steel, improving the quality of steel products in steel production, is the improvement of the technology of the smelting process to obtain an equilibrium, most uniform melt before crystallization [6]. Crystallization of the metal occurs from a nonequilibrium state, which enhances the chemical and physical heterogeneities of the solid metal, reduces its service characteristics and leads to significant deviations of its quality from melting to melting.

There are different ways to transfer the melt into equilibrium. The most affordable way to obtain equilibrium melt is a high-temperature effect. Classical heat treatment does not always allow castings with the necessary structure to be obtained. The authors of [6, 7] revealed that the preparation of a melt for crystallization by heat-time treatment has an effect on the solidification process and helps to improve the structure and properties of solid metal.

Extensive research conducted at Zlatoust and other metallurgical plants showed that the use of heat-time treatment for the production technology of a number of alloyed steel grades gives good results, leading to improved mechanical properties, reduced scrap, etc. [7].

Heat-time treatment of the melt leads to a decrease in its chemical and physical heterogeneity, a reduction in the size and number of shrinkage defects. Thermal treatment allows the modes of heat treatment to be significantly reduced and simplified and the energy costs for heat treatment of steel castings to be reduced. The mentioned above phenomenon, as a rule, is accompanied by an increase in ductility, corrosion resistance and other characteristics of the quality of the alloys. The structure becomes more uniform, and the hardness is maintained at the same level, which allows higher mechanical properties of the castings to be obtained. There is a recrystallization and ordering of metal grains. This structure helps to increase the hardness and wear resistance of steel, which are practically not inferior to those obtained with standard hardening. The yield of suitable products increases, the metal consumption of products decreases.

In order to improve the quality of products and reduce their cost in the conditions of LLC "Zavod Gidromash", melting was performed using thermal heat-time treatment on DSP-1.5 furnace with acid lining. Quite interesting results were obtained by remelting the "returning" low-alloy steel grades (40KhL, 35 KhGSL), etc. With the melting technology without an oxidizing period, with a carbon content 0.24%, in the melt there is 0.26% of manganese (figure 1). And using the melting technology with an oxidizing period, with a decrease in carbon content from 0.52% to 0.34%, the manganese content decreases from 0.34% to 0.08%. This shows that the manganese of the melting period, according to the technology with oxidation, is actually almost completely oxidized.

Moreover, with the technology of melting with an oxidizing period, part of the hydrogen is also introduced by slag-forming additives. Therefore, it is considered advisable to use technology with minimal oxidation of carbon during the melting process, a slight oxidation period and the maximum

IOP Conf. Series: Materials Science and Engineering **866** (2020) 012044 doi:10.1088/1757-899X/866/1/012044

use of slag from the melting period [5]. Part of the hydrogen is removed during the oxidation of carbon by the oxidized slag of the melting period. This technology ensures a minimum loss of useful impurities, as a result of which, the consumption of ferroalloys for a number of melts is reduced by almost two, and in some cases the addition of ferromanganese is completely eliminated, while acceptable mixing of the metal, which is one of the main tasks of the oxidation period, is nevertheless carried out, due to the fact that the metal in the furnace in some cases corresponds to either boiling or semi-quiet (the silicon content in the metal is up to 0.11%), which ensures its sufficient mixing in the bath volume of the furnace [8].

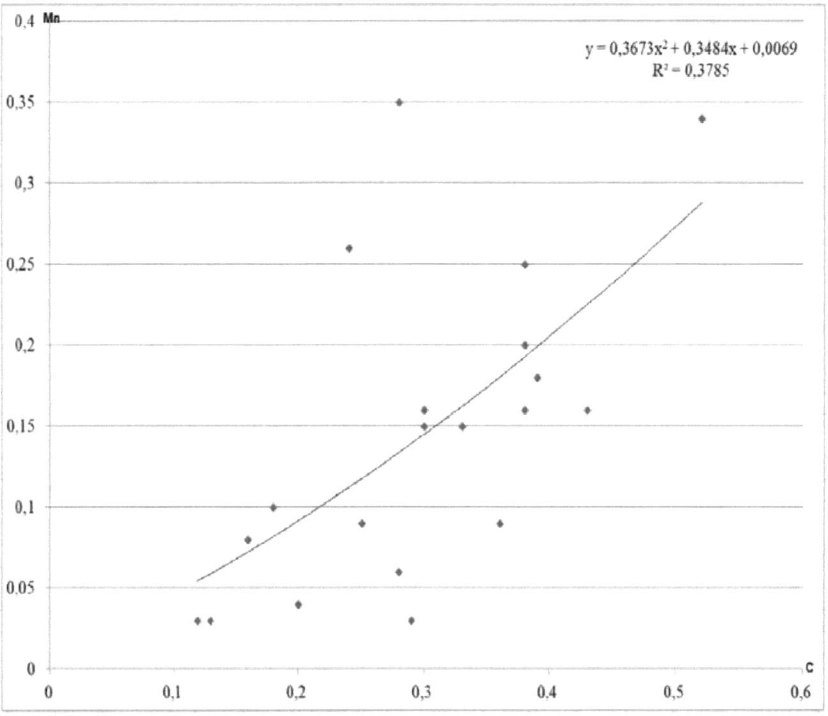

Figure 1. Dependence of Mn content on the C content by melting.

In the period from December 2017 to February 2018, comparative tests of steelmaking technology without an oxidizing period with heat-time treatment and smelting technology usind a fresh batch with full oxidation were conducted. From the conducted melts, we see that there is a decrease in the consumption of cast iron and ferroalloys during smelting of steel 35L in DSP–1.5 (figure 2).

From the histogram, the saving in consumption by the new technology can be seen:

- cast iron 12%;
- ferromanganese 14.6%;
- ferrosilicon 12.7%.

Also, the exclusion of holding time for express analysis and the duration of the smelting of "fresh" carbon charge with oxidation reduces significantly (up to 1 hour or more) the duration of the entire melting.

In addition to effective metal deoxidation (figure 3), when using the studies performed, the optimum quality of the castings is achieved by conducting thermal treatment of the melts, in which the melt is overheated to a temperature of 1650-1700 ^{0}C (for hydro-castings 1680-1710 ^{0}C) with a holding time in the furnace 10 -15 min, which is an effective heat-time treatment and improves the quality of castings. Aluminum, at the end of the heat, is fed to the bottom of the ladle, thereby producing the final deoxidation [8, 9, 10, 11, 12].

IOP Conf. Series: Materials Science and Engineering **866** (2020) 012044 doi:10.1088/1757-899X/866/1/012044

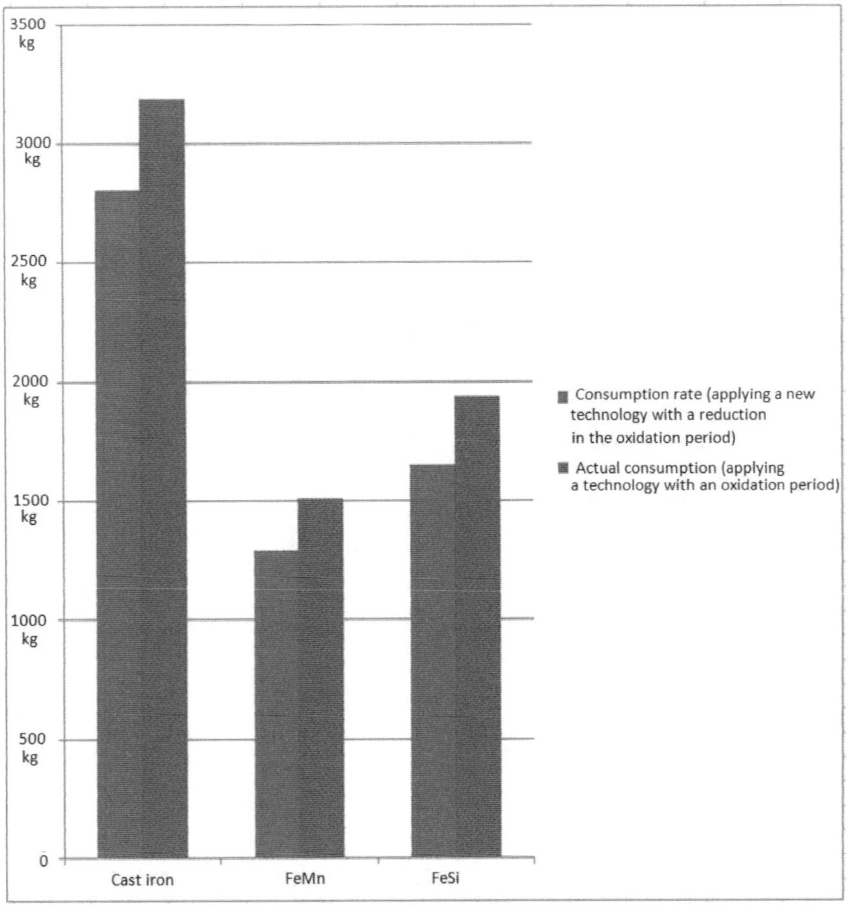

Figure 2. Differences in the technology of smelting without an oxidizing period from the technology of smelting with an oxidizing period.

Figure 3. Casting (bearing housing) with effective metal deoxidation and heat-time treatment.

Figure 4. Responsible casting (rotor disk) without blown holes.

3. Conclusion

Implementation of this metal smelting technology allows obtaining acceptable quality of steel castings with minimal formation of blown holes (figure 4). The described technology has been tested and implemented at LLC "Plant Gidromash" in Novokuznetsk.

References

[1] Povolotsky D Ya and Gudim Yu A 1987 *Alloy Steel Smelting in Arc Furnaces* (Moscow: Metallurgy) p 136
[2] Kramarov A D 1958 *Steel Production in Electric Furnaces* (Moscow: GNTI CHITSM) p 439
[3] Todorov R P and Peshev P Ts 1984 *Defects in Ferrous Alloy Castings* (Moscow: Mashinostroyeniye) p 184
[4] Soyfer V M and Kuznetsov L N 1989 *Arc Furnaces in a Steel Shop* (Moscow: Metallurgiya) p 176
[5] Lubyan D A, Karashkevich B N et al 2017 *Deformation and Destruction of Materials* 504-505
[6] *Collection of Scientific Papers: Ferrous Metallurgy of Russia and the CIS Countries in the XXI Century* 1994 vol 3 (Moscow)
[7] Kudrin V A 2003 *Theory and Technology of Steel Production: Textbook for University* (Moscow: Mir, ACT) p 528
[8] Lubyanoy D A, Karashkevich B N et al 2018 *II Int. Sci. and Pract. Conf. in Modern Scientific Achievements of Metallurgical Heat Engineering and Their Implementation in Industry* (Ekaterinburg: UrFU) pp 122–125
[9] Mamedov R O, Sokolov B M et al 2018 *Int. Sci. and Pract. Conf. in Fundamental Studies of the Main Directions of Technical and Physics and Mathematics* (Volgograd) pp 34–36
[10] Mamedov R O, Lubyanaya D A at al 2018 *II Int. Sci. and Pract. Conf. in Questions of Modern Science: Problems, Trends and Prospects* (Novokuznetsk) pp 40–42
[11] Cheprasov A I, Knyazev S V et al 2016 *IOP Conference Series: MSE* **150** 012026
[12] Antipenko V I and Knyazev S V 1987 *Liteinoe Proizvodstvo* **7** 34–37

IOP Conf. Series: Materials Science and Engineering **866** (2020) 012045 doi:10.1088/1757-899X/866/1/012045

Influence of samarium and europium on Al+1.9%Mn alloy phase composition and microstructure

T M Umarova and O Normahmedov

Lomonosov Moscow State University in Dushanbe, 35/1 Bokhtar Street, Dushanbe, Tajikistan

E-mail: umarova04@mail.ru

Abstract. The experimental results were proposed on the base of the research of the X-ray phase and microstructural analysis of eutectic structured aluminum-manganese alloy, modified by samarium and europium (up to 0.10 wt. %). The optimal concentration of europium was determined by the metallographic microscopy in order to obtain a fine dispersed structure. The results of microstructural analysis indicate the effectiveness of modifying the aluminum-manganese alloy of the eutectic composition with samarium and europium, especially Eu, in the existence of which the primary precipitates become compact, leading to a spherical shape, which allows us to consider the synthesized Al-1.9% Mn- 0.05% Eu alloy in the future as a more promising (compared with the Al + 1.9% Mn alloy doped by samarium) structural material with improved physicochemical properties.

1. Introduction

Alloys of the Al-Mn system are the most important for welded structures and cold stamping. The high ductility, good weldability (by all types of welding), high corrosion resistance, and plastic deformation hardens are characterized by the most famous AMn and AMn1 brands industrial alloys of the alloys almost twice [1].

For a long time this system had been used in industry, however, nowadays this alloys continue to attract the attention of the scientists to interesting patterns of changes in structure and properties: because of supersaturated solid solutions deformation at a process of fast crystallization rates and the solid solution the decomposition upon subsequent heating, etc.

Accelerated cooling of aluminium-manganese alloys, creates the crystallization conditions of sharp supercooling, which causes the eutectic point to shift to the region of high manganese concentrations. By change the period of crystalline and metallographic analysis was found that the most of the manganese in continuous casting remained in the supersaturated solid solution. It was found that the supersaturated solid solution formed during crystallization along manganese with subsequent heating to 450 - 500°C disintegrates, as this temperature corresponds to the tempering temperature of alloys of this system [2].

It is known that the alloys of high electrical conductivity could be used in electrical products as well as aluminum alloys having low electrical conductivity (high resistivity). These alloys are used for motor rotors because the requiring lower conductivity than aluminum and its conductor alloys. They are casting alloys containing 1-2% Mn and/or 0.8 to 13% Si [2].

The aluminum alloy with 0.8- 2.0% Mn containing cerium and nickel for finefilm conductors and microcontact pads could be f practical used [3].

IOP Conf. Series: Materials Science and Engineering **866** (2020) 012045 doi:10.1088/1757-899X/866/1/012045

Not only Cerium, but other rare-earth metals (REM) are increasingly used in aviation make a certain contribution to changing the structure and properties of Aluminum alloys. Since the solubility of REMs metals in aluminum is extremely small, but even that alloys no more than 0.5% (by weight), significantly affect the properties of aluminum alloys. In this case, it is necessary to note the dual role of REM alloying additions: on one hand, lanthanides are effective refiners, because, due to their high chemical activity, they neutralize the harmful effects of impurities, forming refractory chemical compounds, and on the other hand, as surface-active elements, they located on the phase interfaces (grain boundaries, block boundaries, interphase boundaries), strengthening these surfaces and inhibiting the development of diffusion processes.

Therefore, REMs have a positive effect on the structural stability of alloys, reduce their segregation heterogeneity, and prevent the formation of harmful structural components.

In the melting process, microadditives grind grain neutralize impurities, linking them into finely dispersed compounds, while improving the manufacturability of the alloy during subsequent processing by pressure [4]. The versatile positive effect of lanthanide microadditives makes it possible to improve existing and create new alloys for producing working blades of modern aircraft engines with a unique combination of various properties.

Thus, a promising way to improve the properties of structural materials is microalloying them with a metal of the lanthanide group, which, by modifying the structure, significantly increases the mechanical and technological properties of the alloys. Therefore, the use of REMs will be economically justified.

The aim it is to study the effect of REMs (samarium and europium) on the physicochemical properties of eutectic aluminum-manganese system alloys. The subject is a part of general research concerning the effect of REMs on aluminum alloys [5-7].

2. Research methods

In that research work, the alloys of the eutectic Al-Mn system are synthesized, modified with samarium and europium (alternately) in the content of: 0.01; 0.02; 0.05 and 0.10 wt.% Under vacuum at a temperature of T = 1000°C.

For that research alloys were obtained from aluminum grades: A995 (GOST 11069-74), since the influence of impurities (Fe and Si) are quite substantial and the solubility of manganese in aluminum sharply decreases existing iron, electrolytic manganese MP00 (GOST 6008-82) with an impurity of 99.95%, samarium - 99.9%, the latter were introduced ligatures Al-Mn (10%) and Al-Sm (3%), metal europium with a purity of 99.9%.

The metal charge was sealed in a quartz ampoule.

The following equipment and instruments were used for the synthesis and study of alloys: KS 400/10 furnace (Germany); vacuum post; NIKON SMZ 670 microscope, D2 Phaser diffractometer (Bruker, Germany) with CuKα (λ = 1.5418 A). Processing of the obtained data was carried out using the Diffract.eva software products.

Figure 1 shows the diffraction patterns of powders of an alloy of the Al-Mn system (1.9%) modified with europium (up to 0.10 wt.%). A similar X-ray phase analysis was carried out of Al + 1.9% Mn powders, modified with samarium (up to 0.10 wt.%).

According to the results, by the increase of the doping component (Sm and Eu) the diffraction peaks increase could be. The diffraction patterns of all samples are similar to all diffraction peaks from the PDF database.

For preparation the alloys samples: the Al-Mn-Eu and Al-Mn-Sm systems for microstructural analysis we have used the recommendations of the author [5]. Before analysis should be done the mechanical processing of samples (grinding and polishing to a mirror surface).

IOP Conf. Series: Materials Science and Engineering **866** (2020) 012045 doi:10.1088/1757-899X/866/1/012045

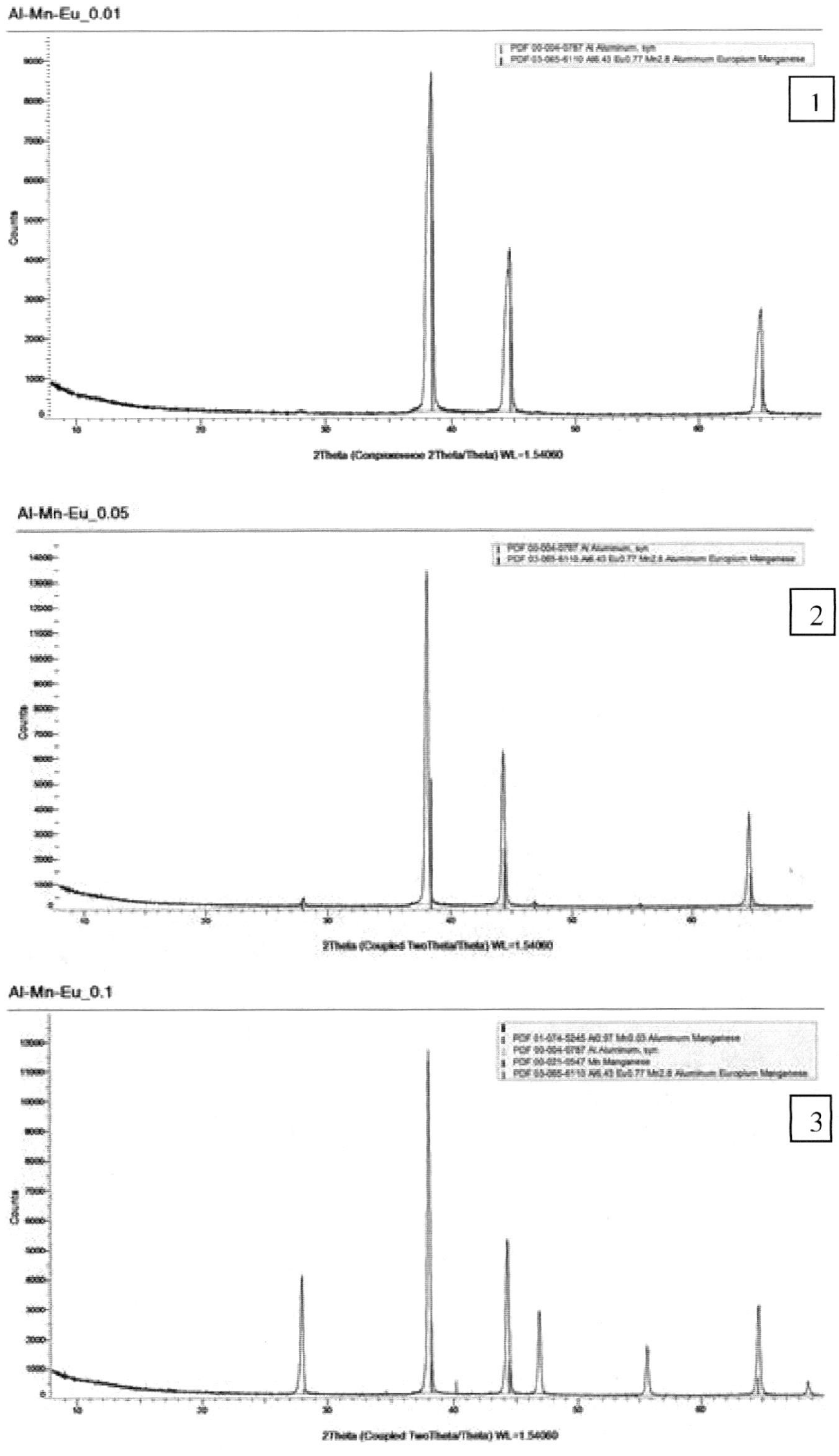

Figure 1. Diffractograms of Al + 1.9% Mn powders modified with europium in the contents of: 0.01 (1), 0.05 (2) and 0.10 wt.% (3).

To reveal the microstructure of the studied alloys, the samples were etched with various reagents, depending on the composition of the alloy and the presence of the phase present.

For example, reagent No. 53 (0.5% HF) was used for the studied alloys for 30–40 seconds (at a temperature of 500 °C). Microstructural analysis was carried out by Nikon optical microscope at a metallographic increase of 200 times, which made it possible to observe a change in the microstructure of the samples, depending on the composition. Figures 2 and 3 show the microstructures of the synthesized alloys of the Al-Mn-Sm and Al-Mn-Eu systems (respectively).

The aluminum-manganese microstructure alloy study by the eutectic composition modified with samarium to 0.05% (figure 2) indicated that it is a branched dendrite. The effect of europium on aluminum-manganese eutectic was more affective. Comparing the photographs of the microstructures of the Al-Mn-Sm and Al-Mn-Eu alloys, could be noted that, unlike samarium, the addition of 0.05% europium leads to the fine grained structure. The twofold increase in the concentration of europium (from 0.05 to 0.10%) the structure of the ternary alloy, where the fine-grained structure changes its shape, that in that sphere of the plate looks like needles.

Thus, by analyzing the obtained results, we noted that the interpretation of the diffraction patterns of the Al-Mn-Sm and Al-Mn-Eu alloys showed in both cases the presence of intermetallic compounds: Al 0.97 Mn 0.03 and Al 6.43 Eu 0.77 Mn 2.8, respectively.

A ternary compound was not found in the Al-Mn-Sm system in the selected concentration range (1.9% Mn and up to 0.10% Sm).

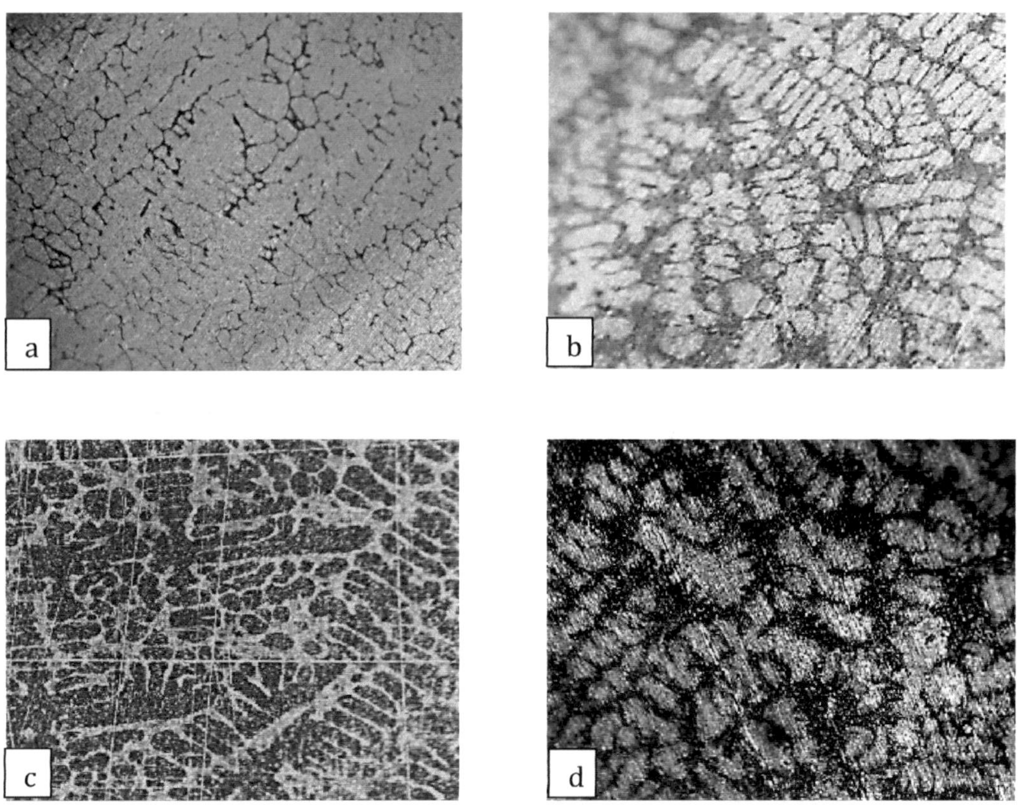

Figure 2. Microstructure (x 200) of Al + 1.9Mn modified alloy:
a – 0%; b – 0.01%; c – 0.02%, d – 0.05% wt. Sm.

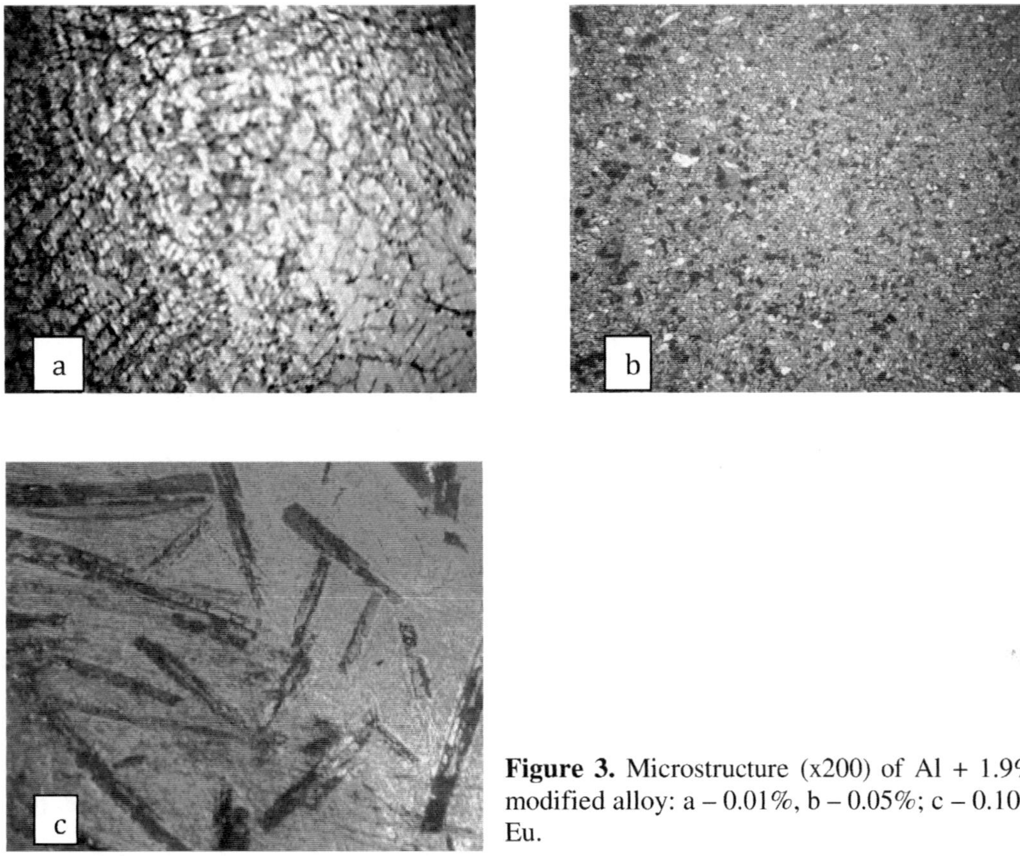

Figure 3. Microstructure (x200) of Al + 1.9% Mn modified alloy: a – 0.01%, b – 0.05%; c – 0.10% wt. Eu.

3. Conclusion

The results of microstructural analysis indicate the effectiveness of modifying the aluminum-manganese alloy of the eutectic composition with samarium and europium, especially Eu, in the existence of which the primary precipitates become compact, leading to a spherical shape (figure 3b), which allows us to consider the synthesized Al-1.9% Mn-0.05% Eu alloy in the future as a more promising (compared with the Al + 1.9% Mn alloy doped by samarium) structural material with improved physicochemical properties.

References

[1] Fomin B, Moskvitin V and Makhov S 2004 *Metallurgy of Secondary Aluminum* (Moscow: Metallurgiya) p 240
[2] *Industrial Aluminum Alloys* 1984 ed by S G Aliev, M B Altman and S B Hambartsumyan (Moscow: Metallurgiya) p 528
[3] Bessonov V A, Bogatov P N and Bochvar N R 1977 *Aluminum based alloy*. Copyright Certificate No. 544703. Description of the invention to copyright Certificate 980454 A
[4] Ospennikova O G 2012 *Aviation Materials and Technologies* (Moscow: VIAM) pp 19–36
[5] Umarova T M 2016 *Materials Science and Engineering Technology* vol 890 331–338
[6] Umarova T M and Ganiev I N 2008 *Anode Alloys of Aluminium with Manganese, Iron and Rare Earth Metals. Monograph* (Dushanbe) p 273
[7] Umarova T M, Jalolov F N and Ganiev I N 2019 *Reports of the Academy of Sciences of the RT* vol 62, No 3-4 203–207
[8] Kovalenko V S 1981 *Metallographic Reagents. Handbook* 3rd ed. p 120

IOP Conf. Series: Materials Science and Engineering **866** (2020) 012046 doi:10.1088/1757-899X/866/1/012046

Experience of ceramic production from silicon carbide with the addition of SiC nanopowder

Yu Nepochatov[1], A Bogayev[1], O Khasanov[2], E Dvilis[2], I Kuchumova[3,4] and R Khabirov[3]

[1]LLC "NANOCERAMICS", 220 Krasniy Ave., Novosibirsk, 630049, Russia
[2]National Research Tomsk Polytechnic University, 30 Lenin Ave., Tomsk, 634050, Russia
[3]Novosibirsk State Technical University, 20 K. Marx Ave., Novosibirsk, 630073, Russia
[4]M.A. Lavrentyev Institute of Hydrodynamics SB RAS, 15 Academician Lavrentyev Ave., Novosibirsk, 630090, Russia

E-mail: ivannakz@mail.ru

Abstract. The effect of silicon carbide nanopowder additives on the properties of armor ceramics from this compound has been investigated, and the results are presented. The study reveals that the powder introduction in an amount from 5 to 10 wt. % leads to an increase in hardness and crack resistance when density is maintained. It enabled us to obtain the material which can be used in structural and armored ceramics.

1. Introduction

Oxygen-free refractory compounds are characterized by high melting points, chemical bond strength, thermal conductivity, dielectric properties and chemical resistance. Silicon carbide (SiC) occupies an important place among them. Consequently, SiC is a promising material finding application as structural ceramics to manufacture parts of gas turbine engines, cutting tools and ceramic bearings. High strength of the chemical bonds in this compound allows it to be also used as lightweight armor, because most of the kinetic energy is consumed on breaking the chemical bonds of the armor material under dynamic impact.

Due to the covalent structure of the chemical bond in silicon carbide, ceramics based on this compound with high relative density can be obtained only by means of the solid-phase sintering mechanism at elevated temperatures (from 2050 to 2200 °C) [1]. To obtain dense ceramic from SiC at lower temperatures, it is necessary to apply liquid phase sintering with the use of modifying additives such as oxides in sufficiently large quantities, which leads to porosity formation and a decrease in strength [2, 3]. Another method appears to be an alternative to this ceramics manufacturing technique. This method considers reaction-sintered ceramics (Si-SiC) fabrication for a short period of time and at a lower temperature [4] via penetration of the liquid phase (liquid silicon) into the porous crude SiC preform. The most common current methods of dense SiC ceramics production are: pressureless sintering (PS-SiC), reaction sintering (Si-SiC), and hot pressing (HP-SiC). Silicon carbide-based ceramics are mainly obtained by reaction sintering (Si-SiC) and liquid phase sintering (LPS-SiC) methods.

The use of LPS-SiC ceramics as armor materials with the mechanical properties outweighing advantages of Si-SiC materials allows either increasing the armor resistance class of such ceramics with the same armor thickness or reducing the armor weight under maintaining the armor-defeating performance. However, all these methods have certain disadvantages. For instance, S-SiC materials are resistant to oxidation when being exposed to high temperatures, but their crack resistance is rather low (from 3 to 4 $MPa\cdot m^{1/2}$) [5, 6]. Moreover, an abnormal growth of SiC grain is observed in ceramics owing to high-temperature sintering and relatively long exposures at these elevated temperatures. Si-SiC fabricated by reactive sintering cannot be used at high temperatures due to the low melting point of its silicon constituent. Along with this, the hot pressing method (HP-SiC) is energy-consuming, expensive, and inefficient, that's why it has not found wide application. The complete compaction can be achieved via high-temperature sintering during short exposure periods, which is accomplished by the method of spark plasma sintering (SPS). The use of additives from metal and ceramic nanopowders as modifiers is considered to be one of the ways to the problem solution. SiC nanopowder modifier is known to be the most promising additive that allows the efficient control over the ceramics structure based on silicon carbide.

The study objective is to propose approaches to increasing strength, hardness, and crack resistance of silicon carbide ceramics by means of SiC nanopowder introduction into its composition and the use of the SPS method to consolidate experimental samples.

2. Materials and experimental procedure

To conduct the experiments, we used 64C green silicon carbide produced by Volzhsky Abrasive Works. Its specific surface was 7.5 $m^2\cdot g^{-1}$ after grinding in a jet mill followed by sieving on a classifier. The average particle size of the base powder was equal to 1 μm after grinding. Silicon carbide nanopowder was obtained by mechanochemical activation in the AGO-2C ball planetary mill. Deagglomeration of the nanopowder was carried out by means of ultrasound with 20 kHz oscillation frequency and the processing time of 120 min. Its specific surface was 30 $m^2\cdot g^{-1}$. The average particle size of the silicon carbide nanopowder ranged from 75 to 100 nm after deagglomeration. The presence of predominantly hexagonal modification of silicon carbide in the sample (α-SiC) was determined by X-ray diffraction analysis with the use of an ARL X`TRA diffractometer and CuKα radiation. A Technai G2 FEI transmission electron microscope was used to determine the particle size of nanopowders, which ranged from 12 to 200 nm, but the particles with sizes up to 90 nm make up the largest volume fraction.

A powder mixture from base powder particles and a silicon carbide nanopowder was ground in a ball planetary mill. The mass ratio of the dry processed material (the nanostructured powder mixture) to the mass of the planetary mill balls was 1:1, the number of revolutions - 900 rev and the grinding time - 1 min. After grinding, the average particle size of the powder mixture was 0.9 μm. A suspension of a nanostructured mixture of silicon carbide powder was prepared with ethanol used for mixing substances with 1:5 volume-to volume ratio of "ethanol-mixture". When the silicon carbide powder was mixed with the silicon carbide nanopowder, the mass ratio of the dry processed material to the balls mass was 1:1. To provide the uniform distribution of nanopowder particles, the number of the planetary mill revolutions was 900 rpm, and the duration of the suspension preparation was 5 minutes. The suspension was dried in an oven at 100 °C for 10 hours, then, the mixture was rubbed through a sieve. The relative humidity of the nanostructured powder mixture amounted to 0.2 wt. % after drying. To determine the optimal amount of the introduced nanopowder, its content had to vary from 1 to 10 wt. % at a constant sintering temperature of 2200 °C. To determine the optimal quantitative content of the SiC nanopowder additive in the powder-based matrix of silicon carbide, model samples of ceramic armor elements with different content of the nanopowder additives were made in the form of 2.5 mm-thick disks with a diameter of 30 mm. The beams with standard sizes were also made for testing of bending strength. The method of spark plasma sintering (SPS) was applied to fabricate the model samples. SPS was carried out in the synthesis mode at a temperature of 2200 °C, under prepress pressure of 85 MPa, and 5 min exposure time at a maximum temperature. X-ray phase analysis and an energy dispersive elemental analysis were jointly applied with scanning electron microscopy (SEM) to control the phase composition of the powders and ceramics and to check the presence of extraneous phases.

3. Results and discussion

The obtained results confirm that the initial powder additives and fabricated ceramics satisfy the requirements for phase composition and chemical purity. The analysis of the porosity nature was carried out by SEM images (see figure 1) and showed the presence of small intergranular pores in the control sample (without nanopowder additives). Rather large pores are observed in the ceramics with the introduction of 1 % nanopowder (figure 1b), on the contrary, the samples with 5 % (figure 1c) and 10 % (figure 1 d) nanopowder additives practically do not contain pores. Based on the research results given in table 1, we can come to conclusion that the use of SiC nanopowder additives leads to a noticeable increase in the intergranular boundaries strength of ceramics. A significant decrease in residual porosity is apparent for samples with 5 wt. % and 10 wt. % nanopowder additives.

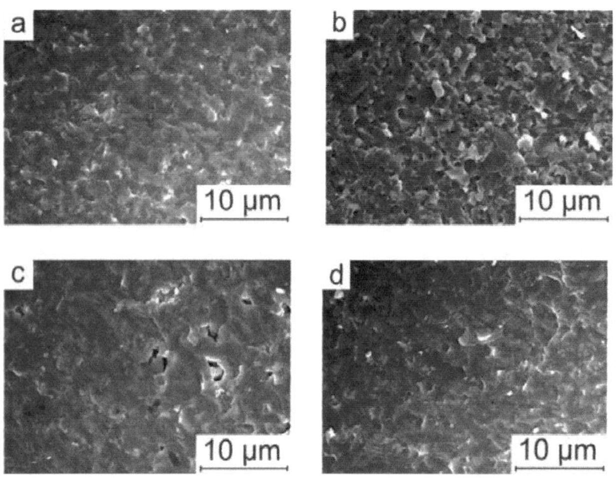

Figure 1. Morphology of cleaved SiC ceramics: (a) control sample; (b) 1 % nanopowder; (c) 5 % nanopowder; and (d) 10 % nanopowder samples.

Table 1. Physico-mechanical characteristics of the ceramics based on silicon carbide.

The content of the SiC nanopowder additive (wt. %)	Relative density (%)	Fracture toughness, K_IC (MPa·m$^{1/2}$)	Vickers hardness (GPa)
1	97.6	5.0	28.0
5	98.8	5.0	24.5
10	98.8	6.4	28.0

4. Conclusion

The use of silicon carbide nanopowder additives can significantly improve the strength characteristics of SiC-based ceramics. These properties are significantly enhanced for the ceramics containing 5 wt. % and 10 wt. % SiC nanopowder additives. Meanwhile, basic ceramic properties remain almost unchanged. The use of silicon carbide nanopowder additives leads to the formation of a less porous and more uniform microstructure of the material that, consequently, allows increasing the crack resistance of ceramics. Further optimization should be carried out with the ceramics containing 5 wt. % to 10 wt. % nanopowders.

References

[1] Riedel R and Chen L-W 2015 *Ceram. Sci. and Techn.* **2** 3–37
[2] Baud S, Thévenot F, Pisch A and Chatillon C 2003 *J. Eur. Ceram. Soc.* **23** 1–8
[3] Samanta A K, Dhargupta K K and Ghatak S 2001 *Ceram. Int.* **27** 123–33
[4] Grinchuk S *et al.* 2018 *J. Eur. Ceram. Soc.* **38** 4815–23
[5] Swain M 1994 *Materials Science and Technology* vol (New York: VCH Publishers Inc.) p 842
[6] Minori M, Said J, Hamid M, Makio N, Keizo U, Rolf W and Roger M 2002 *Improved Ceramics through New Measurement, Processing and Standarts* (American Ceramic Society. Westerville, OH) p 282

IOP Conf. Series: Materials Science and Engineering **866** (2020) 012047 doi:10.1088/1757-899X/866/1/012047

Effect of carbon nanotubes on ballistic armour performance of ceramics from boron carbide

Yu Nepochatov[1], V Kuznetsov[2], A Bogayev[1], A Bandin[1], A Abraamyan[1] and I Kuchumova[2,3]

[1]LLC "NANOCERAMICS", 220 Krasniy Ave., Novosibirsk, 630049, Russia
[2]Novosibirsk State Technical University, 20 K. Marx Ave., Novosibirsk, 630073, Russia
[3]M.A. Lavrentyev Institute of Hydrodynamics SB RAS, 15 Academician Lavrentyev Ave., Novosibirsk, 630090, Russia

E-mail: ivannakz@mail.ru

Abstract. The structure, phase composition and properties of hot-pressed mixtures based on boron carbide (B_4C) and carbon nanotubes (CNT) were studied. Used boron carbide powder was produced by LLC PIChT and CNTs were made in Russia and Korea. Sintering was carried out in vacuum at 2100 °C and under the pressure of 30 MPa. The samples of ceramics have been obtained with the ballistic armour performance, which is ~ 26% higher than that of ceramic samples without additives.

1. Introduction

One of the global trends in armor-plating of ground and aircraft equipment is the use of ceramic armor, which is lighter than steel and not inferior to it in terms of strength. Elements of ceramic armor find wide application in production of personal protective equipment, military automobiles, airplanes and helicopters. Boron carbide (B_4C) appears to be a promising ceramic material used for such armor. B_4C has extremely high hardness (38 GPa) and low specific gravity (compared with armored steel and aluminum oxide), and the Young modulus (460 GPa), which is close to the corresponding values for diamond [1].

The disadvantage of ceramics based on B_4C is a low value of crack resistance. One of the effective ways to increase the fracture toughness of ceramics is the formation of various heterogeneous structures that contribute to the deviation of the crack trajectory, its branching and lead to an increase in energy dissipation during fracturing. Therefore, the initial ceramic matrix is reinforced either with fibers or with dispersed solid particles. As a result of a dissimilar substances combination, we can obtain the materials that combine the initial strength of the ceramic matrix and increased crack resistance owing to reinforcing components.

One of the most promising approaches to improving the mechanical properties of B_4C-based ceramics is the introduction of reinforcing elements, i.e. particles and fibers into its matrix, which can significantly improve its strength characteristics such as the elastic modulus, crack resistance, strength, hardness, and increase of resistance to wear. For these purposes, the use of carbon nanotubes (CNTs) is the most promising, due to the uniqueness of their properties: high tensile strength (up to 1500 GPa), high Young's modulus (~1 TPa), high thermal and corrosion resistance, high thermal conductivity (> 3000 W·m^{-1}·K^{-1}), high electrical conductivity (> 105 S·m^{-1}) and low density (1.4 g·cm^{-3}) [2].

In [3], data are presented on the study of composite ceramics based on B_4C multi-wall CNTs (MWCNTs) fabricated by hot pressing. The isothermal exposure time was 60 min, pressing was carried out at 1950 °C, and pressure reached 60MPa. Modification of boron carbide MWCNTs (0.5…10 vol. %, diameter ~ 150 nm) results in high values of bending strength (~ 400 MPa), hardness (11 GPa), crack resistance (4.2 MPa·m$^{1/2}$) and change in the density of the obtained material. According to [4], MWCNTs (d ~ 150 nm) introduction into a boron carbide matrix leads to an increase in crack resistance (up to 7.9 MPa·m$^{1/2}$) of B_4C ceramics obtained by the method of spark plasma sintering. The increase of following parameters is also noted by the authors: resistance to bending (710 MPa) and microhardness (36 GPa). According to [5], modification of boron carbide with carbon nanotubes (average diameter 15 nm) tends to improve the mechanical properties of the fabricated composite compared to the initial material, in particular, the ultimate tensile strength.

The research was conducted on a single carbon nanotube coated with boron carbide. An abnormally high value of the flexural strength was observed (~ 25000 MPa), which is several orders of magnitude higher than in unmodified B_4C ceramics. The experimental data [6] also indicate a significant improvement in strength characteristics of carbon modified B_4C ceramics. Ceramic samples were obtained by hot pressing method (the duration of isothermal exposure was 120 minutes, temperature 2025 °C, pressure 200 MPa). It is shown in this study that crack resistance reaches a high value (4.2 MPa·m$^{1/2}$). The density of the resulting composite is 2.51 g·cm^{-3}, which is close to the maximum theoretical density of boron carbide [1]. The flexural strength of modified ceramics also increases and amounts to 564 MPa. The results obtained in [7] indicate the improvement of physical and mechanical characteristics of the ceramics obtained by hot pressing (duration of isothermal exposure 60 min, temperature 1850 °C, pressure 60 MPa) when a MWCNT (diameter of 40…60 nm) and a silicon activating additive (8 wt. %) are introduced into the boron carbide matrix. The data analysis shows that B_4C–Si–MWCNTs composites are characterized by high values of bending strength (506 MPa) and fracture toughness 5.7 MPa·m$^{1/2}$.

The objective of this research is to study the effect of carbon nanotube additives on the ballistic efficiency of boron carbide ceramics.

2. Materials and experimental procedure

During the experiments, boron carbide powder fabricated by Polygon of Innovative Chemical Technologies (PIChT), LLC and carbon nanotubes (CNTs) manufactured by OCSiAl (Russia) and Kumho Petrochemical Co., Ltd. (Korea) were used as feedstock material. The average particle size of boron carbide powder was 3.55 µm.

Its elemental composition was (wt. %): B 72.29, C 25.51, O 1.16, Al 0.06, Si 0.55, Fe 0.19 and Cu 0.24. To remove oxygen impurities (in the form of boron oxide) during hot pressing, an additional isothermal exposure was carried out at 1300 °C in vacuum for two hours. B_2O_3 evaporates at this temperature, and thereby the starting material is purified. The distinctive characteristics of OCSiAl CNTs are: 89 ± 2 wt. % carbon content (the rest is metal impurities); one layer; length ≈ 5 µm; average diameter 1.5 ± 0.4 nm; specific surface is ~ 510 m^2·g^{-1}. TEM image of a carbon nanotube manufactured by OCSiAl is shown in Figure 1.

Modified boron carbide ceramics were manufactured by the following techniques. A calculated sample of a CNTs aqueous suspension was added to an aqueous solution of organic binders and the resulting mixture was subjected to ultrasonic treatment for 10 minutes. Then, the dispersed CNTs solution was sputtered onto the initial powder in a XLB-5 dryer to obtain a granular boron carbide powder. Next, 59.5x59.5x7 mm plates and 59.5x7x7 mm standard samples were pressed from the granular powder, the binders were removed, the plates and standard samples were placed in a graphite mold, and hot pressing was performed on an FCT HP W-400 System at 2100 °C and under the pressure of 30 MPa. The obtained sintered samples were studied by X-ray phase analysis (XRD) and scanning electron microscopy (SEM) method. Density, porosity, bending and compression strength were measured on the standard samples. Additionally, micro-hardness was determined by indentation with the Vickers diamond pyramid. Ballistic testing of the elements was carried out with 7.62 mm caliber bullets fired out of the SVD rifle on flat samples being up to 9 mm thick.

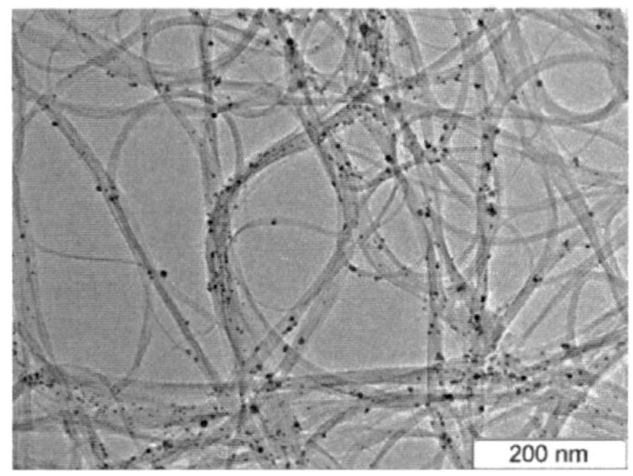

Figure 1. TEM image of CNTs manufactured by OCSiAl.

3. Results and discussion

When the samples were fabricated, the content of CNTs in ceramics was equal to (wt. %): 0.1, 0.3, 0.5, 3 and 5. In case the CNTs content amounted to 0.1, 0.3, 3 and 5 wt. %, low values of wet density were determined for pressed samples, and hot-pressed samples were characterized by low density and reduced strength. Consequently, in that case the optimal content of CNTs is 0.5 wt. %. X-ray analysis shows that the following phases are present in the sintered samples: B_4C, FeB and graphite. The presence of the latter could be explained by the reaction conditions:

$$4Fe+B_4C=4FeB+C$$

The presence of iron could be attributed to grinding of the grinding media (steel balls). With regard to the laws of thermodynamics such a reaction is quite possible. To avoid this undesirable reaction, it is advisable to use the ceramic grinding media. Figure 2 shows the micrographs of the sintered samples after electrochemical etching. It can be seen that boron carbide-based ceramics are a homogeneous materials without introduction of CNTs. However, ceramics appear to be composite materials with the addition of CNTs fabricated by OCSiAl. Similar results were obtained with the use of carbon nanotubes (CNTs) manufactured by Kumho Petrochemical Co., Ltd. (Korea).

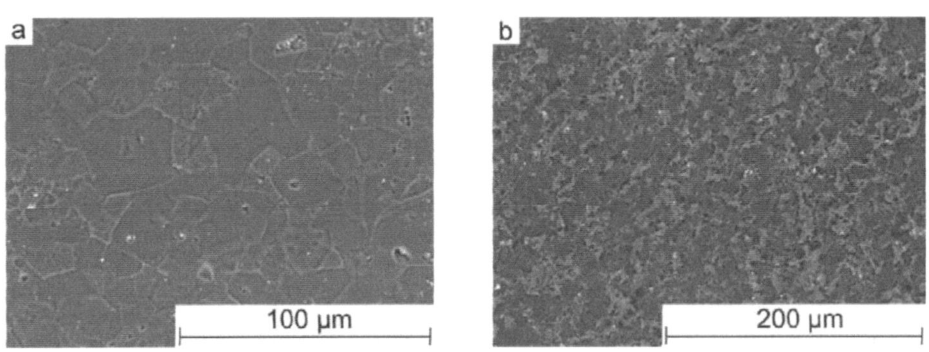

Figure 2. Micrographs of microstructures: (a) the sample of boron carbide; (b) the sample of boron carbide with the addition of CNTs (0.5 wt. %).

The indentations made with a diamond indenter on the samples are shown in Figure 3. The micrographs show the significant difference between fracture and crack growth patterns in the samples. With the addition of CNTs, a branched fracture pattern appears in the sample by means of the diamond pyramid. However, the indentation left with the diamond pyramid is clearly defined in the sample without introduction of CNTs, and it has clear boundaries in the form of a rhombus, from whose vertices straight cracks propagate along the sample. The fracture pattern in the sample with the addition of CNTs is more preferable from the point of view of the armor-defeating performance since the kinetic energy of the bullet will be more efficiently dissipated.

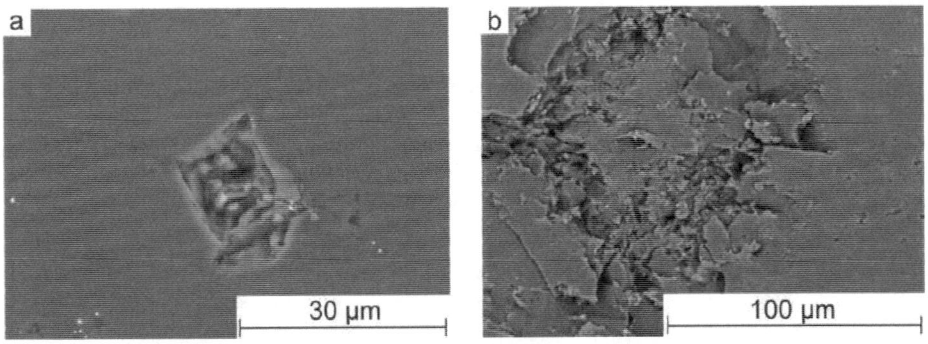

Figure 3. The indentations: (a) on the surface of the boron carbide sample; (b) on the surface of boron carbide sample with the addition of CNTs (0.5 wt. %).

Samples of 59.5×59.5×7 mm armored plates were made for ballistic tests without and with the addition of CNTs (0.5 wt. %). Armor-piercing bullets of 7.62 mm caliber were fired out of the SVD rifle on the armored plates at a distance of 10 meters, the speed of the bullet was 801 m/s. The ballistic armor performance was calculated based on the test results by the DoP method. The test results are given in table 1.

Table 1. Results of ballistic tests.

Parameter	Armoured ceramics based on Al_2O_3	Armoured ceramics based on B_4C	Modified armored ceramics based on B_4C	
			CNTs made by Kumho Petrochemical Co., Ltd. (Korea)	CNTs fabricated by OCSiAl (Russia)
Density ($g{\cdot}cm^{-3}$)	3.90	2.51	2.51	2.51
Porosity (%)	0.0049	0.0031	0.0032	0.0031
Hardness (GPa)	16.8	31.0	31.1	31.3
Bending strength (MPa)	354	481	458	450
Compressive strength (MPa)	2300	2562	2577	-
The speed of sound propagation in ceramics, $m{\cdot}s^{-1}$	11348	15548	14526	14623
Ballistic armor performance, (BAP)	5.13	6.93	8.55	8.77

With consideration for the data presented in the table, it is assumed that the ballistic armor performance (BAP) of the samples with the addition of 0.5 wt.% CNTs is 1.23 and 1.27 times higher for Korean and Russian CNTs, respectively, compared to BAP of boron carbide sintered without CNTs additives. It is related to the fact that fracturing of boron carbide modified with CNTs results in a branched fracture formation which is responsible for the efficient kinetic energy dissipation of the armor-piercing bullet.

IOP Conf. Series: Materials Science and Engineering **866** (2020) 012047 doi:10.1088/1757-899X/866/1/012047

4. Conclusion

A composite material based on boron carbide with the addition of 0.5 wt. % CNTs was fabricated. The material exhibits high ballistic armour performance compared to boron carbide without CNTs addition. BAP growth accounts for 26%. It was found that fracturing of boron carbide modified with CNTs leads to a branched fracture pattern which is a cause of efficient kinetic energy dissipation of the armour-piercing bullet.

References

[1] Kislyy P, Kuzenkova M et al 1988 *Boron Carbide* (Kiev: Naukova Dumka) p 216
[2] Rakov E 2006 *Nanotubes and Fullerenes* (Moscow: Logos) p 376
[3] Kobayashi T, Yoshida K and Yano T 2014 *Key Engineering Materials* **616** 27–31
[4] Hirota K, Nakayama Y et al 2009 *Int. J. Appl. Ceram. Techn.* **6** 607–16
[5] Li K, Yang Y et al 2014 *Adv. Eng. Mater.* **63** 286–9
[6] Schwetz K, Sigl L and Pfau L 1997 *J. Solid State Chem.* **133** 68–76
[7] Wei H, Zhang Y and Gong H 2009 *Mater. Res. Innovations.* **13** 70–3

IOP Conf. Series: Materials Science and Engineering **866** (2020) 012048 doi:10.1088/1757-899X/866/1/012048

Hardness of low-alloy steel obtained by cold metal transfer in the magnetic field

V Shlyarov[1,2], D Zagulyaev[1,2], X Chen[1] and V Gromov[1]

[1]Siberan State Industrial University, Novokuznetsk, Russia
[2]Wenzhou University Institute of Laser and Optoelectronic Intelligent Manufacturing, Wenzhou, China

E-mail: zagulyaev_dv@physics.sibsiu.ru

Abstract. In recent years, the technology of creating metal materials by the Cold Metal Transfer method has been actively developed, which helps to obtain products of various shapes and sizes. The hardness of 9CrMoV-N low-alloy steel obtained by the Cold Metal Transfer method in the field of a permanent magnet was investigated. The magnetic field induction did not change in the entire series of tests and was equal to 200 mT. As a result of the studies, it was found that the hardness of the obtained sample is more than 2 times higher than the hardness of the starting material and is 590.97±14.87 HV in comparison with the hardness of the wire before surfacing 265 HV. It was found that the distribution of hardness over the volume of the material is heterogeneous and depends on the place of testing. It was suggested that changes in hardness are associated with the structural features of the deposited material.

1. Introduction

The development of industry requires constant modernization of existing technologies in connection with the need to reduce costs, increase the operational properties of finished products, reduce technological operations to bring products to the final stage. This is especially true of high-tech industries, the products of which are subject to increased requirements. The development of metal materials by the Cold Metal Transfer (CMT) method has been actively developed. The technological process for creating products using the CMT method is as follows: the wire automatically moves in the direction of the deposited substrate, between the electrode (deposited wire) and the substrate a difference of potentials is applied.

At a certain distance from the substrate, a short circuit occurs, which is detected by the automatic process control system and the wire is removed from the zone of deposition, thereby reducing the time of heat supply to the weld (the process of wire approach and its removal is cyclic up to 70 times per second). As a result, the average process temperature is much lower in comparison with the conventional welding, since there is no constant arc burning process. For this reason this method is called Cold Metal Transfer. Summarizing, we can say that controlled processes of short circuit, metal separation from the electrode and its transfer to the pool of the weld (solder bead) literally drop by drop take place during CMT. The result is a metal transfer process, characterized by the absolute absence of splashes.

Surfacing using the CMT process is extremely effective, since the required quality of the metal layer is provided in just one or two passes instead of two or three with traditional methods. The main focus during surfacing is given to the following areas: minimum penetration of the base metal, minimum mixing of the deposited layer with the base metal, minimum value of residual stresses and deformations of the metal in the surfacing zone [1]. However, the production of metallic materials by this method is accompanied by some problems, among which are: obtaining of an inhomogeneous microstructure of the material, gradient of thermal stresses over the cross-section of the resulting part, the high cost of units for the production of parts, etc.

The analysis of the current state of the problems of product formation by additive technology methods showed that the use of cryogenic temperatures [2] and magnetic fields, both transverse and longitudinal, in the process of deposition and the formation of welded joints is a very relevant research area [3-10].

The influence of the magnetic field during arc welding and deposition with wire can increase the coefficient of fusion of the electrode metal, strength characteristics of the weld (deposited metal), control the penetration depth of the base metal, grind the structure of the weld metal [11-15]. Mostly, researchers use small magnetic fields with induction up to 200-500 mT [16]. Based on the foregoing, the purpose of these studies was to study the effect of an external magnetic field on the mechanical properties of 9CrMoV-N steel obtained by the Cold Metal Transfer method. The material for research was not chosen by chance, a steel wire of this chemical composition is used for welding equivalent steels of type 91 "P91 9CrMo", modified with small additives of niobium, vanadium and nitrogen, to improve the long-term creep properties. Modified 9CrMo steels are currently widely used for components such as manifolds, main steam lines and turbine housings in fossil fuel power plants.

2. Methods of experiment

The wire was surfaced using a numerically controlled robot and a multifunctional wire feed device for argon-arc welding FANUC Robot M-10iA. Surfacing parameters were set using software and RSU 5000i controller, which made it possible to select the optimal wire feed speed to the melt pool, as well as current and voltage parameters. Additive production of samples of specified sizes was carried out under the action of a constant transverse magnetic field. A highly homogeneous magnetic field was created using WD-175 electromagnet and RA-3KW programmable direct current source, and the field induction was measured using CH-1500 gauge. Figure 1 schematically shows the process of surfacing 9CrMoV-N steel in the permanent magnet field.

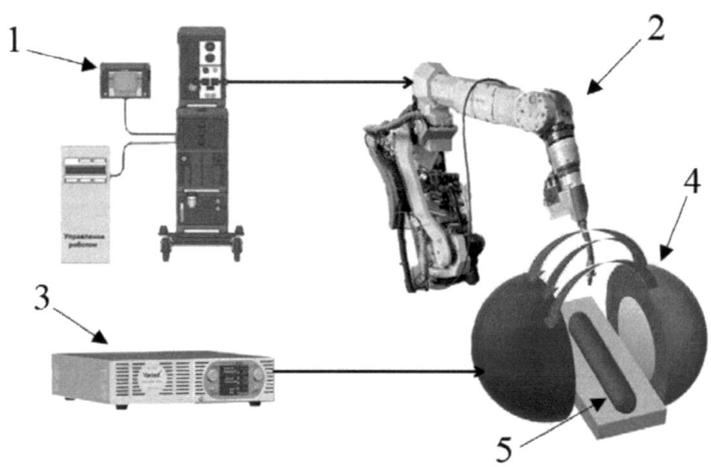

Figure 1. Schematic representation of material deposition in the permanent magnet field: 1 – RSU 5000i controller; 2 – robot for argon-arc welding FANUC Robot M-10iA; 3 – direct current source RA-3KW; 4 – electromagnet WD-175; 5 – deposited sample.

The surfacing parameters (with application of software) were initially set using RSU 5000i controller (1), as well as the magnetic field parameters, the field induction was changed by adjusting the current flowing through the coils. In this work, the magnetic field induction did not change in the entire series of tests and amounted to 200 mT. Then low-alloy steel (5) was surfaced by means of FANUC Robot M-10iA (2) in the field of a permanent magnet (4). Deposition of 9CrMoV-N was carried out in argon. The wire feed speed into the molten bath was 5 m/min at a welding current 160 A and arc voltage 14.1 V, the surfacing speed was 8 mm/s.

The evaluated mechanical characteristic showing the strength of the resulting material was hardness. The measurement was carried out according to the Vickers method, in accordance with the requirements of the standard ISO 6507-1:2005 "Metallic materials. Vickers hardness test. Part 1 Test Method "[17]. Microhardness tester HXD – 1000TM/LCD was used in the work, the indenter load was 9.8 N, the application time and load retention time was 15 s, and the removal of test load – 10 s. To assess the hardness of the material, samples of 10×6×30 mm in size were cut by EDM cutting, and measurements were made on transverse sections.

3. Results and discussion

As a result of electric arc deposition in the field of a permanent magnet, 9CrMoV-N steel samples were obtained. The sample was a parallelepiped, consisting of 30 layers of weld metal with dimensions of 60×6×100 mm. According to AWS A5.28 technical documentation, the 9CrMoV-N steel wire has the chemical composition shown in table 1 and a hardness approximately 265 HV.

Table 1. The chemical composition of the wire 9CrMoV-N in wt. %.

C	Mn	Si	S	P	Cr	Ni	Mo	Nb	V	N	Cu	Al
0.08-0.13	0.40-0.80	0.15-0.50	⩽ 0.010	⩽ 0.010	8.5-9.5	0.40-0.80	0.85-1.10	0.03	0.15-0.25	0.03-0.07	⩽ 0.10	⩽ 0.40

For greater reliability of the obtained experimental results, the hardness was measured at various distances from the substrate with a step of 5 mm, in different parts of the sample. The quantitative value of hardness was determined by averaging more than 70 measured values. The measurement results showed that the hardness of the obtained sample exceeds the hardness of the original wire by more than 2 times, and is 590.97 ± 14.87 HV. It is also worth noting that the values of hardness, which reflect the mechanical properties of the whole sample, are heterogeneous, and the distribution of hardness by volume of the material is shown in figure 2.

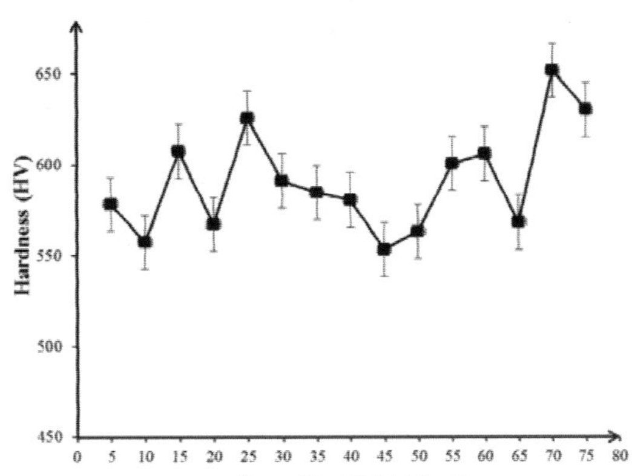

Figure 2. Hardness distribution of the 9CrMoV-N steel obtained by cold metal transfer.

The analysis of experimental data shows that the hardness is not uniformly distributed over the volume of the material and depends on the place of measurement. The minimum hardness values are observed at distances of 10 and 45 mm from the substrate, the maximum hardness takes at 25 and 70 mm. Such behavior is possibly associated with different states of the microstructure of the material in different regions of the samples. According to crystallization kinetics and solidification theory [18], the formation of a fine-grained or larger structure depends on the cooling rate, the heat accumulation of the deposited layer will be less in the upper layers, which will lead to faster cooling and, as a result, grain refinement.

Thus, it can be assumed that in the upper part of the sample the microstructure is finer-grained, and therefore, the hardness value is higher than that of the starting material and the material located closer to the substrate. However, this assumption requires verification by metallographic analysis or scanning electron microscopy, which will be the goal of our subsequent work.

4. Conclusion

The paper analyzes the hardness of low-alloy steel 9CrMoV-N obtained by the cold metal transfer method. As a result of studies, it was found that during CMT, the material is formed, the hardness of which is more than 2 times than the hardness of the original wire. It was suggested that a significant increase in hardness is associated with the structural features of the material obtained as a result of surfacing. All this makes it possible to conclude that the CMT technology can be used as a reliable method for obtaining modern structural materials, however, further study and improvement of technological parameters is required.

Acknowledgments

The research was financially supported by the National Natural Science Foundation of China (Grant No. 51975419).

References

[1] Sun Q, Li J et al 2018 *Journal of Manufacturing Processes* 32 48–56
[2] Liu K, Chen X et al 2020 *Materials Characterization* 159 110011
[3] Chena R, Jiang P et al 2018 *International Journal of Fatigue* 112 308–317
[4] Chen R, Jiang P et al 2018 *Journal of Materials Processing Technology* 254 114–123
[5] Chen R, Jiang, P et al 2017 *Corrosion Science* 126 385–391
[6] Gómez C A, González-Sánchez J and Bilyy O 2019 *Procedia Structural Integrity* 265–272
[7] Liu Y, Sun Q et al 2015 *MaterialsLetters* 152 29–31
[8] Chen X, Luo M et al 2019 *Journal of Manufacturing Processes* 41 111–118
[9] Rosado-Carrasco J, Krupp U et al 2019 *International Journal of Fatigue* 121 243–251
[10] Chen R, Wang C 2016 *Materials and Design* 109 146–152
[11] Wu D, Hu C et al 2019 *Vacuum* 169 108958
[12] Huang M, Zhang Q et al 2020 *Journal of Manufacturing Processes* 50 456–466
[13] Jin P, Liu Y et al 2020 *Journal of Alloys and Compounds* 819 152951
[14] Liu Y, Sun Q et al 2015 *Materials Letters* 152 29–31
[15] Sun Q, Li J et al 2018 *Journal of Manufacturing Processes* 32 48–56
[16] Yan F, Wang X et al 2019 *Optics & Laser Technology* 113 164–170
[17] ISO 6507-1, Metallic materials – Vickers hardness test – Part 1: Test method.
[18] Wang X, Gong X and Chou K 2017 *Proceedings of the Institution of Mechanical Engineers, Part B: Journal of Engineering Manufacture* 231(11) 1890–1903

IOP Conf. Series: Materials Science and Engineering **866** (2020) 012049 doi:10.1088/1757-899X/866/1/012049

Improvement of the working surface resistance of a stamping tool with coating and SSS modeling of the preliminary mechanical activation of the surface layer of the stamp working parts

V P Tabakov[1], V N Kokorin[1], O I Morozov[1], M V Ilyushkin[2], Y A Titov[1] and E L Kornyakov[3]

[1]Ulyanovsk State Technical University, 32 Severny Venets Street, Ulyanovsk, 432027, Russia
[2]JSC Ulyanovsk NIAT, 34 Doctor Mikhailov Street, Ulyanovsk, 432057, Russia
[3]JSC Ulyanovsk Cartridge Plant, 1 Shofyorov Street, Ulyanovsk, 432007, Russia

E-mail: vnkokorin@mail.ru

Abstract. A research technique was developed. The selection of the processed samples (sizes, material, operating conditions) was carried out. The physical and computational model of the process of cold plastic deformation of samples from tool heat-resistant steel X12M was developed. The technique for studying the effect of cold plastic deformation on the adhesive ability of a substrate – coating system is presented.

1. Introduction

The relevance of the study lies in the creation of a new technology for producing wear-resistant coatings on a modified tool working surface, the selection of processing modes and the design of a working tool for sheet, cold volume and hot stamping dies from semi-heat and heat-resistant steels of high viscosity, as well as molds. Currently, surface modification methods are implemented only in the technology of increasing the resistance of the cutting tool (cutters, drills, etc.) and are not effective for increasing the resistance of punch and dies in metal forming processes, where it is necessary to carry out a comprehensive modification of the substrate-coating system. The working tool of dies (punches and dies) works in severe loading conditions: the presence of shock, alternating loads; the presence of features of the friction forces between the processed material and the tool and temperature conditions of processing.

The essence of the proposed technology developed in the Ulyanovsk State Technical University consists in a comprehensive modification of the surface layer: preliminary mechanical activation and deposition of hardening coatings by the ion-plasma method. During its implementation, there is a significant increase in the adhesion of the "substrate-coating" system, a decrease in thermal conductivity and softening of the metal of the stamp surface layer. The observed effect is due to the distortion of the metal crystal lattice, an increase in the density of dislocations during plastic deformation. The use of complex (mechanical activation + hardening coating) surface modification will significantly increase resistance, increase the service life of dies and molds.

2. Research methods

The technology to increase the resistance of the working surfaces of parts of dies and molds made of heat-resistant steels using complex modifications is as follows:

1. Preliminary cold plastic deformation of the machined metal surface is carried out in any traditional way (rolls, impact between the strikers, etc.). At this stage, a favorable texture is formed (macro-, micro-), activation and mechanical hardening (strain hardening) takes place in the area of the working area;
2. The final modification of the surface layer is carried out by applying a wear-resistant multilayer coating to the hardened working surface by ion-plasma spraying.

The main types of materials that can be used in the implementation of the developed technology are a group of heat-resistant and semi-heat-resistant tool steels having a heat resistance of more than 500 °C and retaining their physical and mechanical characteristics at temperatures above 300-400 °C, which makes it possible to use them as a material substrates during coating by ion-plasma coating. The permissible overall dimensions of the samples are usually determined by the dimensions of the working area of the Bulat installation for applying wear-resistant coatings and are 200x200x200 mm.

To increase the efficiency of the adsorption process, in a number of publications by foreign authors it is proposed to use mechanical deformation, which consists in mechanical activation of the metal when implementing the uniaxial compression scheme of a workpiece [1-4]. Based on our experimental work, a physical model was proposed that combines mechanical deformation with the binding energy between the elements of the system under consideration: a substrate (acting as a catalyst) and a coating (acting as a reagent).

It was found that in order to increase the adsorption efficiency, it is necessary to reduce the activation energy in the longitudinal direction of the crystal lattice (its tension), which corresponds to the effect created by uniaxial mechanical compression deformation [1-3].

Figure 1 shows a physical model of the structuring of the substrate-coating system (a fragment of the crystal lattice of the substrate) during complex modification (cold plastic deformation followed by ion-plasma coating), which allows illustrating the increase in the efficiency of the adsorption of particles of the coating material (reagent) onto the substrate.

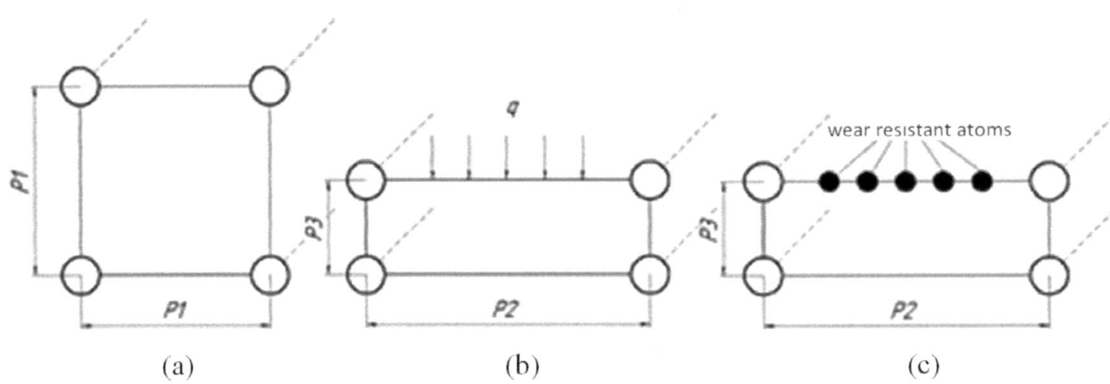

Figure 1. Physical model of the structuring of the "substrate-coating" system with complex modification of the surface layer: a – the original structure; b – deformed structure, – uniaxial compression; c – modified structure of the substrate-coating system.

As was established in the works [1,3] of foreign authors, the arising mechanical stress increases the bonding strength between the elements of the "substrate-coating" system, which is the result of "pulling together" of the atoms of the metal of the substrate at a certain amount of deformation of the surface of the substrate.

In preliminary experimental studies, tool heat-resistant steel X12M GOST 5950-2000 was used as the material of the model and samples for. Based on an analysis of literary sources, the results of mechanical tensile tests of X12M steel were studied and a tensile test diagram was used (maximum tensile strength σ amounted to 1422 MPa, average 1364 MPa).

For a more accurate verification of material parameters when creating the model, the results of A.D. Khvan [8], according to which, according to the results of compression tests on cylindrical specimens for X12M steel, a flow curve was constructed in coordinates to evaluate its mechanical characteristics: stress intensity (σi) - strain intensity (ei).

The experimental tests (sediment of sample-tracks) were carried out on a hydraulic press (nominal press force – 1500 kN), the material of the samples was semi-heat-resistant tool steel X12M, the initial dimensions of the samples were a*b*s0=10x10x10 (mm), the experimental modes are presented in table 1.

Table 1. Modes of experimental studies with uniaxial compression.

No.	Material sample	Initial overall dimensions of the sample			Deformation force F, tf	Pressure deformations R, MPa	Overall dimensions of the deposited samples			The degree of deformation under uniaxial compression ε ,%
		a mm	b mm	S_0, mm			a_1, mm	b_1, mm	S , mm	
1	X12M	10	10	5	9	882.6	11	11	4.2	17
2					15	1470.9	12	12.5	3.6	28
3					20	1961.3	12.5	13	3,1	38
4					25	2451.6	13	13.5	2.7	47
5					30	2941.9	14	14	2.2	59

After mechanical activation, a wear-resistant coating, titanium nitride, was applied to the samples. For the application of wear-resistant coatings, specialized Bulat-type plants were used, which are used in the laboratory of wear-resistant coatings of UlSTU. The mode of applying a wear-resistant coating by ion-plasma spraying included the following parameters: coating material - titanium nitride (TiN); coating thickness - 3 ... 4 μm; temperature - 480 °C.

As the parameter evaluating the increase in the resistance of the surface layer of the samples, we selected the parameter of the adhesive ability of the wear-resistant coating – the peeling coefficient Ko, a decrease in which indicates an increase in the adhesion strength of the coating with the tool base. To evaluate the adhesiveness parameter of a wear-resistant coating based on titanium nitride, it is proposed to use a device for testing preforms with a coating for peeling [9], which consists in fixing the minimum compression force of the springs at the moment of breaking the continuity of the protective coating (peeling).

In order to study the nature of the stress and strain state of the body during preliminary cold plastic deformation, a mathematical model of the process of cold plastic deformation (mechanical activation) using the LS-DYNA software package was developed, and an analysis of the adequacy of the obtained model is presented.

Draft processes were simulated using the LS-DYNA software environment, a multi-purpose program designed to analyze the nonlinear dynamic response of three-dimensional inelastic structures. It includes a fully automated process for solving contact problems, as well as many functions to verify the resulting solution that can successfully solve the most complex problems of impact, destruction and shaping.

IOP Conf. Series: Materials Science and Engineering **866** (2020) 012049 doi:10.1088/1757-899X/866/1/012049

The implementation of the modeling process consisted of 3 main stages: preprocessing; the solution of the problem; postprocessing processing.

When working with the LS-DYNA program, it was required to enter the values of stresses and strains in the form of true stresses and strains. At small strains, the values of the calculated and true parameters are almost identical. However, as the strains increase, these values diverge significantly. To obtain data on the plasticity of the material, the strain curve after uniaxial tension was used [5–7].

As a result of the simulation, a model of the sample sedimentation process was created with the overall dimensions corresponding to the samples precipitated during the full-scale experiment, physicomechanical characteristics, type and parameters of the material, type of contact, precipitation speed, striking motion maps and the time of completion of the precipitation process were determined and set.

The developed model of the precipitation process is presented in figure 2.

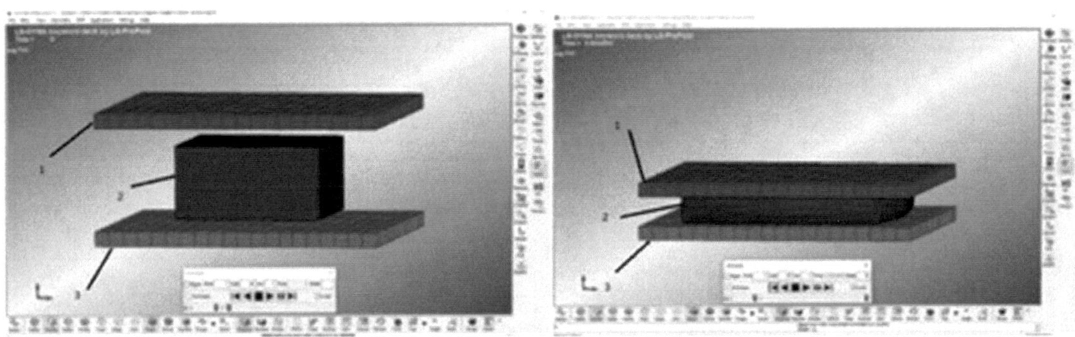

Figure 2. Model of the precipitation process performed in the LS-DYNA program: 1 – top striker; 2 – blank; 3 – lower striker.

As a parameter evaluating the adequacy of the physical model of precipitation with real results, we selected the dependence obtained by the results of modeling and a full-scale experiment, the precipitation effort on the degree of deformation. When processing the results of the experimental values and calculations in MS Excel , the dependencies were approximated, a graphical interpretation of which is shown in figure 3. In this case, the average statistical deviation of the results does not exceed 15%, which indicates the adequacy of the adopted model. Calculation errors can be reduced by increasing the elements involved in the calculation and, accordingly, the model calculation time. The resulting model, as can be seen from the graph (figure 3), quite correctly displays the results of a full-scale experiment (the discrepancy between the curves in the ΔF value does not exceed 5 ... 7 tc). An increase in ΔF for strains exceeding 40% occurs as a result of simplifications of the program calculation caused by the increased scale of the finite element mesh.

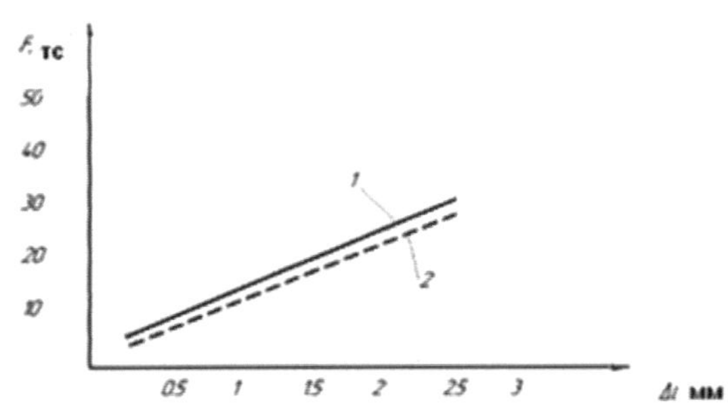

Figure 3. The graph of the dependence of effort on the degree of deformation F = f (Δl): 1– approximated curve obtained from the simulation results in LS-DYNA ; 2 – approximated curve obtained as a result of the experiment.

IOP Conf. Series: Materials Science and Engineering **866** (2020) 012049 doi:10.1088/1757-899X/866/1/012049

Analysis of the simulation allows us to conclude about the nature of loading during the upsetting process, as well as the distribution and magnitude of stresses. The maximum stresses arise at the boundaries of the deformable sample (angular stress concentrators), and with an increase in the loading speed and the upsetting force, the stress growth rate increases, which confirms the correctness of the developed physical model.

3. Conclusion

It is established that during the mechanical activation of the substrate, a change in energy is observed in the workpiece material. On rice nke 4 presents the results of a study of the kinetic and potential energy in the material samples during deformation in uniaxial compression scheme LS-program Dyna . A sharp jump in kinetic energy to the maximum value is characterized by the onset of a sharp metal flow. A further change is due to the gradual compaction of grains in the process of strain hardening. Over time, the metal flow decreases, so the change in the amplitude of the kinetic energy decreases.

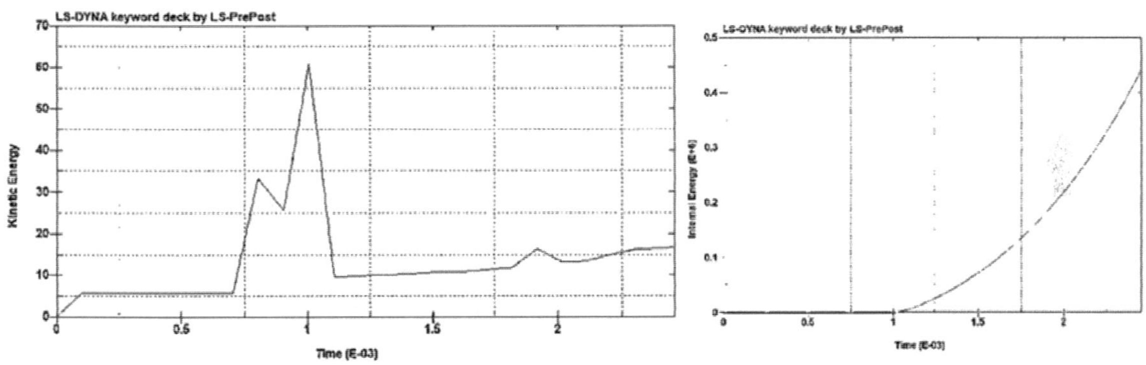

(a) (b)

Figure 4. Graphs of energy changes: a – kinetic energy of the workpiece in the process of deformation; b – potential energy of the workpiece.

Slowing the growth of potential energy is also due to the fact that in order to start the deformation, it is necessary to achieve stresses exceeding the yield strength of the metal. A further increase in energy is due to an increase in the dislocation density of the workpiece material and the force necessary for its deformation.

Analysis of the data obtained in the LS- Dyna program allows us to conclude about the nature of loading during the upsetting process, as well as about an increase in tensile stresses on the contact surface of the workpiece material (up to 1.6 · 103 MPa), which corresponds to the presented physical model, in which there is a decrease activation energy in the longitudinal direction of the crystal lattice (its tension). Maximum stresses arise at the boundaries of the deformable sample (angular stress concentrators), and with an increase in the loading speed and the upsetting force, the growth rate of tensile stresses on the contact surface increases, which confirms the correctness of the developed physical model.

The following stages of the study were completed: a methodology was developed for research and development of technology to increase the resistance of the surface layer; selection of materials and sizes of processed samples; The physical and computational models of the process of cold plastic deformation of specimens from tool heat-resistant steel X12M have been developed; The adequacy of this model and the possibility of using software environments to study the features and modes of cold plastic deformation are established. It is planned to continue research to increase the durability of the working parts of the stamping tool.

Acknowledgements
This work was supported by the Russian Fund for Basic Research – Grant No. 18-48-730-011.

References

[1] Khorshidi A, Vio-let J et al 2018 *Jornal Nature Catalysis* **1** 263–268
[2] Wang H et al 2016 *Science* **354** 1031–1036
[3] Agrawal P M, Rice B M et al 2002 *Surf. Sci.* **515** 21–35
[4] Browell R and Lin G 2000 *Ansys Solutions* vol 2 No 1
[5] Chernyavsky A O 2003 *Finite Element Method. Fundamentals of Practical Application* (Moscow: Mashinostroyeniye) p 24
[6] Kaplun A B et al 2003 *Ansys in the Hands of an Engineer: A Practical Guide* (Moscow: URSS Editorial) p 272
[7] Friedman Ya B 1974 *Mechanical Tests. Structural Strength. Part two* (Moscow: Mashinostroyeniye)
[8] Hwan A D et al 2012 Bulletin of the Voronezh State Technical University vol 8 No 5 131–134
[9] Ilyushkin M V et al 2003 Patent RU 37 221 U1 *Device for Testing the Coating for Peeling* 2003135362/20, 12/9/2003 Published: 04/10/2004 Bull. No. 10

Structure and microhardness of bioinert coatings of Ti-Ta-N system

D A Romanov, K V Sosnin, S Yu Pronin, S V Moskovskii, V E Gromov, A P Semin and E V Protopopov

Siberian State Industrial University, 42 Kirova Street, Novokuznetsk, 654007, Russia

E-mail: romanov_da@physics.sibsiu.ru

Abstract. The coatings of Ti-Ta-N system have been obtained for the first time. Production of bioinert coatings of Ti-Ta-N system was realized by means of promising technique of electroexplosion spraying and subsequent electron-ion-plasma modification with nitrogen ions. Titanium of VT6 grade was used as a substrate for spraying of coatings. Electroexplosion spraying with use of tantalum foil leads to formation of tantalum coating on titanium substrate. Subsequent electron-ion-plasma modification with nitrogen ions results in synthesis of the following phases: TiN, Ta and β-Ti. Average microhardness values of the coatings formed vary from 449 kgf/mm^2 (E = 11.47 %) to 530 kgf/mm^2 (E = 10.02 %). Electroexplosion processing promotes the increase in titanium substrate microhardness near coating – substrate interface. In volume of titanium substrate the microhardness decreases to values corresponding to reference data. Irradiation of electroexplosive tantalum coating by electron beam with subsequent nitriding leads to homogenization of structure. Optimum mode of irradiation should be considered the one at powder density of electron beam of 0.5 MW/cm^2. At lower value of power density the electron beam melts the electroexplosion coating to insufficient depth. At higher values of power density the electron beam leads to boiling of the melted coating and formation of larger quantities of pores .These phenomena are caused by thermal and physical characteristics of titanium substrate.

1. Introduction

In modern medicine the metal implants for different purposes are widely used. Nowadays, metals and alloys free from toxic alloying elements of Al and V [1, 2] are employed as base materials for implants. About 70 – 80 % of medical implants are manufactured from bioinert metallic materials which are presented by stainless steels, cobalt alloys, commercially pure niobium, tantalum, titanium and their alloys [3, 4]. Metallic biomaterials are extremely important for recovery of damaged hard tissue, for improvement of patients' life quality [5, 6].

Material for bone implants should meet the requirements of biomechanical and biomedical compatibility with bone tissue [7, 8]. It is necessary for it that modulus of elasticity of the material being implanted to be approximate to that of bone tissue (30 GPa) that permits one to avoid a manifestation of so-called' stress-shielding effect' which shows itself in dissolution of bone in region of absence of mechanical load [9, 10]. Biomaterial should contain only safe components allowed for medical application and possess high corrosion resistance in organism's media [11, 12]. Among metallic biomaterials titanium and its alloys are the most preferable that is caused by their unique biocompatibility, weak toxicity, low coefficients of thermal expansion and heat conductivity, non-magnetization and high specific strength relative to stainless steels [13, 14].

Titanium and its alloys possess moderate modulus of elasticity (100 – 120 GPa) in contrast to steels (190 – 200 GPa) as well as low cost as compared to pure niobium and tantalum [15]. Disadvantage of titanium alloys is their corrosion in biological media being ac-companied by emission of heavy metals' ions which cause unfavourable electro-chemical reactions in organism and affect the compatibility of implant with surrounding tissues [16]. Applying the coatings to surface of metal implants allows this problem to be solved effectively [17]. Decrease in elastic modulus of products' surface made of titanium alloys is possible by means of coatings based on Ti-Ta-N system.

However, the attempts to synthesize the coatings of the system on titanium surface were not successful so far. In recent years the method of electroexplosion spraying of coatings with different systems is developed by Russian [18] and Chinese scientists [19]. To the present moment of time we have succeeded in obtaining the bioinert coatings of Ti-Zr system [20] by electroexplosion method. The search is concerned with electron-ion-plasma modification with nitrogen ions to strengthen surface layer of electroexplosive tantalum coating by formation of nitrides as well as supersaturated solid solutions.

2. Materials and methods

The object of the research was titanium pin for intramedullar osteosynthesis of ulna (elbow bone) on which electroexplosion coating from tantalum was formed by electroexplosion method [18]. Such implants with electroexplosion tantalum coating need a reliable protection from corrosion effect of biological liquids of living organism. Therefore, to protect the implants' surface it is enough to form coating of Ti-Ta-N system with thickness of not more than 300 μm on their surface. Tantalum foil with mass of 500 mg was used as current-carrying exploded material. Time of plasma effect on sample's surface was ~ 100 μs, absorbed power density on jet axis ~ 1.5 GW/m^2, pressure in shock compression layer near surface being irradiated ~ 12.5 MPa, residual pressure of gas in working chamber ~ 100 Pa; temperature of plasma on nozzle cut ~ 10^4 K, thickness of thermal effect zone ≈ 50 μm.

After electroexplosion spraying the electron-ion-plasma modification with nitrogen ions was performed in order to strengthen the surface layer of electroexplosion tantalum coating by formation of nitrides as well as supersaturated solid solutions. Irradiation of electroexplosion tantalum coating by electron beam was carried out in atmosphere of nitrogen at the following values of power density of electron beam: 0.45, 0.5, 0.55 and 0.6 MW/cm^2. Structure and morphology of the coating were analyzed by methods of light microscopy (device OLYMPUS GX71 with digital camera DP70 and software Image Scope M) and X-ray structural analysis (X-ray diffractometer ARLX' TRA). Microhardness measurement was taken using microhardometer HVS-1000.

3. The results and discussion

Structural studies of transverse metallographic section of electroexplosion coating based on tantalum after electron-ion-plasma modification with nitrogen ions (figures 1 and 2) were performed by light microscopy method. Characteristic images of general view of transverse sections' structure of Ti-Ta-N system samples are shown in figure 1. These images demonstrate clearly the structure obtained with metallographic microscope. According to data in figure 1 the thickness of coating and characteristic constituents of structure are well recognized.

Coatings' thickness of Ti-Ta-N system formed after electron-ion-plasma modification with nitrogen ions at values of power density of electron beam 0.45 (figure 1, a), 0.5 (figure 1,b), 0.55 (figure 1,c) and 0.6 MW/cm^2 (Figure 1, d) amounts to 128 – 232, 94 – 166, 81 – 161 and 279 – 375 μm, respectively. Such a broad distinction in coating thickness, on the one hand, may be explained by choice of corresponding power density of electron beam and on the other hand, by the fact that with growth of internal mechanical stresses the increase in thickness of coatings and pore dimensions takes place. The coating obtained under irradiation in the mode of 0.45 MW/cm^2 is characterized by weak homogenization of structure. In this mode the thickness of melting by electron beam reaches less than 50 μm. The most homogeneous pore-free coating is that obtained in mode of 0.5 MW/cm^2.

This mode of electron-ion-plasma modification with nitrogen ions provides melting of coating throughout thickness of 94 – 166μm, homogenization of structure and saturation with nitrogen ions. With increase in power density of electron beam to 0.55 MW/cm^2 the melt of coating begins to boil and pores are formed in it. Low values of temperature conductivity (9.3·10^6 m^2/s at temperature 300 K) and heat conductivity (22.3 W/(m·K) at temperature 300 K) of titanium substrate [21] contribute to slow heat removal into volume of integrally cold sample. With temperature growth the temperature and heat conductivities of titanium substrate decrease [22] that induces even greater overheating of coating's melt and its boiling. It is the explanation for increase in quantity of pores (dark areas in figure c, d) in coatings obtained in modes of 0.55 and 0.6 MW/cm^2. The coatings obtained in all modes of processing have highly dispersed structure.

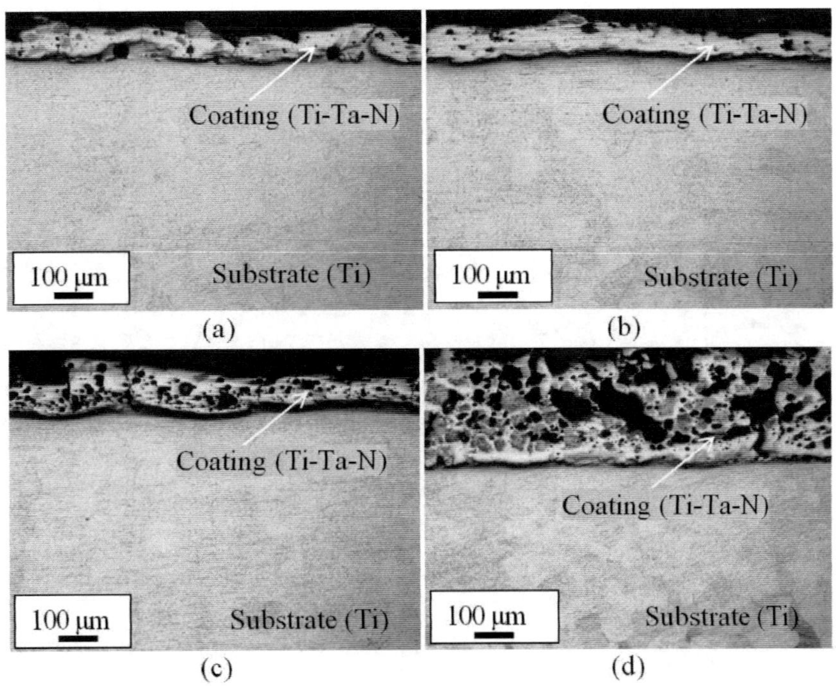

Figure 1. General view of electroexplosion coatings of Ti-Ta-N system after electron-ion-plasma modification with nitrogen ions at values of power density of electron beam 0.45 (a), 0.5 (b), 0.55 (c) and 0.6 (d) MW/cm^2. Coating of Ti-Ta-N system is designated by arrow.

In spite of described features of structure, and relying on the examination results of figure 1 the boundary between coating and substrate (figure 2) contains no pores. The boundary has a wave-like character with wavelength of about 10 μm. It is in good agreement with obtained results of electroexplosion spraying of bioinert coatings of Ti-Zr system [20]. Scientists of Lanzhou University of Technology [19] note the analogous features in structure at boundary of coaling with substrate at electroexplosion spraying of high-melting Ta10W and non refractory Ni60A coatings. It is explained by general feature of electroexplosion methods of applying the coatings. At coating-substrate boundary the thickness of diffusion zone is less than 10^{-3} mm with diffusion coefficient D = 10^{-5} cm/s and pulse action time of 10^{-2} – 10^{-7} s. At coating-substrate interface the initiation of mechanical is inescapable and their stresses whose value is the higher the stronger is the difference of coating properties from those of substrate material.

Low adhesion strength of coatings obtained by conventional methods of spraying [23–25] is caused, first, by jump of properties at interface because of absence or weakly expressed diffusion zone. Interface, essentially, is a macroscopic defect-source of stresses of the first kind. Therefore, features of structure at boundary of electroexplosion coatings with substrate contribute to greater adhesion compared with coatings being obtained by traditional method of spraying. The increase in quality of coatings by forming the broad diffusion zone at coating-substrate interface and rise in efficiency of coating's formation process is possible due to use of electroexplosion spraying technique.

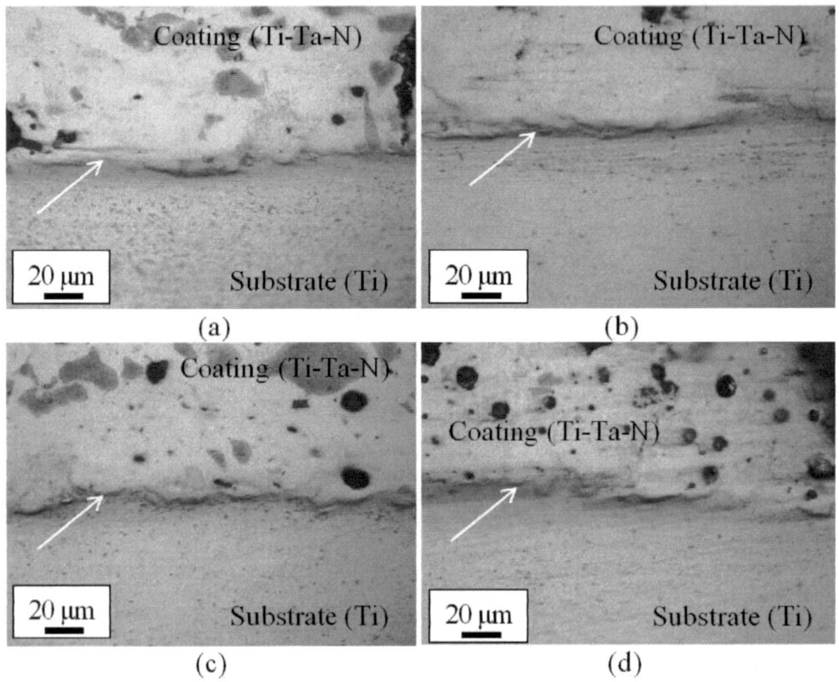

Figure 2. Boundary features of electroexplosion coatings of Ti-Ta-N system with titanium substrate after electron-ion-plasma modification with nitrogen ions at values of power density of electron beam 0.45 (a), 0.5 (b), 0.55 (c) and 0.6 (d) MW/cm^2. Boundary of coating with substrate is designated by arrow.

Analysis of metastable state diagram of titamium alloys indicates that increase in quantity of beta-isomorphous stabilizer should result in the following evolution in phase composition of material [26]: $\alpha' \rightarrow \alpha'' \rightarrow \alpha'' + \beta\ (\omega) \rightarrow \beta\ (\omega) \rightarrow \beta$. In initial state VT6 alloy is a binary alloy containing $\alpha + \beta$ phases. Polycrystalline and homogeneous, in lattice parameter, material gives narrow diffraction peaks while amorphous or noncrystalline inhomogeneous material-wide and low ones. X-ray diffraction patterns of coatings under study (X-ray structural analysis was done from surface layer of coatings formed) are shown in figure 3. Analysis of X-ray diffraction patterns of electroexplosion tantalum coatings after electron-ion-plasma modification with nitrogen ions (figure3) has shown that principal phases being present in the coating are β-titanium with cubic lattice and lattice parameter a = 0.3311nm, tantalum (lattice parameter a = 0.3302nm) and titanium nitride with cubic lattice and lattice parameter a = 0.4241 nm.

As a part of the study the quantitative analysis of X-ray diffraction patterns was carried out and percentage of phases in coatings under examination was determined. Table 1 shows phase composition of samples under study. Diffraction patterns contains reflections at 18.50°, 23.00°, 32.55°, 39.90°,48.52° corresponding to titanium nitride. Reflections at 35.50°, 45.50°, 48.55°, 55.60° correspond to β-titanium phase, and reflections at 24.85°, 27.00°, 35.90°, 52.00° belong to tantalum (figure 3).

IOP Conf. Series: Materials Science and Engineering **866** (2020) 012050 doi:10.1088/1757-899X/866/1/012050

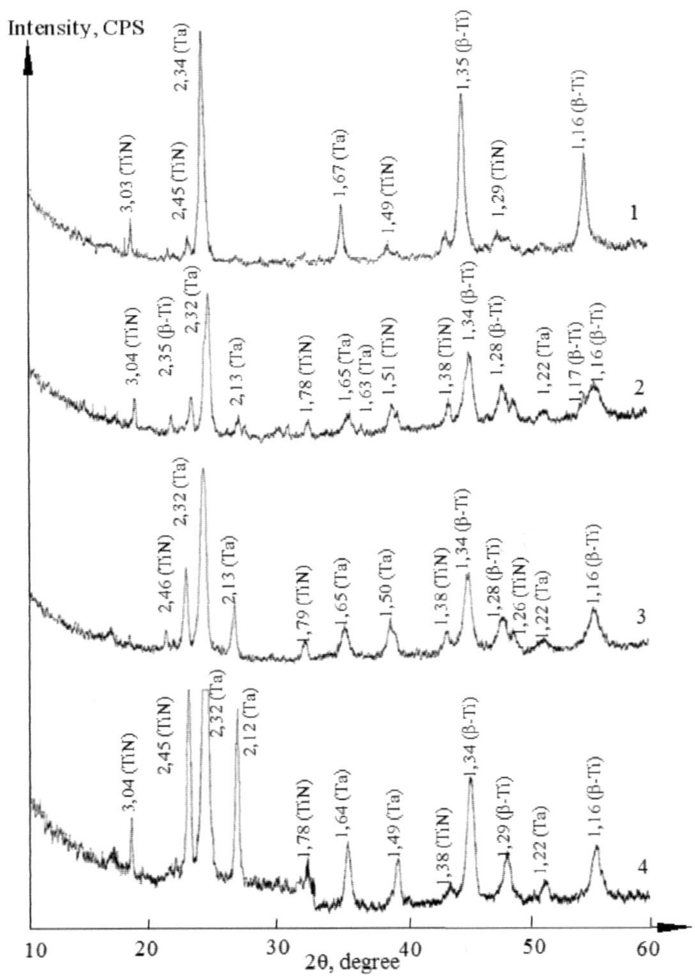

Figure 3. Portions of X-ray diffraction patterns of electroexplosion coatings of Ti-Ta-N system after electron-ion-plasma modification with nitrogen ions at values of power density of electron beam of 0.45 (1), 0.5 (2), 0.55 (3) and 0.6 (4) MW/cm^2.

Table 1. Phase composition of bioinert coatings of Ti-Ta system formed by electroexplosion spraying method.

Value of powerdensity of electron beam, MW/cm^2	Phase composition
0.45	0.34 vol.%TiN; 47.24 vol.%Ta; 52.41 vol.%β-Ti
0.5	75.77 vol.%TiN; 7.42 vol.%Ta; 16.80 vol.%β-Ti
0.55	10.15 vol.%TiN; 42.14 vol.%Ta; 47.71 vol.%β-Ti
0.6	21.03 vol.%TiN; 52.65 vol.%Ta; 26.32 vol.%β-Ti

When performing a series of tests for microhardness measurements (figure 4) the microhardness values of electroexplosion tantalum coating after electron-ion-plasma modification with nitrogen ions at different thickness of coating were obtained. Microhardness of coating amounts to 532.24 ± 53.25 kgf/mm^2 (E = 10.02 %), 499.56 ± 28.33 kgf/mm^2 (E = 5.68 %), 466.47 ± 49.77kgf/mm^2 (E = 10.55 %), and 439.26 ± 50.44 kgf/mm^2 (E = 11.47 %) for samples irradiated in modes of 0.45, 0.5, 0.55 and 0.6MW/cm^2, respectively. Average microhardness value of coating adjacent to substrate is 466.47 ± 77.89 kgf/mm^2 (E = 13.08 %), 400.23 ± 37.22 kgf/mm^2 (E = 9.30%), 466.47 ± 32.52 kgf/mm^2 (E =

6.97 %), and 363.40 ± 29.22 kgf/mm^2 (E = 8.04 %) for samples irradiated in modes of 0.45, 0.5, 0.55 and 0.6MW/cm^2, respectively.

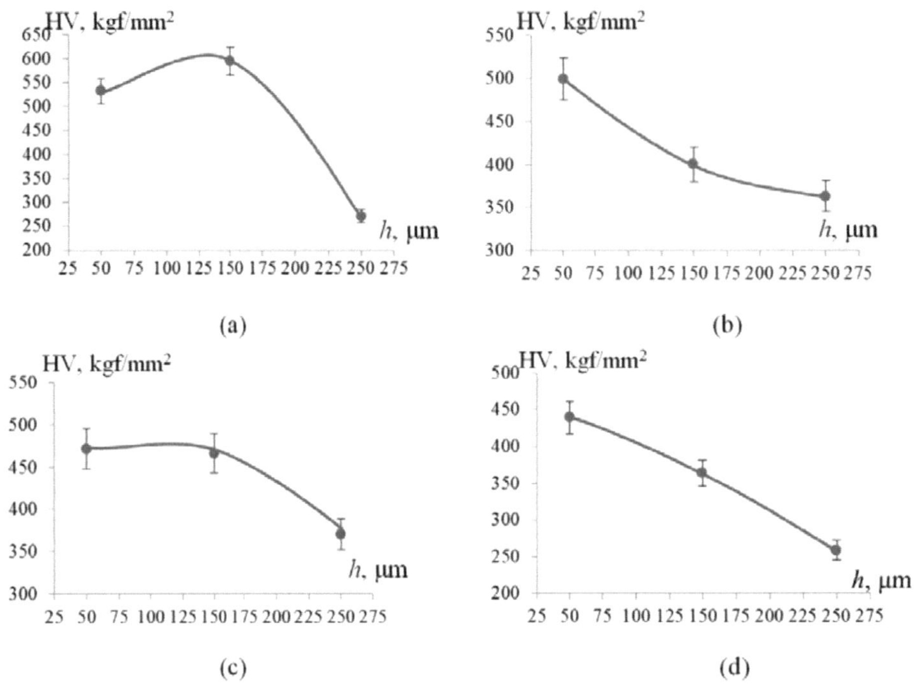

Figure 4. Dependences of change in microhardness on coating's thickness of Ti-Ta-N system after electron-ion-plasma modification with nitrogen ions at values of power density of electron beam 0.45, 0.5, 0.55 and 0.6 MW/cm^2.

Average microhardness values of substrate adjacent to electroexplosion coatings were determined as well. According to studies performed these amounted to 270.81 ± 24.38 kgf/mm^2 (E = 9.00 %), 363.34 ± 53.74 kgf/mm^2 (E = 14.79 %), 370.35 ± 46.23kgf/mm^2 (E = 12.48 %), 258.76 ± 21.24 kgf/mm^2 (E = 8.21 %). According to reference data the microhardness of VT6 titanium amounts to 81 kgf/mm^2 [27]. Thus, electroexplosion processing promotes the increase in microhardness of titanium substrate near coating- substrate interface. In volume of substrate the titanium microhardness amounts to 81 kgf/mm^2.

4. Conclusion

At surface of titanium pin for intramedullar osteosynthesis of elbow bone the bioinert coatings of Ti-Ta-N system were obtained for the first time. Formation of Ti-Ta-N system coating was performed by means of processing of tantalum foil. Subsequent to it, electron-ion-plasma modification of electroexplosion tantalum coating with nitrogen ions was carried out. Combined processing results in formation of coating from 94 to 375 μm in thickness depending on mode parameters of processing. Coatings have highly dispersed structure .The most homogeneous pores-free coating is that obtained in mode of 0.5 MW/cm^2. At lower value of power density of electron beam the coating's structure homogenizes weakly but at higher one –pores form in coating. Wave-like boundary of coating with substrate (wavelength is about 10 μm) contributes to greater adhesion compared to coatings obtained by traditional techniques of spraying. Phase composition of coating is presented by the following phases: TiN, Ta and β-Ti. Combined processing promotes the increase in microhardness of coatings to values of 439 – 532 kgf/mm^2. In total, the studies performed showed prospects for formation of bioinert coatings of Ti-Ta-N system by combined method.

Acknowledgments
The present work was performed within Russian Science Foundation project No. 19-79-00016.

References
[1] Zhu M, Liu X, Tan L et al 2020 *J. Hazard. Mater.* **383** 121122
[2] Vo M T, Nguyen T T et al 2020 *IFMBE Proceedings* **69** 319–322
[3] Zhang Y, Chu K, He S, Wang B, Zhu W and Ren F 2020 *Mater. Sci. Eng., C* **106** 110165
[4] Kim D G, Kim K H, Jo Y et al 2020 *J. Mechanical Behavior of Biomed. Mater.* **101** 103416
[5] Lubov Donaghy C, McFadden R et al 2020 *Opt. and Laser Technol.* **121** 105793
[6] Saeed F, Muhammad N, Khan A S et al 2020 *Mater. Sci. Eng., C* **106** 110167
[7] Zhou W, Huang O, Gan Y et al 2019 *Artif. Cells, Nanomedicine and Biotech.* **47** 290–299
[8] Zhang R, Xu N, Liu X et al 2019 *Artif. Cells, Nanomedicine and Biotech.* **47** 2838–45
[9] Yuan Z, Tao B, He Y et al 2019 *Biomaterials* **223** 119479
[10] Lee J S, Lee S J, Yang S B et al 2019 *Appl. Surf. Sci.* **496** 143675
[11] Ghosh S, Abanteriba S et al 2019 *J. Mechanical Behavior of Biomed. Mater.* **100** 103412
[12] Alfaro M F, Rossman P K et al 2019 *J. Bio- and Tribo-Corrosion* **5** 81
[13] Saha S, Pramanik K and Biswas A 2019 *Mater. Sci. Eng., C* **105** 109982
[14] Shemtov-Yona K, Özcan M et al 2019 *Eng. Failure Anal.* **106** 104160
[15] Jeon C, Oh K C, Park K H and Moon H S 2019 *Sci. Rep.* **9** 2581
[16] Kunrath M F, Leal B F, Hubler R, de Oliveira S D and Teixeira E R 2019 *AMB Express* **9** 51
[17] Stewart C, Akhavan B, Wise S G and Bilek M M M 2019 *Prog. Mater Sci.* **106** 100588
[18] Romanov D A, Moskovskii S V et al 2019 *Mater. Res. Express* **6** 055042
[19] Han F, Zhu L, Liu Z H and Gong L 2019 Surface and Coatings Technol. **374** 44–51
[20] Romanov D A, Sosnin K V, Gromov V E et al 2019 *Materials Letters* **242** 79–82
[21] Wei K, Ma B, Huang X, Xiao Y and Liu H 2019 *Renewable Energy* **141** 431–443
[22] Rahman M S, Schilling P J et al 2019 *J. Eng. Mater. Technol., Trans. ASME* **141** 041006
[23] Luo X, Smith G M and Sampath S 2019 *J. Therm. Spray Technol.* **27** 296–307
[24] Chen Y, Shang J, Liang X, Wang H and Zhou Z 2019 *Surf. Coat. Technol.* **367** 135–147
[25] Ollik K, Rybarczyk M, Karczewski J and Lieder M 2020 *Appl. Surf. Sci.* **499** 143914
[26] Chikwanda H K and Mahlatji L 2018 *Key Eng. Mater.* **770** 95–105
[27] Meng W, Xiaohui Y, Zhang W et al 2020 *J. Mater. Process. Technol.* **275** 116368

IOP Conf. Series: Materials Science and Engineering **866** (2020) 012051 doi:10.1088/1757-899X/866/1/012051

Electromechanical installation based on a powerful current pulse generator for materials treatment

V A Kuznetsov[1], E S Kuznetsova[1], V E Gromov[1], O A Peregudov[2] and Yu A Rubannikova[1]

[1]Siberian State Industrial University, 42 Kirova Street, Novokuznetsk, 654007, Russia [2]Omsk State Technical University, 11 Prospekt Mira Street, Omsk, 644050, Russia

E-mail: gromov@physics.sibsiu.ru

Abstract. The electromechanical device containing an electromagnetic solenoid with a piston operating in the reciprocating motion mode was developed. Technologies with similar operating modes are widely used in various industries. Economical generator of powerful unipolar current pulses with unique systems is used in the electromechanical device. It allows its main parameters to be regulated in a wide range and with high speed: pulse reproduction frequency and amplitude. The principle of generator operation is based on the periodic discharge of pre-charged capacitors to a low-resistance active-inductive load. The parameters of the mechanical and electrical parts of the electromechanical device are calculated: the initial coordinate of the piston, the magnetomotive force resulting from a change in the inductance L(x), the spring elasticity, the resistance force of the piston proportional to its speed, amplitude, duration and frequency of current pulses. A simulation model of the device in the environment of Matlab-Simulink was developed. Plots of transients during the device idling and under load were constructed. The analysis of the device operating modes was performed. The developed electromechanical installation allows the process parameters to be adjusted.

1. Introduction

Progress in the development of modern technology requires new structural materials with high technological and operational characteristics. Such materials and products from them cannot be obtained and processed by traditional methods. Therefore, one of the most important tasks of scientific and technical policy is the creation and implementation of qualitatively new technological processes, including the use of external energy effects on the strength and ductility of materials [1]. Experimental studies of high-density electric currents, high-energy electromagnetic fields, plasma currents, laser radiation and combined effects on electrically conductive materials have led to the creation of a new class of high-performance technological methods for materials treatment.

For the domestic and foreign metallurgical and machine-building industries, the technology using powerful pulsed electric currents in the processing of difficultly deformed metals and alloys has been disseminated [1]. These are the processes of forging and rolling [2, 3], drawing [4-7], metalworking [8-10], sintering [11-13], materials joining [14-16] and sheet forming [17-19] and others. The detailed review of these electrical technologies is given in [20, 21]. They are based on the electroplastic effect, discovered more than 60 years ago and consists in decreasing the resistance to deformation when current pulses are transmitted [22]. Its complex physical nature has not yet been clearly established, since it consists of a number of interconnected ones – the thermal effect, electron-dislocation interaction, the ponderomotive effect, and others [1].

The basis of industrial electrical technologies is the use of powerful sources of electric current [22-25]. Generators of powerful ($\approx$ 15 kA) current pulses created in recent years have been successfully used for electrically stimulated wire drawing [26]. Such installations can be implemented not only in the processing of metals by pressure, but also in the mining industry (crushing of ore, coal, rock), as well as mechatronics, robotics.

The aim of this work is to create, on the basis of a generator of powerful current pulses, an electromechanical device with a reciprocating piston for electrical stimulation of forging and stamping and to study its parameters.

2. Results and discussion

The main part of the device is an electromagnetic solenoid with a piston (figure 1). The device comprises the magnetic circuit with stacked steel on which a coil with a wire wound around is rigidly fixed. A piston is connected to the two springs in the steel casing of the magnetic circuit, as shown in the figure. The piston reciprocates along the vertical coordinate x. The first (pushing) spring with coordinates x = 0), and the second spring with a stiffness coefficient k = 20000 during the deformation process performs the function of a smooth stop of the piston at the extreme lower point of the coordinate x = x1.

In the coordinate range from x = 1 to x = 0.7 (the working area of the piston), the piston experiences resistance from the load side (for example, coal seams), and depending on the force opposing the piston, an equilibrium state point appears at a distance of x = xequil. The transient motion of the piston under the action of electromagnetic forces occurs in accordance with the known equations [22].

$$U = \left(R \cdot i + L(x)\, {}^{di}\!/_{dt} + i \cdot {}^{dL(x)}\!/_{dx} * {}^{dx}\!/_{dt} \right) \tag{1}$$

where: U – the voltage of the power source, V; i – the current flowing through the winding of the electromagnet, A; R– active equivalent resistance of busbar and coil winding, Ohm; L(x) – inductance as a function of the coordinate x of the piston.

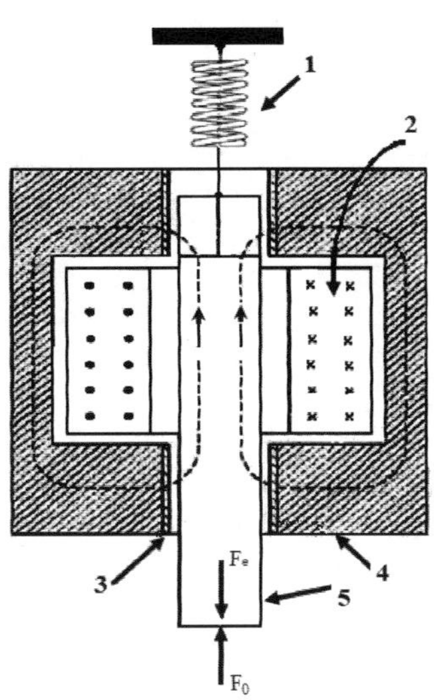

Figure 1. Design of the electromechanical device. 1 – return spring, 2 – solenoid coil; 3 – a directing ring; 4 – steel casing with lined iron; 5 – a cylindrical steel piston; 6 – force impact on the piston, F_0.

IOP Conf. Series: Materials Science and Engineering **866** (2020) 012051 doi:10.1088/1757-899X/866/1/012051

The main variable component of the mechanical system is the coordinate of the piston x. The equation of the mechanical part of the system is determined on the basis of Newton's second law:

$$Fсум = M \cdot a = M\frac{d^2 l}{dt^2} \tag{2}$$

where: Fsum – the resultant of forces on the mechanical part of the system, N; M is the mass of the rod, kg; a – piston acceleration, m/s.

The equation for changing the variable x has the form:

$$M \cdot \frac{d^2 x}{dt^2} = Fe - k(x - x_0) - b\frac{dx}{dt} - f_0 \tag{3}$$

where: x_0 is the initial coordinate of the piston, m; k is the resistance coefficient, N/m; f_0 – external force on the piston, N.

Based on these equations, the model of the mechanical part of the electromechanical device is presented in the Matlab – Simulink environment (figure 2).

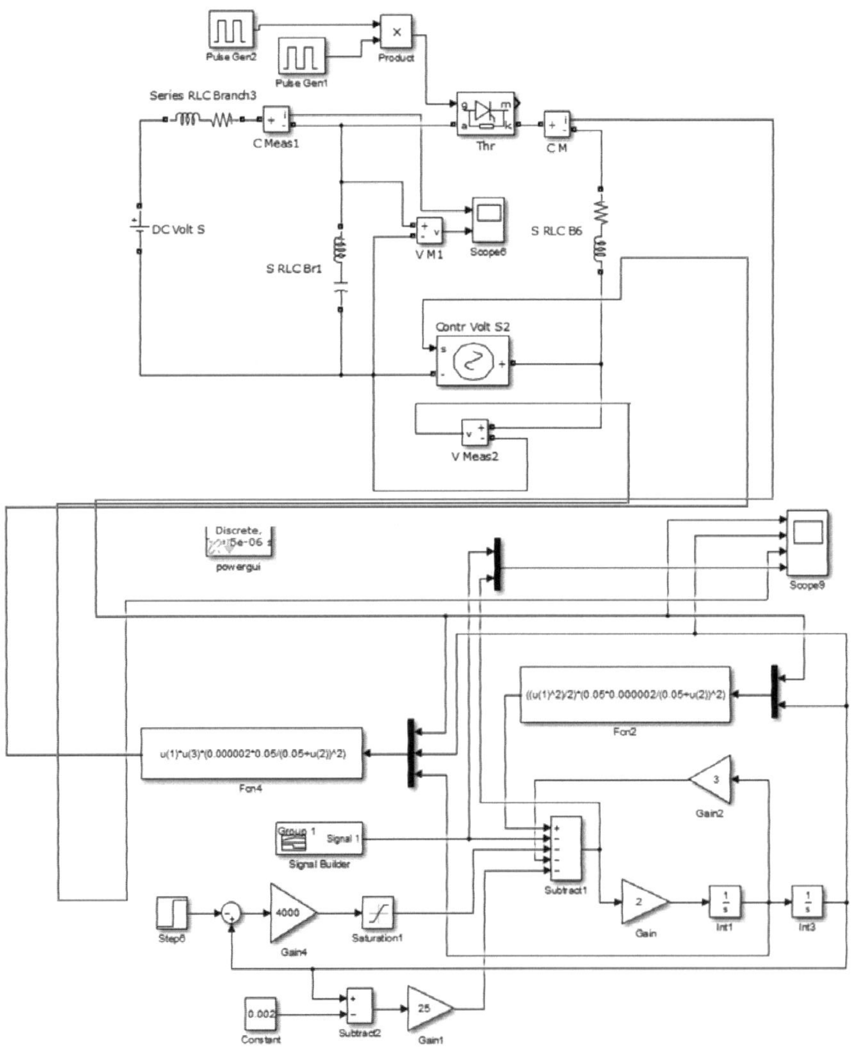

Figure 2. Model diagram of the electromechanical device.

The articles are known in which similar electromechanical devices are described using a battery as a power source [23, 24]. In this work, to implement such setup, a powerful current pulse generator was used [25, 26].

The principle of operation of the generator is based on a periodic discharge to a low-impedance load of pre-charged capacitors. The shape of the current pulse of such generator is close to sinusoidal, the pulse duration depending on the capacitance and inductance can be 100 – 10 000 μs, and the reproduction frequency up to 1000 Hz. The block diagram of the generator is shown in figure 3.

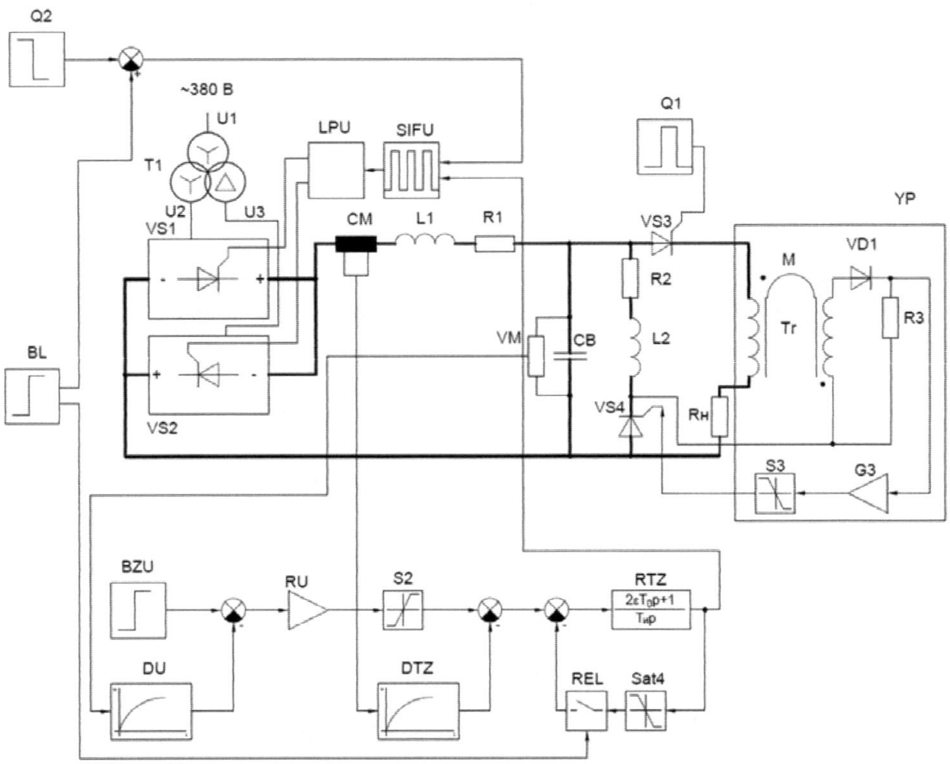

Figure 3. Structural diagram of a powerful current pulse generator.

The generator contains a charge unit with a reversible thyristor converter, consisting of counter-parallel connected thyristor bridges SV1, SV2, which is connected to the capacitors CB via series-connected equivalent resistance R1 and inductance L1, a capacitor discharge block for the load (CB capacitors, thyristor switch VS3, load RH), recharge unit YP (transformer M, diode VD1, resistor R3, amplifier G 3, restriction unit S3), the principle of operation of which is described in [13], as well as the automatic regulation unit (system) ACS of generator parameters (BZU block voltage reference, RNZ regulators, RU, sensors of basic parameters DTZ, DU).

The automatic control systems formed in the generator allow the main parameters in a wide range to be adjusted: the pulse reproduction frequency and its amplitude. To create a reciprocating motion of the piston, device B2 (figure 2) generates a packet of pulses of adjustable duration, while the piston makes a working movement "forward" until it touches a solid body that needs to be destroyed or deformed. After the termination of the pulses, the return spring returns the piston to its original position.

Figure 4 shows the transient graphs of the electromechanical device model with a reciprocating piston.

IOP Conf. Series: Materials Science and Engineering **866** (2020) 012051 doi:10.1088/1757-899X/866/1/012051

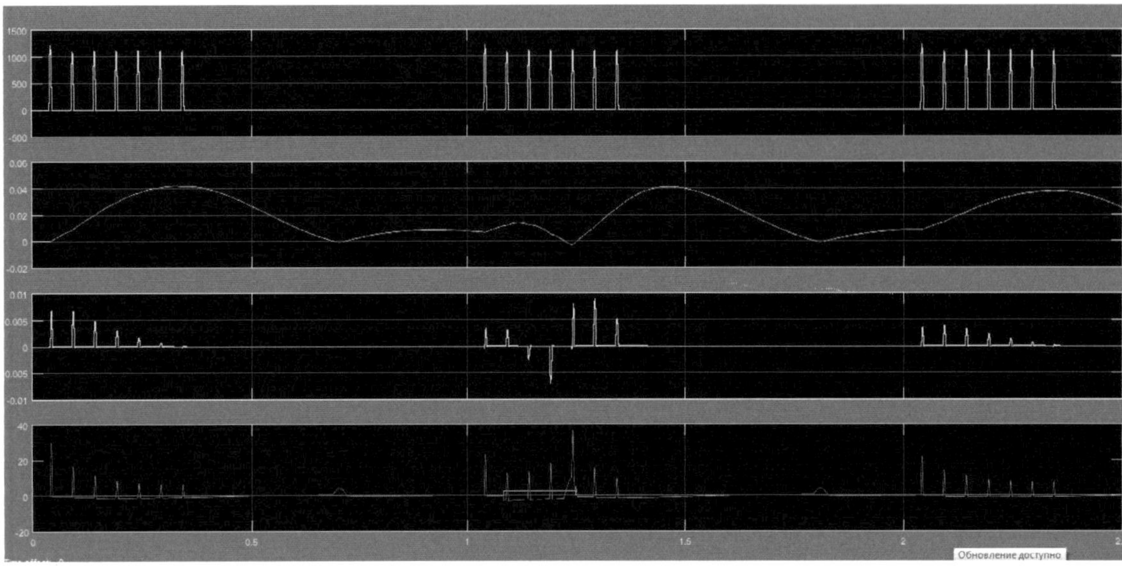

Figure 4. Transient graphs of the electromechanical device.

The graphs show the following electromagnetic accelerator parameters:
1 – generator current pulses, A;
2 – distance traveled by the piston, m;
3 – counter-EMF formed by the inductance of the electromagnet coil, V;
4 – working force on the piston side and external load, kg.

3. Conclusion

The electromechanical device with a reciprocating piston and the influence of the latter on the load to deform it was developed. The model of the device in the Matlab – Simulink environment was implemented. The use of a powerful current pulse generator with an automatic parameter control system allows high-speed control of process parameters: force and piston distance.

References

[1] Gromov V E, Zuev L B et al 1996 *Electrostimulated Plasticity of Metals and Alloys* (Moscow: Nedra) p 293
[2] Salandro W, Jones J et al 2014 *Electrically Assisted Forming: Modeling and Control* (Basel. Switzerland: Springer) p 355
[3] Jones J J, Mears L and Roth J T 2012 *ASME Journal of Manufacturing Science and Engineering* **134(3)** 034504
[4] Gromov V E, Erilova T V et al 1994 *Izv. Vuzov. Ferrous Metallurgy* 456–458
[5] Gromov V E, Kozlov E V et al 1994 *Problems of Mech. Eng. and Machine Reliability* 246–251
[6] Gromov V E, Zuev V I et al 1996 *Izv. Vuzov. Physics* 366–396
[7] Kozlov A, Mordyuk B and Chernyashevsky A 1995*Mater. Sci. Eng. A.***190(1)** 75–79
[8] Hameed S, Rojas H A G et al 2016 *Int. J. Adv. Manuf. Technol.* **87(5-8)** 1835–41
[9] Zhang D, To S et al 2012 *Metall. Mater. Trans. A.* **43(4)** 1341–46
[10] Ji R, Liu Y et al 2011 *J. Mech. Sci. Technol.* **25(6)** 1535–42
[11] Langer J, Hoffmann M J and Guillon O 2009 *Acta Mater.* **57(18)** 5454–65
[12] Grasso S, Sakka Y and Maizza G 2009 *Sci. Technol. Adv. Mater.* **10(5)** 053001
[13] Munir Z A, Quach D V and Ohyanagi M *J.* 2011*Am. Ceram. Soc.* **94(1)** 1–19

[14] Liu X, Lan S and Ni J 2015 *J. Mater. Process. Technol.* **219**112–123
[15] Santos T G, Miranda R and Vilaca P 2014*J. Mater. Process. Technol.* **214(10)** 2127–33
[16] Santos T G, Lopes N et al 2015 *J. Mater. Process. Technol.* **216** 375–380
[17] Jeswiet J, Micari F et al 2005*CIRP Ann. Manuf. Technol.* **54(2)** 88–114
[18] Fan G, Sun F, Meng X, Gao L and Tong *G* 2010 *Int. J. Adv. Manuf. Technol.* **49(9-12)** 941–947
[19] Fan G, Gao L, Hussain G and Wu Z 2008 *Int. J. Mach. Tools Manuf.* **48(15)** 1688–92
[20] Nguyen-Tran H, Oh H et al *2015 Int. J. Precis. Eng. Manuf. Green Technol.* **2(4)** 365–376
[21] Guan L, Tang G and Chu P K 2010 *J. Mater. Res.* **25(7)** 1215–24
[22] Troitskii O A and Rozno A G 1970 *Solid State Physics* **12(1)** 203–210
[23] Zhmakin Yu D, Romanov DA et al 2011 *Industrial Energy* 622–26
[24] Zhmakin Yu D, Zagulyaev D V et al 2010 *Industrial Energy* 639–42
[25] Zhmakin Yu D, Zagulyaev D V et al 2011 *Industrial Energy* 128–32
[26] Kuznetsov V A, Gromov V E et al 2017 *Izv. Vuzov. Ferrous Metallurgy* **60(2)** 157–163

IOP Conf. Series: Materials Science and Engineering **866** (2020) 012052 doi:10.1088/1757-899X/866/1/012052

Boron electroexplosive alloying of austenite steel

Yu F Ivanov[1], E A Petrikova[1], A D Teresov[1], D V Romanov[2], O V Ivanova[3], V E Gromov[2], A A Klopotov[3] and A P Semin[2]

[1]Institute of High-Current Electronics, SD RAS, Tomsk, Russia
[2]Siberian State Industrial University, Novokuznetsk, Russia
[3]Tomsk State University of Architecture and Building, Tomsk, Russia

E-mail: yufi55@mail.ru

Abstract. Studies of phase and elemental composition, defect substructure of surface layer of $Cr_{18}Ni_{10}Ti$ stainless high chromium steel subjected to alloying by electroexplosion method were performed by techniques of X-ray structural analysis, optical and electron microscopy. Titanium and boron were chosen as alloying elements. It has been stated that electroexplosion alloying of steel results in the formation of surface layer having a multiphase submicro-nanocrystallyne structure being characterized by the presence of micropores, microcracks and microcraters. Dependence of phase composition and microhardness of surface layer on relation of titanium and boron masses at electroexplosion alloying has been revealed. It has been found that microhardness of the modified layer is determined by relative masses of titanium borides' fractions in the surface layer of steel and it may increase 18-fold that in the initial (prior to electroexplosion alloying) state.

1. Introduction

Nowadays, boriding is one of the most promising methods of chemical and thermal treatment of metals and alloys, including steels [1]. The most common types for industrial application are the boriding techniques including saturation in gaseous, liquid and solid media. The process of liquid nonelectrolyzed boriding which permits one to design one-or multi-phase structure of surface layer and give the prescribed properties to the core (from annealed to hardness state) [2] is perspective for use. The methods of thermocycle boriding [3], electron beam boriding [4] and boriding with application of laser beams [5] are being at the stage of development. The given technique of boriding, by its mechanism, is a modification of boriding from solid phase saturating coatings under chemical and thermal treatment [4].

The most uniform and homogeneous penetration of boron into the surface layer of metal is realized when method of gas boriding. Boriding of parts is performed in medium of gases containing oxides, halogenides and hydrogen compounds of boron. Atomic boron, being liberated in thermal decomposition of gases, deposits on the surface of parts and diffuses deep into metal. Nevertheless, boriding mixtures of gases being formed are dangerously explosive that imposes restrictions to the application of this technique. The limited use of boriding in powders and pastes is caused by high labour consumption and high costs of the processes and difficulty associated with formation of definite structure and phase composition of the hardened layer and properties of core.

In the past few decades the modification methods of metals and alloys based on application of concentrated fluxes of energy [6] are actively developed. One of these methods is electroexplosive alloying. Electroexplosive alloying (EEA) of surface of metals and alloys is based on the phenomenon of failure of conductor under the action of powerful pulses of electric current [7]. Plasma stream, formed from products of electric explosion of a conductor, melts the surface being treated, while saturating a melt by components of plasma jet. Subsequent high-speed cooling caused by removal of heat into the bulk of material results in self-hardening of modified layer. Powdered weighed samples of various substances may be placed into the area of conductor explosion (current carrying foil). They are transported to surface being hardened by plasma stream providing additional modification of structure and properties of a material [8]. EEA is characterized, in most cases, by substantial overheating of surface being hardened, above boiling temperature due to high pressure of jet on a surface that is accompanied by formation of high developed relief with a large number of microdroplets, microcraters, microcracks, metal burrs. Electroexplosive alloying, due to little time of thermal effect, is characterized by high rate of cooling of material's modified layer. In consequence of it, a submicro-nanocrystalline structure [9] is formed in a surface layer.

The paper is concerned with the analysis of results and establishing the regularities of structure and properties, formation of Cr18Ni10Ti stainless high chromium steel subjected to electroexplosive alloying with titanium and boron.

2. Material and methods

Stainless steel $Cr_{18}Ni_{10}Ti$ (Standard 5632-72) [10] was used as material under study. The samples had the shape of plates 10x10x5 mm in size. At the first stage the processing of steel surface was performed by method of electroexplosive alloying (EEA) at device EVU 60/10 [11]. Foil of commercially pure titanium VT1-0 was used as current-carrying material. The weighed sample of boron powder was placed on surface of the foil. Powder of amorphous boron (B-99D- Technical Specifications 1-92-1549, B>99%, diameter of particles 0.5-5 μm). Parameters of electroexplosive alloying were the following: power density and duration of plasma pulse effect 2.2 GW/m^2 and 100 μs. Four modes of EEA were used that differed in foil mass of titanium and powder mass of amorphous boron (table 1).

Table 1. Relations of foil mass of titanium (m(Ti)) and weighed sample of boron powder (m(B)) used for EEA of $Cr_{18}Ni_{10}Ti$ steel.

mode	m (Ti), mg	m (B), mg	m (Ti) / (B)
1	360.7	50	7.214
2	392.2	62.5	6.275
3	423.7	75	5.649
4	455.2	87.5	5.202

Study of steel structure in the initial state and after modification was performed by the methods of X-ray structural analysis (difractometer XRD 600), optical (device Microvisor Viso – MET – 221) and scanning electron (device SEM 515 Philips) microscopy. Examination of samples' elemental composition was carried out by the methods of microspectral analysis. The properties of the modified layer were characterized by determining microhardness (microhardometer PMT-3, indenter load 1 N).

3. Results and discussion

As it was shown in research carried out earlier [7-9, 11] electroexplosion processing of metals and alloys including steels is accompanied by formation of highly developed relief with a large number of microdroplets, microcraters, microcracks and metal burrs. Characteristic image of surface structure of steel being formed in electroexplosion alloying with titanium and boron is shown in figure 1. Diverse defect structure of modification surface is clearly seen: metal burrs (figure 1, a), microcracks (figure 1, b), micropores (figure 1, c). Electroexplosion alloying, due to little time of thermal effect, is characterized by high cooling rate of material's modified layer. As a consequence, the submicro-nanocrystalline structure whose characteristic image is presented in figure 1, d is formed in the surface layer. The structural analysis of steel surface being formed in electroexplosion alloying clarifies the formation of the material with crystallites' sizes of 50-70 nm (figure 1, d).

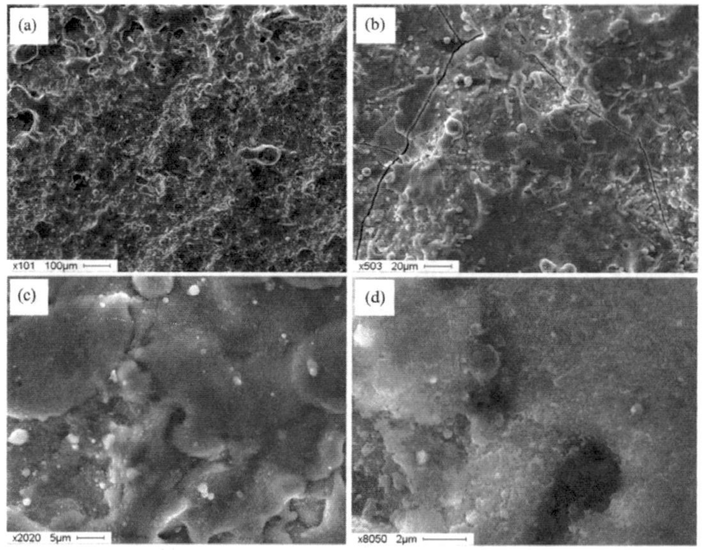

Figure 1. Electron microscopic image of $Cr_{18}Ni_{10}Ti$ steel surface structure after EEA according to mode No. 1. Scanning electron microscopy.

The studies of elemental composition of steel layer subjected to electroexplosion alloying were performed by methods of micro-X-ray spectral analysis. The characteristic image of steel surface portion and energy spectra obtained from it is depicted in figure 2. It is clearly seen (figure 2, b) that on the picture of energy spectra the maximums of both main elements of steel (Fe, Cr, Ni) and alloying elements (B, Ti) are present. The latter lends support to the statement that electroexplosion alloying is accompanied by saturation of steel surface layer with elements of electroexploded foil (Ti) and powder (B) placed on it.

The quantitative analysis results of elemental composition of $Cr_{12}Ni_{18}Ti_{10}$ steel surface layer subjected to electroexplosion alloying are shown in table 2. When analyzing the results presented in table 2 it may be noted that boron and titanium are the main elements of steel surface layer; the elements forming steel are present in negligible quantity. Maximum quantity of boron and titanium atoms is revealed in steel surface layer modified by electroexplosion method according to mode 3. The relative content of steel elements (iron, chromium, nickel) is little in surface layer. Therefore, electroexplosion processing with parameters selected in the research is accompanied by formation of not only steel alloyed layer but also a thin coating based on titanium and boron.

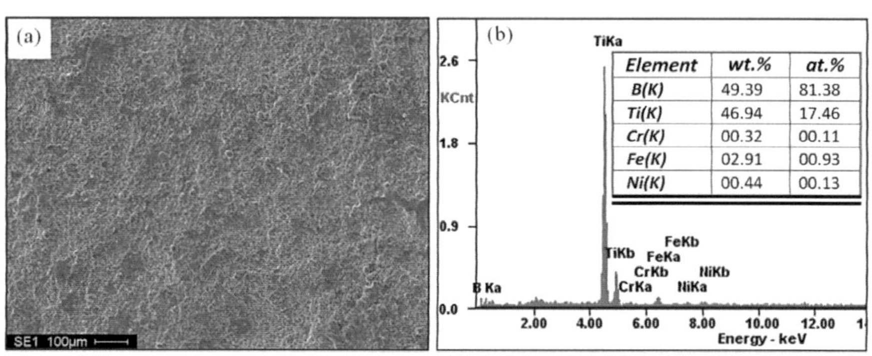

Figure 2. Electron microscopic image of $Cr_{18}Ni_{10}Ti$ steel surface layer after EEA (a); b – energy spectra obtained from sample's surface portion presented in (a); the quantitative analysis results of

elemental composition of sample's portion are set out in table 1; EEA according to mode No. 1; scanning electron microscopy.

Table 2. Micro-X-ray spectral analysis results of $Cr_{18}Ni_{10}Ti$ steel surface layer subjected to EEA with titanium and boron. EEA models are designated by numbers in Table (corresponding to table 1).

Element	Weight %			
	No. 1	No. 2	No. 3	No. 4
B(K)	49.39	41.35	21.80	18.81
Ti(K)	46.94	32.97	43.96	71.03
Cr(K)	0.32	4.59	5.84	0.73
Fe(K)	02.91	18.31	25.0	8.43
Ni(k)	0.44	2.78	3.41	0.00

Phase composition of steel layer modified by electroexplosion method was studied by methods of X-ray structural analysis. The portion of X-ray diffraction pattern from steel surface layer subjected to electroexplosion alloying according to mode No. 1 is depicted in figure 3.

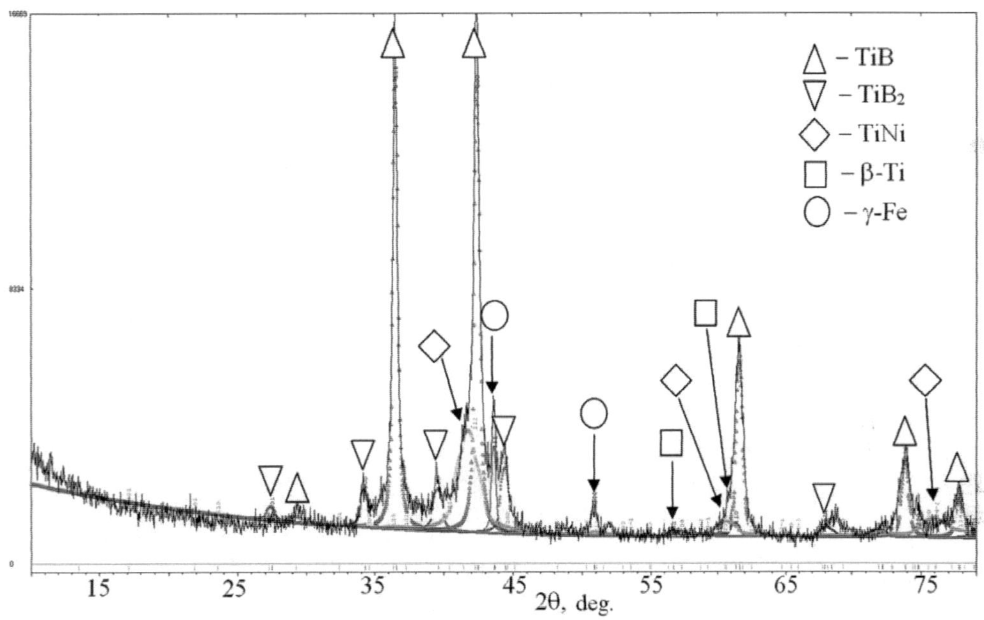

Figure 3. Portion of X-ray diffraction pattern obtained from steel surface layer subjected to EEA according to mode №1.

The quantitative X-ray phase analysis results of steel surface layer subjected to electroexplosion alloying with different relations of titanium and boron masses are presented in table 3.

Analysis of the results shown in table 3 testifies that electroexplosion alloying of steel (independent of alloying parameters) is accompanied by formation of multiphase surface layer whose main phases are titanium borides TiB and TiB_2, the total content of which is maximum in the sample processed according to mode No. 1. These results are in good agreement with micro-X-ray spectral analysis results of steel surface layer presented in table 2 and demonstrating the maximum quantity of titanium and boron precisely in the sample processed according to mode No. 1.

Mechanical properties of steel surface layer modified by electroexplosion alloying were characterized by microhardness. The obtained results are shown in table 4. When analyzing the results of table 4 it can be noted that EEA is accompanied by multiple (5.2-18.1-fold) increase in microhardness of steel surface layer, independent of treatment mode. Maximum increase in microhardness of steel surface layer is obtained for samples modified according to mode No. 1. When

IOP Conf. Series: Materials Science and Engineering **866** (2020) 012052 doi:10.1088/1757-899X/866/1/012052

comparing the results of elemental and phase composition of the modified layer presented in table 2 and table 3 with the results of mechanical tests (table 4) it can be stated that maximum hardness of steel surface layer corresponds to maximum concentration of boron in the layer and maximum relative content of titanium borides TiB and TiB_2.

Table 3. X-ray phase analysis results of $Cr_{18}Ni_{10}Ti$ steel surface layer subjected to electroexplosion alloying with titanium and boron.

EEA mode	Phase	Mass. %	Parameters of lattice, nm			D(coherent scattering region), nm	Δd/d, 10^{-3}
			a	b	c		
No. 1	TiB	61.25	0.42516			29.3	1.213
	TiB_2	13.69	0.30193		0.32448	16.7	0.597
	γ-Fe(Ni)	2.02	0.35866				
No. 2	TiNi	18.87	0.75437		0.51849	10.5	7.747
	β-Ti	4,17	0.32136			23.28	6.443
	TiB	44.07	0.42587			24.9	1.153
	TiB_2	1.78	0.30318		0.32231	10.2	4.992
	$TiFe_2$	41.17					
	$Ti_{0.95}Cr_{0.05}$	12.98					
No. 3	TiB	23.73	0.42804			42.9	2.133
	$TiFe_2$	29.34	0.48626		0.79119	31.4	2.317
	FeNi	1.72	0.35840	0.35368	0.35552	9.3	6.687
No. 4	TiNi	43.46	0.30148			86.6	32.27
	β-Ti	1.75	0.32808			398.5	1.992
	TiB	51.59	0.42513			27.1	2.443
	FeNi	0.82	0.34975	0.35081	0.36870	37.8	1.23
	α-FeNi	0.44	0.28776			30.2	0.91
	γ-Fe(Ni)	0.35	0.35056			33.6	3.21
	$TiCr_2$	9.96	0.49610		0.80199	61.9	0.916
	$Ti_{0.95}Cr_{0.05}$	12.42	0.46076		0.27584	37.8	2.059
	α-Ti	24.41	0.29528		0.47549	10.6	6.561

Table 4. Microhardness of $Cr_{18}Ni_{10}Ti$ steel in initial state and after electroexplosion alloying (δ(HV) – measurement error).

Treatment mode	HV, MPa	δ(HV), MPa
Initial	1952.0	970
No. 1	35398.4	15564.4
No. 2	10243.4	2638.4
No. 3	11144.0	3373.9
No. 4	18993.9	1361.4

4. Conclusion

Surface alloying of $Cr_{18}Ni_{10}Ti$ high chromium stainless steel with boron and titanium atoms by electroexplosion method has been performed. Studies of phase and elemental composition, defect substructure of steel modified layer by methods of modern materials science (scanning electron microscopy, X-ray structural and micro-X-ray spectral analysis) have been carried out. It has been stated that electroexplosion alloying of steel results in formation of surface layer having the multiphase submicro – nanocrystalline structure being characterized by presence of micropores, microcracks and microcraters. Dependence of phase composition and microhardness of surface layer on relation of boron and titanium masses at electroexplosion alloying has been detected. It has been suggested that electroexplosion alloying with parameters selected in the research is accompanied by not only steel alloying but also formation of thin coating enriched by titanium and boron atoms. It has been found that microhardness of the modified layer is determined by relative mass of titanium borides' fractions in steel surface layer and may increase by more than 18 times that of steel in the initial (prior to electroexplosion alloying) state.

Acknowledgements
The research was financially supported by the RSF grant (project No. 19-19-00183).

References

[1] Voroshin L G, Lyakhovich L S 1978 *Boriding of Steel* (Moscow: Metallurgiya) p 274
[2] Tsikh S G, Martynov V N et al 2015 *Repair – Innivations – Technologies – Modernization* **8** 26–28
[3] Gur'ev A M, Vlasova O A et al 2007 *Polzunovsky Almanac* **1-2** 85–88
[4] Semenov A P, Sizov I G et al 2020 Patent RF №20186872 BI No. 22
[5] Kotkowiak M, Piasecki A and Kulka M 2016 *Archives of Mechanical Technology and Materials* **36** 7–11
[6] Kadyrzhanov K K, Komarov F F, Pogrebnyak A D et al 2005 *Ion-beam and Ion-plasma Modification of Materials* (Miscow: MSU) p 640
[7] Bagautdinov A Ya, Budovskikh E A et al 2007 *Physical Fundamentals of Electroexplosive Alloying of Metals and Alloys* (Novokuznetsk: SibSIU) p 301
[8] Konovalov S, Gromov V and Ivanov Yu 2018 *Material Research Express* **5** 116520
[9] Romanov D A, Gromov V E et al 2015 *Advances in Physical Sciences* **16** 119–157
[10] *Reference Book of Steel and Alloys' Grades* 1989 ed. by V G Sorokin (Moscow: Mashinostroenie) p 640
[11] Budovskikh E A, Sarychev V D et al 2002 *Fundamentals of the Technology of Surface Treatment of Materials with Pulsed Heterogeneous Plasma* (Novokuznetsk: SibSIU) p 170

IOP Conf. Series: Materials Science and Engineering **866** (2020) 012053 doi:10.1088/1757-899X/866/1/012053

Structure and properties of differentially hardened 100-m rails after long-term operation

V E Kormyshev[1], E V Polevoy[2], A V Golovatenko[2], V E Gromov[1], Yu F Ivanov[3] and A P Semin[1]

[1]Siberian State Industrial University, 42 Kirova Street, Novokuznetsk, 654007, Russia
[2]JSC "Evraz-Integrated West-Siberian Metallurgical Combine", Novokuznetsk, 654043, Russia
[3]Institute of High Current Electronics SB RAS, 2/3 Academicheskii Ave., Tomsk, 634055, Russia

E-mail: gromov@physics.sibsiu

Abstract. Using methods of modern materials science the analysis of structural phase states of defect substructure and mechanical properties of tread surface at different distance (from 0 to 22 mm) along central axis and fillet of differentially hardened 100 meter rails of DT 350 category manufactured at JSC "Evraz-Integrated West-Siberian'' metallurgical combine after long term operation (passed tonnage 1411 mln. t. brutto) at experimental test ring has been performed. Microstructure of rails' metal is presented by fine-dispersed lamellar pearlite (grain confidence index 2) with inclusions of excess ferrite along grain boundaries estimated by 2 numbers of scale No. 7 of State Standard 8233. The value of interplate distance in rail head varies within 0.10-0.15 μm. Long term operation of rails is accompanied by the formation of gradient structure being expressed in regular change in hardness, microhardness, impact strength, destruction degree of cementite plates, scalar dislocation density throughout cross-section of rails head. The failure mechanisms of lamellar pearlite structure have been discussed. It has been shown that microhardness at 2 mm depth from tread surface amounts to 1481-1486 MPa. At depth up to 10 mm the microhardness values decrease up to 1210-1385 MPa that is caused by increase in interplate distance and decrease in level of cold strain hardening of metal in long-term operation of rails.

1. Introduction

In modern conditions of high axial loads and rates of motion the surface layers of rails undergo the intense plastic deformation leading to failures in long-term operation [1, 2]. Analysis of recent papers [3-8] shows that already at a comparative small operating life – 100-500 mln. t. brutto the structural phase states with irregular high microhardness, small size of grain in the interval from 20 to 500 nm are formed. Cementite plates are either bent or fractured with extremely high density of dislocations noted at interphase boundaries. It is clear that different processes (recrystallization, relaxation, phase transitions, formation and decay of phases, amorphisation etc.) may take place in long term operation that result in evolution of structural phase states being accompanied by change in mechanical properties [3-12].

The purpose of the research is to analyze the structure and properties of differentially hardened 100-meter rails being formed in long-term operation.

2. Material and methods of research

The samples of 100-meter differentially hardened rails of DT 350 category made of E76XF vacuum steel removed from the railway at experimental practice ground of Shcherbinka town after passed tonnage of 1411 mln. t. brutto were used as test material. Chemical composition of rail's fragment under study is given in table 1.

Microstructure of metal was revealed by deep etching in 50% hot water solution of hydrochloric acid on incomplete template (head, neck). Estimation of microstructure was done according to RD-14-2R-2001 Classificator of microstructure defects of rails [13]. Metal microstructure was studied on metallographic sections cut from the upper part of head (fillet and treat surface) before and after etching in 4% alcoholic solution of nitric acid. Studies of steel structure, fine and defect substructure were carried out by using the methods of optical microscopy.

Table 1. Chemical composition of rails of DT 350 category from E76XF steel.

Chemical composition	Content of chemical elements, %										
	C	Mn	Si	P	S	Cr	Ni	Cu	V	Al	Ti
tests	0.72	0.77	0.61	0.010	0.009	0.42	0.07	0.14	0.038	0.003	0.003
Specifications TU 0921-276-01124323-2012 for steel E76XF	0.71-0.82	0.75-1.25	0.25-0.60	not more than 0.020	0.020	0.20-0.80	Σ not more than 0.27 % 0.20	0.20	0.03-0.15	not more than 0.004	0.025

Impact strength of steel was determined at test temperature +20°C using two samples of 1 type according to State Standard 9454 cut from rails. Head hardness was measured using Brinell and Rockwell method on tread surface and along cross-section of head. Microhardness was determined at distance of 2 mm and 10 mm from surface at place location of both fillets and central zone of tread surface of sample head.

3. Results and discussion

Visually, treat surface of head rail sample has a smoothed and glitter appearance with some wear shift to one of fillets. In zone of working fillet the cracks of contact fatigue located almost at right angle to rolling axis and slight spalling are observed.

Metal microstructure of fragment being tested along axial segregation, point inhomogeneity, segregation strips and cracks is estimated as satisfactory. No internal defect, as well as continuity violations, was detected. From tread surface a darker etching region whose formation is connected with metal deformation processes, having place in long-term operation, is observed.

When scanning the non-etched metallographic sections through optical microscope, the branched continuity violations passing at acute angle to surface to depth up to 1.09 mm are revealed from surface of working fillet to the place of surface cracks of contact fatigue. Etching of rail metal in zone of non-continuities enabled one to reveal the structure with high level of deformation cold hardening.

On metallographic sections cut from tread surface of head the single fine continuity violations to depth up to 0.03 mm are met. Deformation depth from tread surface is insignificant and amounts to 0.035 mm.

Microstructure is sample's head is represented by fine-dispersed lamellar pearlite with small inclusions of excess ferrite being estimated by grain confidence index 1.5 of scale No. 7 of Standard 8233 (figure1. a, b). Metal microstructure is presented by fine-grained pearlite with small areas of structurally free ferrite (figure 1, c) being met. In pearlite structure there are rather many colonies with failed cementite plates (figure 1 d) besides regular colonies (colonies with regular located cementite plates). Areas of degenerated pearlite are present as well.

IOP Conf. Series: Materials Science and Engineering **866** (2020) 012053 doi:10.1088/1757-899X/866/1/012053

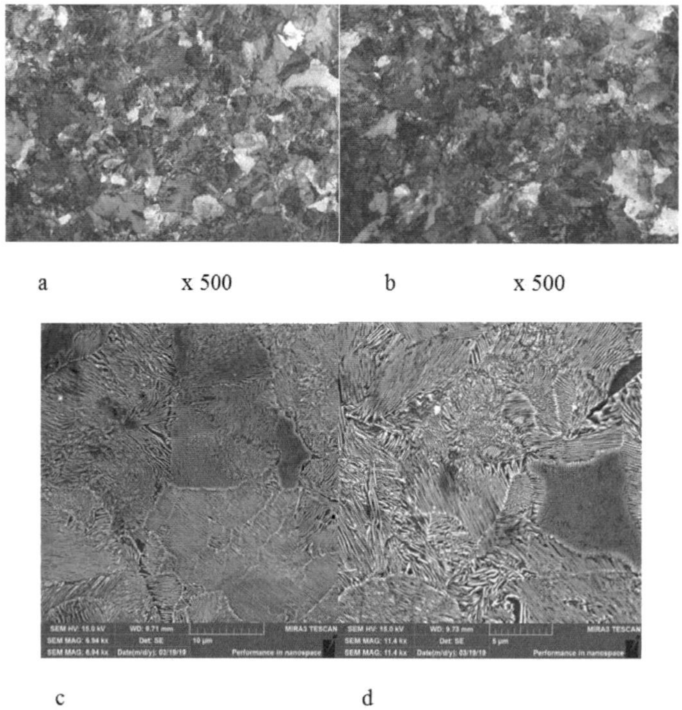

Figure 1. Structure of rail head metal revealed by methods of optical microcopy (a, b) and scanning electron microscopy (c, d) of etched metallographic section at depth of 0.5-1.0 mm.

Degree of failure of lamellar pearlite structure depends on position of volume being analyzed: on tread surface it is twice as high as surface layer of working fillet.

In scientific literature two mechanisms of cementite plate failure are mainly described [1, 5, 6]. The first one involves the intersection of plates by moving dislocations and carrying out carbon atoms by them into ferrite to stress field of dislocations. The second one consists in pulling of carbon atoms by dislocations from cementite lattice due to marked difference in binding energy of carbon atoms with dislocations (0.6 eV) and atoms in cementite lattice (0.4 eV). Dislocation substructure in the form of chaotically distributed dislocations (figure 2 a) and net-like dislocation substructure (figure 2 b) are detected. Scalar density of dislocations on tread surface amounts to $7.5\text{-}8.0 \cdot 10^{10}$ cm^{-2} and it decreases linearly with increase in distance from surface.

When analyzing the results of quantitative microstructure estimate presented in table 2 a more dispersed pearlite structure of tread surface may be noted relative to structure of fillet pearlite of fillet pearlite.

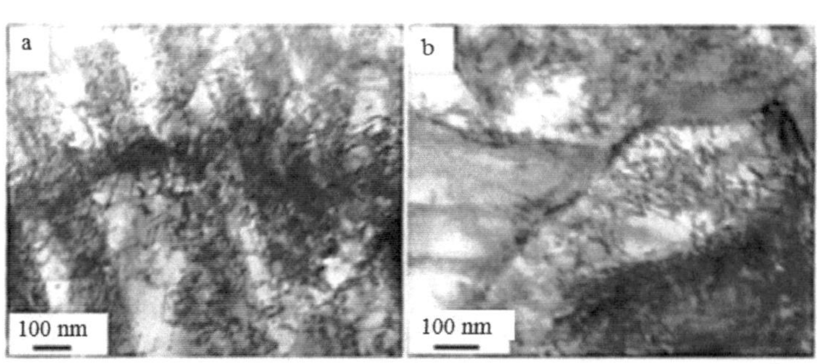

Figure 2. Electron microscopic image of dislocation substructure.

Table 2. Quantitative characteristics of metal structure of rail's head detected by methods of optical and scanning electron microscopy.

Rail index	Interplate distance, μm			Value of pearlite colonies, μm			Grain diameter, μm (grain number)		
	min	max	mean	min	max	mean	min	max	mean
fillet	0.073	0.256	0.132	2.711	12.157	6.17	15.042	51.169	29.8
tread surface	0.073	0.225	0.125	2.634	10.731	5.6	-	-	-

Table 3. Impact strength and hardness of steel on tread surface of head, along its section as well as in upper part of neck.

Specifications	KCU +20°C		Hardness, HB					
	J/cm^2		TSH	10 mm	fillet No. 1	fillet No. 2	22 mm	neck
	30	27	388 399	381	364	362	373	345
Specifications TU 0921-276–01124323-2012 for rails of DT 350 category	not less 15		363-401	not less 341				not more 341

From test results shown in Table 3 it follows that the metal of the test sample meets specifications TU 0921-276-01124323-2012 for rails of DT 350 category for impact strength and hardness on tread surface on head and along its cross-section. Hardness, measured in neck has slightly increased values relative to requirements of specifications.

Analysis of results presented in Table 4 shows that hardness values at 2 mm depth in central zone and working fillet are at a higher level (38.5-37.1 HRC) compared with nonworking fillet (35.3 HRC) that is caused by the presence of dup deformation being accompanied by cold hardening of the material in the zone. At al depth of 10 mm and 22 mm from tread surface the hardness of metal is characterized lower level (by 2-3 HRC) compared to depth of 2 mm and has correlated values (34.8-35.8 HRC).

Table 4. Metal hardness along cross-section of rail's head in transverse direction.

Place of measurement	Hardness HRC, at distance from surface, mm		
	2	10	22
Working fillet	38.5	35.5	34.8
Central zone	37.1	35.8	35.6
Nonworking fillet	35.3	35.5	35.2

Table 5. Rail's microhardness at a distance of 2 mm and 10 mm from surface at place of both fillets and central zone of tread zone.

Zone measurement	Microhardness, MPa, at distance from surface	
	2 mm	10 mm
Working fillet	1475	1385
Nonworking fillet	1486	1274
Tread surface	1481	1210

Averaged (by results of four measurements in each zone) values of microhardness determined at distance of 2 mm and 10 mm from surface at place of both fillets and central zone of head tread surface are shown in table 5. It is seen that microhardness at depth of 2 mm has close values -1481-

IOP Conf. Series: Materials Science and Engineering **866** (2020) 012053 doi:10.1088/1757-899X/866/1/012053

1475 MPa. At depth of 10 mm from surface the microhardness values decreases to 1210-1385 MPa that is evidently caused by increase in interplate distance (decrease in dispersion) and decrease in level of metal deformation hardening taking place in long-term-operation.

4. Conclusion

It has been stated by methods of physical materials science that:

Microstructure of rail metal is presented by fine-dispersed lamellar pearlite of grain confidence index 2 with inclusions of excess ferrite along grain boundaries estimated by grain confidence index 2 of scale No. 7 of Standard 8233. Value of interplate distance in rail head varies within the limits 0.10-0.15 μm. Average value of pearlite colonies in fillet zone amounts to 6.2 μm, on tread surface -5.6 μm. The main value array of real grain dimensions estimated only in zone of nonworking fillet is 7-8 number of Standard 5639-82.

Scalar density of dislocations amounts to $7.5\text{-}8.0 \times 10^{10}$ cm^{-2} and decreases linearly with increase in distance from tread surface.

Impact strength and hardness on tread surface of head and along its cross-section meets specifications TU 0921-276-01124 324-2012 for rails of DT 350 category. Hardness measured by Rockwell method at depth of 2 mm from surface amounted to 38.5-37.1 HRC, at depth of 10 and 22 mm – 34.8-35.8 HRC, respectively.

Microhardness at depth 2 mm from tread surface amounts to 1481-1486 MPa. At depth up to 10 mm microhardness values decrease to 1210-1385 MPa that is caused by increase in interplate distance and decrease in level of metal deformation level in long-term operation of rails.

Acknowledgments

The research is financially supported by RFBI project № 19-32-60001.

Authors are grateful to A.M. Yunusov for help in performing experiments and discussing the results.

References:

[1] Gromov V E, Yuriev A B, Morozov K V and Ivanov Y F 2016 *Carbide* (CISP Ltd) p 153
[2] Shur E A 2012 *Damage of Rails* (Moscow: Intext) p 192
[3] Ivanisenko Yu and Fecht H J 2008 *Steel Tech* **3** 1 19–23
[4] Gavriljuk V G 2003 *Mater.Sci. and Eng. A.* **345** 81–9
[5] Li Y J, Chai P et al 2011 *Acta Mater* **59** 3965–77
[6] Gavriljuk V G 2001 *Scripta Mater* **45** 1469–72
[7] Gromov V E, Peregudov O A et al 2015 *J. of Surface Inv. X-ray, Synchrotron and Neutron Techniques* **9** 6 1292–98
[8] Gromov V E, Ivanov Yu F et al 2015 *Materials and Electronics Engineering* **2** 4 1–4
[9] Ivanov Yu F, Gromov V E et al 2015 *Steel in Translation* **45** 4 254–257
[10] Ivanov Yu F, Gromov V E et al 2016 *Bulletin of RAS: Physics* **80** 12 1483–88
[11] Peregudov O A, Morozov K V et al 2016 *Russian Metallurgy* **4** 371–74
[12] Sheiman E 2012 *Friction and Wear* **33** 4 413–422
[13] Tyfour W R, Beynon J H and Kapoor A 1995 *Wear* **180** 79–89
[14] Singh U P, Singh R 1993 *Wear* **170** 93–99
[15] Specification' Defects of Rails. Classification, Catalogue and Parameters of Defect and Sharply Defect Rail's. OJSC Russian Railways (Moscow – Order No. 2499 from 23.10.2014)

IOP Conf. Series: Materials Science and Engineering **866** (2020) 012054 doi:10.1088/1757-899X/866/1/012054

Testing experimental parts of metallurgical units for wear and impact in industrial conditions

A Z Issagulov, Sv S Kvon, V Yu Kulikov, D R Aubakirov, Ye P Chsherbakova and A M Dostayeva

Karaganda State Technical University, 56 N. Nazarbayev Ave., Karaganda, 100027, Republic of Kazakhstan

E-mail: mlpikm@mail.ru

Abstract. The paper presents the results of industrial testing parts made of experimental wear-resistant alloys. Cast iron was smelted with a high content of nickel and vanadium and steel with alloying elements, which vary widely over from 15% to thousandths. The prototypes wear resistance was increased by about 30-34%. The work was performed at Karaganda State Technical University. These studies were carried out as a part of the grant of the Science Committee of the Ministry of Education and Science of the Republic of Kazakhstan BR05236295 "Designing, Developing and Implementing Technologies of Producing and Machining New Generation Wear-Resistant Materials for Obtaining Parts of Metallurgical Units".

1. Introduction

One of the existing trends in developing wear-resistant materials is improving the composition and properties of the alloy based on improved steels [1-4]. In previous studies, steel 30HN2MA deoxidized with ferrosilicon was considered as an analogue of the alloy with high wear-resistant properties [5-7]. The results of the experiments show that the composition and properties of 30HN2MA steel after deoxidation with ferrosilicon are close to the composition and properties of Hardox steel due to introducing micro amounts of boron.

Another analogue can be considered steel 30H3MF, which is used for manufacturing diesel parts, has wear-resistant properties and heat resistance up to 450 ^{0}C. Compared to Hardox steel, 30H3MF steel has a higher chromium content and does not contain boron. However, vanadium is present in the composition of 30H3MF steel, forming solid and resistant carbides of the MeC type; in addition, the presence of vanadium contributes to the grain grinding, which further strengthens the structure. Taking into account these features of the effect of vanadium, it can be assumed that the presence of vanadium in the composition compensates for the absence of boron.

The disadvantage of 30H3MF steel in this aspect is also a low content of nickel and manganese in the composition, which play an important role in the formation of alloy properties. Nickel does not form carbides and, therefore, does not affect hardness, but at the same time increases the matrix toughness and partially lowers the temperature threshold of cold brittleness. Manganese forms cementite-type carbides, thus hardening matrix. Thus, the presence of nickel and manganese in the composition of steel in the given amount (0.5-0.7% and 1.5-1.7%, respectively) is necessary, because these elements form the alloy matrix properties.

There are also carried out the studies [8-10], where it is proposed to use low- and medium-carbon steels that are microalloyed with strong carbide formers, such as vanadium and niobium, as wear-resistant materials. Microalloying with elements such as vanadium, titanium and niobium contributes to grain refinement, formation of stable carbides of the MeC type, and increasing hardness and wear resistance [11-15]. Introducing additives in micro quantities (the total content of vanadium, niobium and titanium does not exceed 0.1%) practically does not affect the cost of steel but significantly improves their properties.

2. Experimental studies

In the industrial conditions there was smelted an experimental alloy which composition is presented in table 1 [16-17].

Table 1. Chemical composition of the experimental alloy, %.

Experimental alloy	C	Si	Mn	Cr	Ni	Mo	V
2	0.32	0.65	1.55	2.1	0.62	0.26	0.12

The complexity of this task is caused by the "fine composition" of the alloys. The number of alloying elements varies over a wide range from 15% to thousandths. It should be borne in mind that alloying is carried out by the same ferroalloys, which must be taken into account when introducing and subsequent processing of alloys.

The total weight of the smelt, taking into account losses and waste, amounted to 418 kg. The smelt was poured into sandy clay molds. In total there were obtained 5 "jaw plate" castings, the weight of one casting was about 83 kg. One casting was obtained with an obvious casting defect, runout and was re-melted into a laboratory sample weighing 5.7 kg. The resulting samples (figure 1) were tested for wear and impact in industrial crushers.

Figure 1. Jaw plate made of the experimental alloy.

Cast iron with a high content of nickel and vanadium was also smelted [18–20]. During the melting process and after completing the melting, the chemical analysis of the alloy was controlled. The results of the final composition of the smelted alloy are shown in table 2.

Table 2. Chemical composition of the preset and smelted alloys.

Sample No.	C	Si	Mn	Cr	Ni	Ti	Mo	V
1 preset	2.4-2.6	2.0-2.2	3.0-3.5	15-18	3.0-3.5	0.5-1.0	-	0.1-0.15
2 smelted	2.5	2.2	3.4	16	3.4	0.9	-	0.13

As it is seen from the data in table 2, the specified alloy composition (cast iron with a high content of nickel and vanadium) is achieved.

The smelt weight was 416 kg, taking into account losses and waste. The smelt was poured into sandy-clay molds for the grinding body "asterisk" (figure 2). There were obtained a total of 500 pieces, the weight of one casting was 0.8 kg. The losses during casting made 16 kg.

(a) (b)

Figure 2. The grinding body (asterisk): *a* – after testing; *b* – in the course of testing.

3. Discussion of the results

The grinding bodies melted of test alloy 1 in an ore-thermal furnace were loaded into the UZT 4/1 crusher in the conditions of the SPA Marganets LLP (Karaganda city) and poured into a sand-clay mold (figure 3). After complete cooling and heat treatment, the bodies were loaded into the UZT 4/1 crusher. For loading and unloading, the ZL 50C loader was used.

The bodies were loaded in batches of 100 pcs. and were used to crush manganese ore. The cycle of each test lasted within 18 hours. In total, 2 tests of grinding bodies were carried out, the total duration was 36 hours. The prototypes of each batch after testing were transferred to the customer for further studying.

Figure 3. UZT 4/1 crusher for testing grinding bodies.

IOP Conf. Series: Materials Science and Engineering **866** (2020) 012054 doi:10.1088/1757-899X/866/1/012054

The testing results are given in table 3.

Table 3. Results of testing grinding bodies made of the experimental alloy 1 in semi-industrial conditions.

No.	Sample	Number of splits, pcs.	Number of cleavages, pcs.	Surface wear over 15%, pcs.
1	IHCh28N2 (reference)	14	43	22
2	Experimental alloy 2	8	29	21

In the UZT 4/2 crusher (figure 4), two jaw plates (a crusher part operating under conditions of increased wear and impact) smelted of experimental alloy 2 were delivered for each cycle.

Smelting alloy 2 was also carried out under the conditions of the SPA Marganets LLP in an ore-thermal furnace. Pouring was carried out in sand and clay molds. The obtained prototypes were placed in the crushers and tested within 18 hours of each cycle. A total of 2 cycles were carried out, the total duration of the work was 36 hours. Loading, unloading and transporting from the smelter to the crusher was carried out by a ZL 50C loader.

The prototypes from each batch after testing were transferred to the customer for further studies.

The remaining part of the batch of prototypes (grinding bodies and plates), according to the transfer act, was left to the contractor as working parts in his ownership, subject to the provision of an act on the experimental parts condition after starting the operation. The customer has no further complaints on this issue.

Figure 4. UZT 4/2 crusher for testing the plate.

The results of testing in semi-industrial conditions are presented in Table 4. Wear was determined by means of changing the sample mass after testing according to the formula:

$$L = \frac{m_1 - m_2}{m_1} * 100\%,$$

where m_1 is the initial mass, kg;

m_2 is the mass after testing, kg.

Table 4. Results of testing samples made of experimental alloy 2 in semi-industrial conditions.

No.	Sample	Initial mass, kg	Mass after testing, kg	Wear, %
1	Operating at present (reference)	83.0	74.2	10.6
2	Experimental alloy Upper plate series 1	83.2	77.1	7.3
3	Experimental alloy Lower plate series 1	83.1	78.1	6.1
4	Experimental alloy Upper plate series 2	83.0	76.9	7.3
5	Experimental alloy Lower plate series 2	83.1	77.1	7.2

The grinding bodies made of cast iron with a high nickel content and additionally microalloyed with vanadium showed high wear resistance with good impact strength. The prime yield (the number of intact parts without splits and ckleavages) after 36 hours of testing increased by 33%. The jaw plates wear resistance under the same test conditions increased by about 34% compared to the reference.

4. Conclusion

Thus, the semi-industrial tests showed the possibility of smelting experimental alloys of a preset composition on existing production equipment and using traditional charge materials. The prototypes wear resistance increases by about 30-34%. However, it should be noted that this characteristic is "floating" and depends on the properties of the mineral raw materials being treated.

References

[1] Zou Xiaodong, Sun Jincheng and Matsuura Hiroyuki 2018 *Metallurgical and Materials Transactions B: Process Metallurgy and Materials Processing Science* **5(49)** 2168–73
[2] Li Hui-rong, Sun Li-gen and Zhu Li-guang 2018 *Metals* **8(8)**
[3] Wang Yan, Zhu Liguang and Zhang Qingjun 2018 Metals **8(8)**
[4] Zou Xiao-dong, Sun Jin-cheng and Zhao Da-peng 2018 *Journal of Iron and Steel Research International* **2(25)** 164–172
[5] Mascaraque-Ramirez C, Franco P 2018 *Ships and Offshore Structures* **7(13)** 750–758
[6] Chaudhari V, Kulkarni D M and Rathi S 2017 *ASME International Mechanical Engineering Congress and Exposition* (Tampa, FL) vol 9
[7] Sekban D M, Akterer S M and Saray O 2018 *Journal of Materials Science & Technology* **1(34)** 237–244
[8] Nuesse G 2017 *Stahlbau* 9(86)
[9] Mert T and Ekinci S 2017 *Acta Physica Polonica* A **3(13)** 495–499
[10] Yu Yan-chong, Li Hao and Wang She-bin 2017 *Metallurgical Research & Technology* **4(114)**
[11] Filippov M A, Gervasyev M A et al 2017 *Diagnostics, Resource and Mechanics of Materials and Structures* **1** 43–54
[12] Orlov V V, Malyshevsky V A et al 2014 *Steel* **9** 79–88
[13] Issagulov A Z, Kim V A et al 2014 *Metalurgija* **53(4)** 685–688
[14] Dub A V, Barulenkova N V et al 2004 *Metalurgija* **4** 67–73
[15] Kulikov V Yu, Issagulov A Z et al 2017 *Metalurgija* **56(3-4)** 409–411
[16] Arinova S K, Issagulov A Z et al 2017 *Proc. of the XX Int. Sci. Pract. Conf. In Metallurgy: Technologies, Innovations, Quality* 222–227
[17] Kvon Sv S, Kulikov V Y et al 2016 *Metalurgija* **55(2)** 206–208
[18] Issagulov A Z, Kulikov V Y et al 2016 *Metalurgija* **55(3)** 426–428
[19] Kvon S S, Kulikov V Y et al 2019 *Metalurgija* **58(3-4)** 315–318
[20] Kovalev P, Riaboshuk S et al 2019 *Metals* **9(2)** 203–208

IOP Conf. Series: Materials Science and Engineering **866** (2020) 012055 doi:10.1088/1757-899X/866/1/012055

Features of the structure formation in the process of obtaining cast aluminum matrix nanocomposites using ultrasonic melt treatment

E S Prusov[1], V B Deev[2,3,4], A A Panfilov[1] and M V Temlyantsev[4]

[1]Vladimir State University n.a. Alexander and Nikolay Stoletovs, Vladimir, Russia
[2]Wuhan Textile University, Wuhan, China
[3]National Research Technological University "MISiS", Moscow, Russia
[4]Siberian State Industrial University, Novokuznetsk, Russia

E-mail: deev.vb@mail.ru

Abstract. The paper presents modern representations about the effect of ultrasonic treatment on the behavior of nanoscale reinforcing particles in aluminum melts. Advantages and potential directions of application of the ultrasonic melt processing method for producing cast aluminum matrix nanocomposites are presented.

1. Introduction

To date, traditional metal-based alloys have practically exhausted their potential from the standpoint of a breakthrough change in the level of properties and characteristics of critical products [1]. A promising direction for improving the mechanical and operational properties of alloys is their reinforcement with dispersed particles of high-modulus refractory compounds, which leads to the formation of heterophase structures characteristic for metal-matrix composites [2].

Such approach opens up wide opportunities for improving the functional characteristics of products and is relatively easy to adapt to the conditions of existing industrial enterprises. Metal matrix composites (primarily based on light alloys) have a unique combination of such valuable properties as high specific strength, low density, low coefficient of thermal expansion and high heat conductivity, high wear resistance and many others [3].

However, a significant limiting factor for industrial applications of metal matrix composites with micro-sized reinforcing phases is the relatively low ductility and impact strength, high abrasion on the counterbody, and high wear of the cutting tool during machining [4]. One way to overcome these limitations is to use nanoscale reinforcing phases [5].

The introduction of nanoparticles into metal melts presents significant technological difficulties due to their strong tendency to form agglomerates because of the action of the van der Waals forces, as well as rapid oxidation even at relatively low temperatures and poor wettability with liquid metals [6-8].

In this regard, the well-known liquid-phase methods for producing exogenously reinforced aluminomatrix composites with micro-sized particles (mechanical mixing of particles into the melt using an impeller, injection of an inert gas, etc. [9]) are low-efficient in the preparation of nanostructured composite materials. Practice shows that the production of high-quality foundry metal-matrix nanocomposites in most cases requires the use of external physical effects on the melt, one of which may be ultrasonic treatment [10-12]. At present, the use of ultrasound in metallurgy has already become one of the traditional solutions for degassing melts and modifying treatment [13]. The possibility of a fundamental change in the structure of alloys using ultrasonic treatment in acoustic cavitation mode is confirmed by many years of research by prof. Eskin G.I. [14].

In these fundamental works it was found that ultrasonic treatment of liquid and crystallizing melt destroys the branches of growing dendrites, fragments of which act as additional centers of crystallization. The resulting excess of crystallization nuclei ensures the formation of a finely ground non-dendritic structure with a grain size approximately equal to the size of the dendritic parameter. The aim of the work is to assess the prospects of using the method of ultrasonic melt treatment for the preparation of aluminomatrix nanocomposites by liquid-phase methods based on an analysis of modern ideas about the processes of their structure formation under conditions of cavitation influences.

2. The technological process and equipment for producing cast aluminomatrix nanocomposites using ultrasonic melt treatment

The synthesis of nanomaterials based on the use of ultrasound is currently considered by researchers as one of the most promising [15]. When metal matrix nanocomposites are prepared by liquid-phase methods, the melt treatment by ultrasonic vibrations is carried out, as a rule, in the frequency range 18–25 kHz directly during the introduction of the reinforcing phase or immediately after it. A diagram of a typical installation for ultrasonic melt treatment in the preparation of nanocomposites is shown in figure 1 (*a*) and includes a radiating waveguide 1, magnetostrictive transducer 2 and ultrasonic oscillation generator 3. In some cases, nanoparticles (4) are supplied in the inert gas protective atmosphere (5 – balloon with gas).

The waveguide is usually made of a titanium alloy, stainless steel, or niobium-based alloys because of their effective transmission of ultrasound and dimensional stability at elevated temperatures [16]. Niobium waveguides have higher chemical resistance in contact with the matrix melt, withstand high processing temperatures with minimal cavitation erosion, but are the most expensive. The destruction of agglomerates of nanoparticles during melt processing is associated with the phenomenon of acoustic cavitation (figure 1, b), which is the process of formation and growth (stage *I*), pulsation and collapse (stage *II*) of tiny bubbles during cyclic exposure to high-intensity ultrasonic waves.

Foci of cavitation appearing and disappearing near agglomerates of nanoparticles create excessive pressure, which allows the melt to pass into the capillary spaces and pores between the particles even when they are poorly wetted by a liquid metal [17]. The saturation of the agglomerates by the melt leads to their decomposition into individual particles (stage *III*) due to the force of viscous friction at the leading edge of the ultrasonic wave. It was shown in [18] that the deagglomeration process will proceed most efficiently during the formation and collapse of cavitation bubbles directly inside agglomerates of nanoparticles.

To create acoustic cavitation in melts of light alloys, the ultrasound intensity should be at least 80 W/cm^2 [13]. Assuming that the surface of the waveguide-emitter is completely wetted by the melt, the intensity I can be expressed as follows [13, 19]:

$$I = \frac{1}{2}\rho_L c_L \left(2\pi f A_0\right)^2,$$

where f is the oscillation frequency, ρ_L is the density of the metal melt, c_L – the propagation velocity of sound waves in the melt, A_0 – the initial amplitude. The calculations performed using the above expression allow us to establish the theoretical nature of the change in the intensity of ultrasonic treatment depending on the frequency of oscillations and the initial amplitude, taking into account the type o alloy being processed.

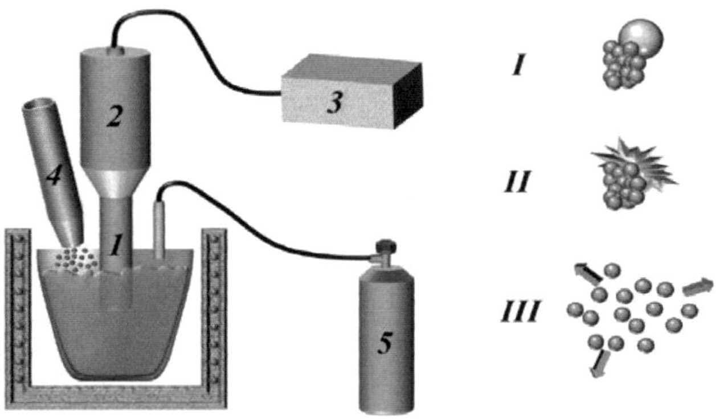

(a) (b)

Figure 1. Schematic diagram of a typical installation for ultrasonic melt treatment (a) and the proposed model of destruction of nanoparticles agglomerates (b)

3. The effect of ultrasonic treatment on the structure formation and formation of properties of cast aluminomatric nanocomposites

Specific examples of the implementation of ultrasonic technologies in the production of cast aluminum matrix composites with exogenous nanoparticles are so far presented in the scientific literature to a limited extent, however, these methods are gradually beginning to attract more and more attention of researchers and specialists. In particular, in [20], ultrasonic melt treatment was used to obtain nanocomposites based on Al-7Si-0.3Mg with the addition of 0.5 wt.% Cu and 1 wt.% of Al_2O_3 nanoparticles.

A niobium probe operating at a frequency of 20 kHz with a maximum power of 600 W was used. Nanoparticles were wrapped in a two-layer aluminum foil before being fed into the melt. The waveguide was immersed into the matrix melt to a depth of 12 mm at a temperature of 700°C and ultrasonic vibrations were applied with a frequency of 20 kHz and an amplitude of 60 μm while feeding heated capsules with nanoparticles. The treatment time was 15 minutes for each capsule administered. The combined modifying effect of nanoparticles and ultrasonic treatment on all phase components, including primary α-dendrites, eutectic silicon, and Al_2Cu inclusions, was noted.

In [21], 1 mass% SiC (50 nm) and Al_2O_3 (20 nm) nanoparticles were blown with an argon stream into the melt of the A356 aluminum alloy with simultaneous ultrasonic treatment at a power of 1.75 kW and a frequency of 18 kHz. When nanoparticles were introduced together with ultrasonic processing of the melt, significant refinement of the average grain size in the matrix alloy and eutectic structure was achieved, as well as a decrease in the distance between the secondary axes of the dendrites by more than 36% compared with the matrix alloy. Ultrasonic treatment was accompanied by degassing of the alloy, providing a significant reduction in the porosity of the castings. The addition of SiC nanoparticles according to the proposed regimes contributed to an increase in tensile strength from 228 to 283 MPa and an increase in the ductility of the material from 4 to 11.2%.

The effect of ultrasonic vibrations on the dispersion and distribution of graphite particles in the preparation of Al – 2 vol.% Graphite composites by mechanical mixing was studied [22]. Graphite powder with an average particle size of 21 μm was introduced into the aluminum melt using an impeller, followed by ultrasonic treatment for 1.3 and 5 minutes. A Ti-6Al-4V alloy emitter was used at a frequency of 20 kHz and a power of 2.5 kW. It was shown that ultrasonic treatment for 1 min sharply reduces the porosity of materials and destroys the accumulation of graphite particles.

A further increase in the treatment time does not significantly affect the change in porosity, however, it affects the size and morphology of graphite inclusions, which are reduced to 100-130 nm during processing for 5 min. The influence of different temperatures of ultrasonic treatment was studied in [23] when producing aluminomatrix composites with 1% mass. Al_2O_3 nanoparticles. Powders of Al (80-100 μm) and Al_2O_3 (65 nm) were mixed in a ball mill in stearic acid to obtain a composite powder and sintered at 400°C for 2.5 h. The matrix melt was superheated to 750°C, then for 10 min the resulting powder was mixed using an impeller. After that, ultrasonic vibrations with a frequency of 20 kHz were applied for 60 s at a power of 300 W and melt temperatures of 650, 670, 680, and 700°C. Processing at 670°C leads to the best microstructure parameters with an average α-grain size of 25 μm and an increase in tensile strength by 37%.

It was confirmed [24] that the presence of Al_2O_3 nanoparticles in an aluminum melt leads to a significant intensification of the process of formation of cavitation centers, as well as to an increase in the melt velocity near the waveguide by an average of 20%. It has been suggested that the noted trends may be due to the nucleating action of alumina nanoparticles with respect to cavitation bubbles. However, an excessive increase in the volume of cavitation bubbles is accompanied by a weakening of the distribution of ultrasonic energy, which can lead to a localized distribution of cavitation effects.

The above examples convincingly indicate the technological feasibility of using ultrasonic treatment when introducing nanosized reinforcing particles into aluminum melts. Available experimental data confirm an improvement in the wettability and uniformity of the distribution of nanoparticles over the melt volume due to the destruction of agglomerates. Modern ideas about the structure formation of aluminomatrix nanocomposites using ultrasonic treatment link the mechanism of deagglomeration of nanoparticle clusters with the sonocapillary effect and the phenomenon of ultrasonic cavitation. Surface activation of nanoparticles during processing increases the thermodynamic stability of the system as a whole. An additional result is a reduction in the porosity of composite castings due to the degassing effect of ultrasonic treatment, as well as a modifying effect on the structure of the matrix alloy. However, the industrial implementation of ultrasonic technologies in the synthesis of cast aluminomatrix nanocomposites is still hampered by the lack of recommendations on the selection of optimal processing conditions depending on the composition of the synthesized materials, taking into account the physicochemical nature of the components.

4. Conclusion

- The advantages and potential directions of the use of ultrasonic melt treatment in the production of aluminomatrix composite materials with nanoscale reinforcing particles in liquid-phase technological processes are presented.
- Modern ideas of the structure formation of aluminomatrix nanocomposites using ultrasonic melt treatment are presented, linking the mechanism of deagglomeration of clusters of nanoparticles with the sonocapillary effect and the phenomenon of ultrasonic cavitation.
- The presented practical examples confirm the improvement in the wettability and uniformity of the distribution of nanoparticles over the volume of the matrix melt, the reduction in the porosity of composite castings, and the combined modifying effect of nanoparticles and ultrasonic treatment on the structure of the material.

References

[1] Vojtech D 2010 *Metalurgija* 49(3) 181–185
[2] Prusov E, Deev V and Rakhuba E 2019 *Materials Today: Proceedings* **11(P1)** 386–391
[3] Prusov E and Panfilov A 2011 *Russian Metallurgy (Metally)* **7** 670–674
[4] El-Labban F E, Abdelaziz M et al 2016 *J. of King Saud Univ.: Engineering Science* **28** 230–239
[5] Casati R and Vedani M 2014 *Metals* **4** 65–83
[6] Min Y, Akbulut M et al 2008 *Nature Materials* **7(7)** 527–538
[7] Sajjadi S A, Ezatpour H R and Beygi H 2011 *Materials Sci. and Eng. A* **528(29-30)** 8765–71
[8] Andrievskii R A, Shul'ga Yu M et al 2016 *Inorganic Materials* **52(7)** 686–693
[9] Panfilov A and Prusov E 2013 *Proc. of 22nd Int. Conf. on Metallurgy and Materials* 1195–99
[10] Vorozhtsov S A, Eskin D G et al 2015 *Metallurgical and Materials Trans. A* **46(7)** 2870–2875
[11] Kannan C and Ramanujam R 2018 *Science and Technology of Materials* **30(2)** 109–119
[12] Deev V, Prusov E and Rakhuba E 2019 *Materials Science Forum* **946** 655–660
[13] Eskin G I and Eskin D G 2014 *Ultrasonic Melt Treatment of Light Alloy Melts* (FL, CRC Press)
[14] Eskin G I 2010 *Metallurgist* **54(7-8)** 505–513
[15] Bang J H and Suslick K S 2010 *Advanced Materials* **22(10)** 1039–59
[16] Li X, Yang Y and Cheng X 2004 *Journal of Materials Science* **39** 3211–12
[17] Kudryashova O and Vorozhtsov S 2016 *JOM* **68(5)** 1307–1311
[18] Eskin D G, Tzanakis I et al 2019 *Ultrasonics Sonochemistry* **52** 455–467

[19] Chen D, Zhao Y et al 2011 *J. of Wuhan Univ. of Tech.-Mater. Sci.* **26(5)** 920–925
[20] Choi H, Jones M, Konishi H and Li X 2011 *Materials Science and Engineering A* **43** 738–746
[21] Liu X, Jia S and Nastac L 2014 *International Journal of Metalcasting* **8(3)** 51–58
[22] Christy Roshini P, Nagasivamuni B et al 2015 *J. of Materials Eng. and Perf.* **24** 2234–39
[23] Su H, Gao W, Feng Z and Lu Z 2012 *Materials & Design* **36** 590–596
[24] Mirihanage W, Xu W et al 2016 *Materials Letters* **164** 484–487

IOP Conf. Series: Materials Science and Engineering **866** (2020) 012056 doi:10.1088/1757-899X/866/1/012056

High-temperature gas corrosion of cast-iron sections of the EcoSøderberg cell collecting bell

M V Temlyantsev[1], E A Pinaev[2], V B Deev[3], E N Temlyantseva[1] and N I Kuvshinnikova[1]

[1]Siberian State Industrial University, 42 Kirova Street, Novokuznetsk, 654007, Russia
[2]JSC "Rusal Krasnoyarsk", 40 Pogranichnikov Street, Krasnoyarsk, 660111, Russia
[3]MISiS National Research Technological University, 4 Leninsky Prospekt, Moscow,119049, Russia

E-mail: uchebn_otdel@sibsiu.ru

Abstract. Investigations of the structure, chemical, mineralogical and phase composition of corrosion products of cast-iron sections of the EcoSøderberg cell collecting bell were carried out. Structural features of the scale were established using electron microscopy. The mechanism of corrosion products formation in the form of loose scale was proposed and scientifically substantiated.

1. Introduction

The gas collecting bell is one of the important structural elements of electrolysis cells. At the Russian aluminum-producing enterprises it is made from cast-iron sections. The transition to a new, more efficient design of the gas collecting bell [1] led to a significant change in the conditions of its service. In particular, the temperature of gases in the central domes of the gas removal system rises to 800 °C and the temperatures of the cast-iron sections of the gas collecting bell to 600–620 °C. Together with the general tendency to increase the sulfur content in the coke used in aluminum production, an increase in temperature led to the intensification of high-temperature gas corrosion of the cast-iron sections of the gas collecting bell, reducing their durability and service life.

Anode gases, in relation to VCh cast iron (high-strength with spherical graphite), from which sections of the gas collecting bell are made, are quite aggressive, since they contain intensifying high-temperature gas corrosion fluorine compounds (HF, $\leq$300 mg/nm^3), elemental sulfur, sulfur oxides (up to 240 – 270 mg/nm^3) and vanadium, resinous substances, the main sources of which are cryolite Na_3AlF_6, products of electrolyte evaporation, anode mass. According to [2], in the general case, during the electrolysis of aluminum, the dust-gas mixture may contain gaseous components: HF, CF_4, C_2F_6, SiF_4, SO_2, SO_3, H_2S, CS_2, COS, CO_2, CO, resinous substances and solid particles: C (carbon black), Al_2O_3, Na_3AlF_6, $Na_5Al_3F_{14}$, K_2NaAlF_6, AlF_3, CaF_2, MgF_2.

The corrosion resistance of the sections largely depends on the structure, chemical and phase composition of the corrosion products (scale) formed on the surface of cast iron. If in the process of operation there is no destruction of the scale layer, then oxidation (metal loss) obeys the square root law.

At the initial stage, when the metal surface is not oxidized, scale formation occurs intensively, almost linearly. Further, with an increase in the operating time, the thickness of the scale increases, it prevents the flow of diffusion processes to a greater extent, and the intensity of scale formation decreases.

In real conditions, the destruction of the scale layer can occur, caused by various factors, which leads to a deviation of the time dependence of the burn from the square root law. It is characteristic that after the destruction of the scale layer or the violation of its continuity, intensification of fumes occurs. The most dangerous from the point of view of metal loss is *catastrophic* oxidation [3].

It occurs in the cases when, during the operation of the section, the scale is removed from the surface of cast iron or practically does not have a protective effect in terms of reducing the intensity of diffusion processes. In such cases, the dependence of oxidation on time obeys an almost linear law and there is no effect of slowing down scale formation.

In this regard, the study of the structure, chemical and phase composition of the corrosion products of the cast-iron sections of the gas collecting bell of aluminum electrolytic cells EcoSøderberg is relevant and of great scientific and practical importance.

2. Methods of research

In the work the study on the structure, chemical, mineralogical and phase composition of VCh cast iron corrosion products selected from the inner surface of the gas collecting bell sections after operation was performed. Investigations (X-ray phase, chemical, spectral analyzes and optical microscopy) were carried out in the collective use center "Material Science" of Siberian State Industrial University.

The microstructure and chemical composition of the corrosion products were investigated using a scanning electron microscope (SEM, TESCAN VEGA 3) and electron microprobe analysis (EMPA, OXFORD AZtec) at MISiS National Research Technological University.

3. Analysis of the research results

The analysis of the study results shows that the corrosion products mainly consist of iron oxides and alloying elements of cast iron, however, a number of samples are characterized by the presence of an increased content of C (up to 4.3%), Al (up to 4.2%), Na (up to 5.62 %), K (up to 3.08%), S (up to 3.32%). X-ray phase analysis confirmed the presence of iron oxides Fe_3O_4 – magnetite, Fe_2O_3 – hematite, revealed the presence of potassium, sodium, aluminum, fluorine-containing compounds Ka_2NaAlF_6 – elpazolite and Na_2MgAlF_7 – weberite.

The scale has a pronounced layered structure, contains a large number of defects in the form of pores and cracks (figure 1). The layers differ in chemical, phase composition and macrostructure (figure 2, table 1). All studied samples are characterized by cyclic alternation of relatively dense layers of iron oxides Fe_2O_3 and Fe_3O_4 and more porous interlayers between them.

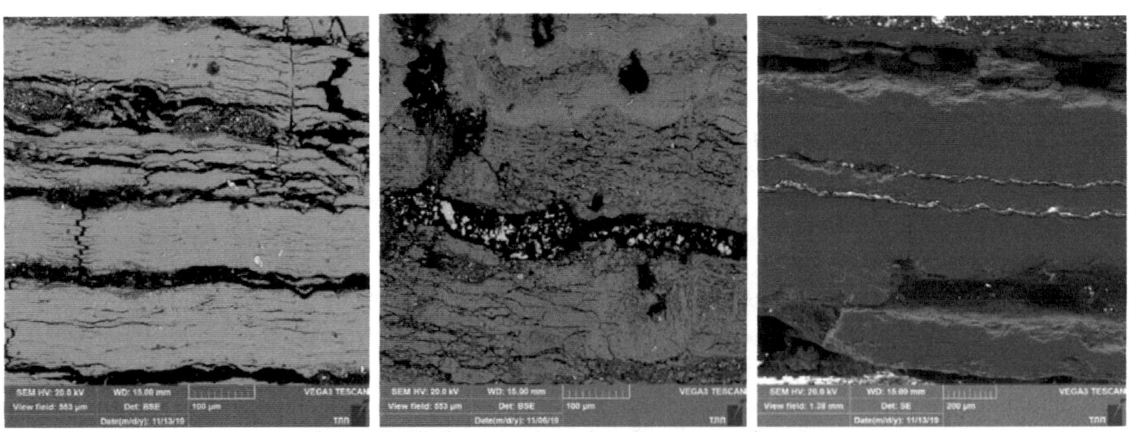

Figure 1. Defects in the microstructure of scale.

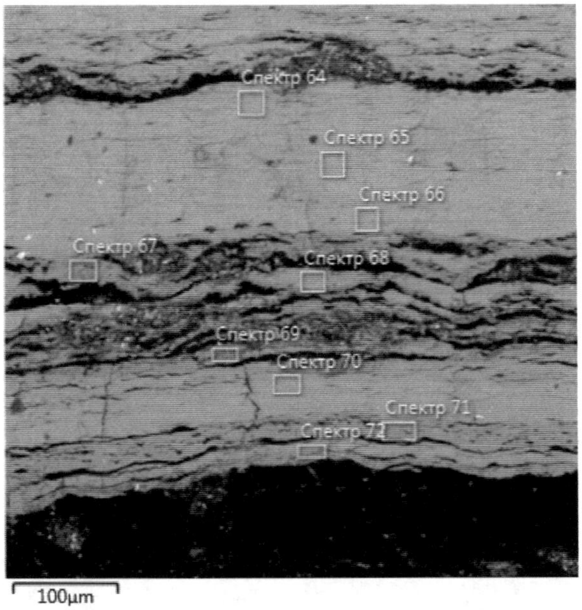

Figure 2. Microprobe analysis points.

The interlayers are characterized by a high content of C and F. Sulfur is distributed fairly evenly over the thickness of the scale. Carbon in the form of small graphite plates is found in scale. A distinctive feature of the investigated samples of scale is high defectiveness, friability, the presence of a large number of pore cracks, discontinuities, low adhesion to the oxidized surface of cast iron. This is a consequence of the content in the scale of various phases and compounds having different coefficients of thermal expansion.

The cyclic changes in the temperature of the section during operation lead to the appearance of structural and phase stresses in the scale contributing to its loosening. It is important that when the section temperature changes from 220 to 620 °C on the surface of cast iron in the scale at temperatures above 560 – 570 °C, a layer of wustite is formed, which decomposes into iron and magnetite at lower temperatures, the cyclicality of this process adversely affects the continuity and protective properties of scale. The source of carbon in the layers of scale is also a gas atmosphere, which as a result of the reaction of interaction of molten aluminum with CO:

$$4Al+6CO = 2Al_2O_3+6C$$
$$2CO=CO_2+C$$

as one of the products contains C in the form of graphite or soot forming coal foam [4].

Table 1. The composition of scale (wt.%).

Spectrum name	C	O	F	Na	Al	Si	S	V	Mn	Fe	Cu	Total
Spectrum 64	**8.51**	31.2	**3.06**	0	0	0.19	**1.6**	0	0	54.98	0.47	100
Spectrum 65	**7.07**	33.06	**2.79**	0.04	0	0.18	**1.38**	0	0.37	55.11	0	100
Spectrum 66	**8.2**	30.41	**5.1**	0	0	0.13	**1.91**	0	0	54.25	0	100
Spectrum 67	**18.16**	14.16	**24.99**	0	0.16	0.14	**2.91**	0.01	0	39.47	0	100
Spectrum 68	**9.74**	28.18	**9.62**	0.03	0	0.12	**1.21**	0.03	0	51.06	0	100
Spectrum 69	**12.24**	12.41	**18.26**	0	0	0.11	**1.21**	0	0	55.35	0.42	100
Spectrum 70	**8.91**	32.1	**3.07**	0	0	0.07	**1**	0	0	54.22	0.64	100
Spectrum 71	**15.3**	22.65	**15.78**	0	0	0.04	**1.71**	0	0	44.51	0	100
Spectrum 72	**10.91**	11.36	**32.01**	0	0	0.09	**2.01**	0.04	0	43.52	0.05	100

The presence of carbon scale in the interlayers is associated, among other things, with parallel processes of decarburization of cast iron in a gaseous medium containing oxidizing and reducing components. According to the data of [5, 6], as a result of decarburization, the oxide film can partially recover, become loose and thin, its expansion and formation of growths filled with soot carbon are observed.

Figure 3 shows the proposed dynamics and mechanism of high-temperature gas corrosion of the cast-iron sections of the gas collecting bell. The initial state (figure 3 a) corresponds to the microvolume of the new section, the surface of which is not covered with scale. During operation, under the influence of a temperature less than 570 °C and an oxidizing medium, a scale of hematite and magnetite forms on the surface of cast iron (figure 3 b).

Further, due to structural, phase, and thermal stresses, cracks, chips, and other discontinuities are formed in the mill scale (figure 3 c). The intensification of their formation is facilitated by the phenomena of cast iron growth, cyclical changes in the temperature of the section, and an increase in the thickness of the scale. Discontinuities open the contact of the surface of cast iron with the atmosphere.

The atmosphere, including solid particles, in particular carbon, for example in the form of soot or graphite, penetrates into the cavity in which these particles are deposited on its walls, and aggressive oxidizing gases form a new layer of hematite and magnetite on the surface of cast iron (figure 3 d), and magnetite surrounding the cavity is oxidized to hematite. In this case, the cavity itself with a different chemical and phase composition, when the section temperature changes, begins to play the role of a strong stress concentrator and the cracks formed in it serve as next channels for the penetration of the oxidizing atmosphere (figure 3 f).

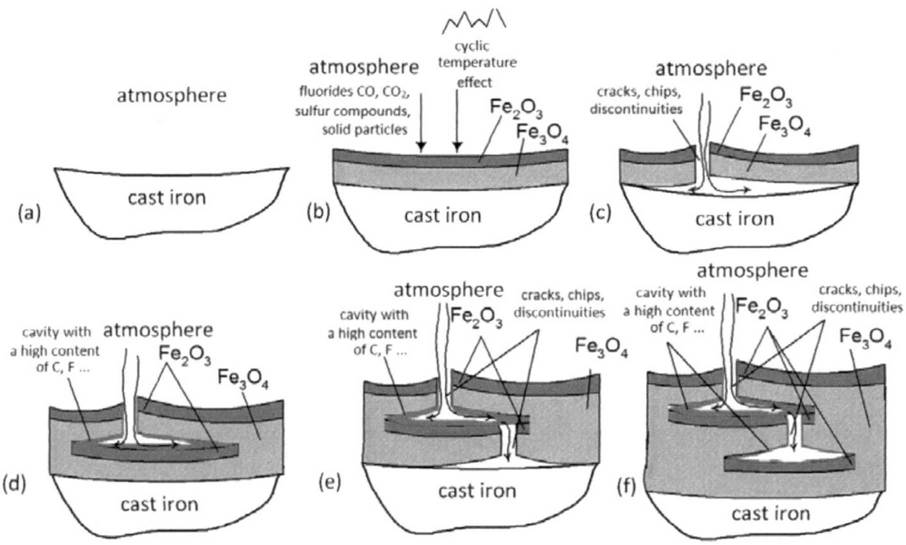

Figure 3. Diagram of the dynamics and mechanism of high-temperature gas corrosion of cast-iron sections of gas collecting bell.

Thus, in the macrovolume, the formation of such cavities and discontinuities repeatedly leads to the formation of loose scale on the surface of the section. Under the influence of gravity and gas flows, partial collapse of loose scale occurs, which further intensifies scale formation.

The transition to a temperature range of more than 570 °C and further cooling are important. Under such conditions, a three-layer scale is formed containing wustite, which decomposes upon cooling. This process causes additional stresses that contribute to the violation of the protective properties of scale [7]. The presence of carbon in the cavities under certain conditions promotes the occurrence of reduction processes, for example, the restoration of hematite to magnetite.

Fluorine-containing and sulfur-containing components of the atmosphere are unconditional intensifiers of high-temperature gas corrosion and an increase in their content in gases in contact with the cast-iron sections of gas collecting bell leads to their more rapid destruction.

4. Conclusion

The study of the structure, chemical, mineralogical and phase composition of the corrosion products of the cast-iron sections of EcoSøderberg cell collecting bell was carried out. The mechanism of the

IOP Conf. Series: Materials Science and Engineering **866** (2020) 012056 doi:10.1088/1757-899X/866/1/012056

corrosion products formation in the form of loose scale was proposed and scientifically substantiated. At the first stage, during operation, under the influence of a temperature of less than 570 °C and an oxidizing medium, a layer of scale from hematite and magnetite is formed on the surface of cast iron. Further, cracks, chips, and other discontinuities are formed in the scale due to structural, phase and thermal stresses. The intensification of their formation is facilitated by the phenomena of cast iron growth, cyclical changes in the temperature of the section, and an increase in the thickness of the scale.

Discontinuities open the contact of the cast iron surface with the atmosphere. The atmosphere, including solid particles, in particular carbon, for example in the form of soot or graphite, penetrates into the cavity in which these particles are deposited on its walls, and oxidizing aggressive gases form a new layer of hematite and magnetite on the surface of cast iron, and magnetite surrounding the cavity is oxidized to hematite.

In this case, the cavity itself with a different chemical and phase composition, when the section temperature changes, begins to play the role of a strong stress concentrator, and the cracks formed in it serve as next channels for the penetration of the oxidizing atmosphere. Thus, in the macrovolume, the formation of such cavities and discontinuities repeatedly leads to the formation of loose scale on the surface of the section.

Under the influence of gravity and gas flows, a partial collapse of the "fur coat" occurs, which further intensifies scale formation. Upon transition to a temperature zone of more than 570 °C and further cooling, a three-layer scale containing wustite is formed, which is destroyed upon cooling. This process causes additional stresses that contribute to the violation of the protective properties of the scale. The presence of carbon in the cavities under certain conditions promotes the occurrence of reduction processes, for example, the restoration of hematite to magnetite.

References

[1] Vinogradov A M, Pinaev A A 2017 *Metallurgy of Non-ferrous Metals* **1** 19 – 30
[2] Mintsis M Ya, Polyakov P V, Sirazutdinov G A 2001 *Aluminum Electrometallurgy* (Novosibirsk: Nauka) p 368
[3] *Oxidation of Metals* 1969 ed Zh Benar (Moscow: Metallurgiya) vol 2 p 444
[4] Foosnas T, Naterstad T, Bruheim M and Grjotheim K 1986 *Light Metals* pp 633–642
[5] Bobro Yu G 1960 *Heat-resistant and Growth-resistant Cast Irons* (Moscow: Mashgiz) p 170
[6] Bobro Yu G 1976 *Alloy Cast Iron* (Moscow: Metallurgiya) p 288
[7] Temlyantsev M V, Mikhailenko Yu E 2006 *Steel Oxidation and Decarburization in Heating Processes under Pressure Treatment* (Moscow: Teplotekhnik) p 200

AUTHOR INDEX

Institute of Physics
Dirac House, Temple Back
Bristol BS1 6BE UK

ISSN: 1757-8981
ISBN 978-1-7138-3618-6

9 781713 836186

CURRAN ASSOCIATES INC.
proceedings.com